PROBABILITY, STATISTICS, AND QUEUEING THEORY

With Computer Science Applications

Computer Science and Applied Mathematics
A SERIES OF MONOGRAPHS AND TEXTBOOKS

Editor
Werner Rheinboldt
University of Maryland

HANS P. KÜNZI, H. G. TZSCHACH, and C. A. ZEHNDER. Numerical Methods of Mathematical Optimization: With ALGOL and FORTRAN Programs, Corrected and Augmented Edition

AZRIEL ROSENFELD. Picture Processing by Computer

JAMES ORTEGA AND WERNER RHEINBOLDT. Iterative Solution of Nonlinear Equations in Several Variables

AZARIA PAZ. Introduction to Probabilistic Automata

DAVID YOUNG. Iterative Solution of Large Linear Systems

ANN YASUHARA. Recursive Function Theory and Logic

JAMES M. ORTEGA. Numerical Analysis: A Second Course

G. W. STEWART. Introduction to Matrix Computations

CHIN-LIANG CHANG AND RICHARD CHAR-TUNG LEE. Symbolic Logic and Mechanical Theorem Proving

C. C. GOTLIEB AND A. BORODIN. Social Issues in Computing

ERWIN ENGELER. Introduction to the Theory of Computation

F. W. J. OLVER. Asymptotics and Special Functions

DIONYSIOS C. TSICHRITZIS AND PHILIP A. BERNSTEIN. Operating Systems

ROBERT R. KORFHAGE. Discrete Computational Structures

PHILIP J. DAVIS AND PHILIP RABINOWITZ. Methods of Numerical Integration

A. T. BERZTISS. Data Structures: Theory and Practice, Second Edition

N. CHRISTOPHIDES. Graph Theory: An Algorithmic Approach

ALBERT NIJENHUIS AND HERBERT S. WILF. Combinatorial Algorithms

AZRIEL ROSENFELD AND AVINASH C. KAK. Digital Picture Processing

SAKTI P. GHOSH. Data Base Organization for Data Management

DIONYSIOS C. TSICHRITZIS AND FREDERICK H. LOCHOVSKY. Data Base Management Systems

WILLIAM F. AMES. Numerical Methods for Partial Differential Equations, Second Edition

ARNOLD O. ALLEN. Probability, Statistics, and Queueing Theory: With Computer Science Applications

ALBERT NIJENHUIS AND HERBERT S. WILF. Combinatorial Algorithms. Second edition.

JAMES S. VANDERGRAFT. Introduction to Numerical Computations

AZRIEL ROSENFELD. Picture Languages, Formal Models for Picture Recognition

ISAAC FRIED. Numerical Solution of Differential Equations

In preparation

This is a volume in
COMPUTER SCIENCE AND APPLIED MATHEMATICS
A Series of Monographs and Textbooks

Editor: WERNER RHEINBOLDT

A complete list of titles in this series appears at the end of this volume.

PROBABILITY, STATISTICS, AND QUEUEING THEORY

With Computer Science Applications

ARNOLD O. ALLEN

IBM Systems Science Institute
Los Angeles, California

ACADEMIC PRESS

New York   San Francisco   London   1978

A Subsidiary of Harcourt Brace Jovanovich, Publishers

ACADEMIC PRESS, INC.
111 Fifth Avenue, New York, New York 10003

United Kingdom Edition published by
ACADEMIC PRESS, INC. (LONDON) LTD.
24/28 Oval Road, London NW1

Library of Congress Cataloging in Publication Data

Allen, Arnold O
    Probability, statistics, and queueing theory.

    (Computer science and applied mathematics series)
    Includes bibliographical references.
    1. Probabilities. 2. Queueing theory. 3. Mathe-
matical statistics. 4. Mathematics--Data processing.
    I. Title.
QA273.A46   519.2   77-80778
ISBN 0-12-051050-2

PRINTED IN THE UNITED STATES OF AMERICA
79 80 81 82   9 8 7 6 5 4 3 2

For BET and JOHN

CONTENTS

Chapter Three   Probability Distributions

Chapter Four   Stochastic Processes

PART TWO   QUEUEING THEORY

Chapter Five   Queueing Theory

Chapter Six   Queueing Theory Models of Computer Systems

PART THREE   STATISTICAL INFERENCE

Chapter Seven   Estimation

Chapter Eight   Hypothesis Testing

*Faith is belief without evidence in what is told by one who
speaks without knowledge of things without parallel.*
Ambrose Bierce

PREFACE

The genesis of this book is my experience in teaching the use of statistics
and queueing theory for the design and analysis of data communication
systems at the Los Angeles IBM Systems Science Institute. After 18 hours
of instruction, spread over a three-week period, with emphasis on how to
apply the theory to computer science problems, students with good mathematical training were able to use statistics and queueing theory effectively
in a case study for a data communication system. Even students whose
mathematical education consisted of high school algebra were able to appreciate how these disciplines were applied. The results were due to the fact
that the applications of the theory were demonstrated by straightforward
examples that could easily be understood and which illustrated important
concepts.

This book contains a great deal more material than could be presented
to students at the IBM Systems Science Institute. The book is designed as a
junior–senior level textbook on applied probability and statistics with computer science applications. It may also be used as a self-study book for the
practicing computer science (data processing) professional. The assumed

mathematical level of the reader is the traditional one year of analytical geometry and calculus. However, readers with only a college algebra background should be able to follow much of the development and most of the practical examples; such readers should skip over most of the proofs.

I have attempted to state each theorem carefully and explicitly so the student will know exactly when the theorem applies. I have omitted many of the proofs. With a few exceptions, I have given the proof of a theorem only when the following conditions apply: (a) the proof is straightforward, (b) reading through the proof will improve the student's understanding, and (c) the proof is not long.

The emphasis in this book is on how the theorems and theory can be used to solve practical computer science problems. However, the book and a course based on the book should be useful for students who are interested not in computer science itself, but in using probability, statistics, and queueing theory to solve problems in other fields such as engineering, physics, operations research, and management science.

A great deal of computation is needed for many of the examples in this book because of the nature of the subject matter. In fact the use of a computer is almost mandatory for the study of some of the queueing theory models. I believe Kenneth Iverson's APL is the ideal choice of a programming language for making probability, statistics, or queueing theory calculations. Short APL programs are presented in Appendix B to aid the student in making the required calculations. In writing these APL programs I have attempted to write as directly as possible from the equations given in the text, avoiding "one liners"; I have not sought efficiency at the expense of clarity. Every APL program referred to in the text can be found in Appendix B.

The excellent series of books by Donald E. Knuth [1–3] has influenced the writing of this book. I have adopted Knuth's technique of presenting complex procedures in an algorithmic way, that is, as a step by step process. His practice of rewarding the first finder of any error with $1 for the first edition and $2 for the second will also be adopted. I have also followed his system of rating the exercises to encourage students to do at least the simpler ones. I believe the exercises are a valuable learning aid.

Following Knuth, each exercise is given a rating number from 00 to 40. The rating numbers can be interpreted as follows: 00—a very easy problem that can be answered at a glance if the text has been read and understood; 10—a simple exercise which can be done in a minute or so; 20—an exercise of moderate difficulty requiring 18 or 20 minutes of work to complete; 30—a problem of some difficulty requiring two or more hours of work; 40—a lengthy, difficult problem suitable for a term project. (All entries with numbers higher than 30 are "virtual.")

We precede the rating number by HM for "higher mathematics" if the problem is of some mathematical sophistication requiring an understanding of calculus such as the evaluation of proper or improper integrals or summing an infinite series. The prefix C is used if the problem requires extensive computation which would be laborious without computer aid such as an APL terminal or some such facility. T is used to indicate an exercise whose solution is basically tedious, even though the result may be important or exciting; that is, the required procedure is too complex to program for computer solution without more frustration than carrying it out manually.

The reader is assumed to have a basic knowledge of computer hardware and software. Thus he or she should be familiar with such concepts as the central processing unit (CPU), main storage, channels, registers, direct access storage devices such as disks and drums, and programming languages such as FORTRAN, COBOL, PL/I, and APL. The reader should have coded and tested several programs, and generally be familiar with the methods whereby computers are used to solve problems.

I have attempted to use the same principles in planning and writing this book that are used in designing and writing good computer programs, namely, the rules of structured programming as advocated by Edsger W. Dijkstra [4] and others. That is, in planning this book, I first decided what my overall objectives were. I then decided what selections from probability and statistics were necessary to achieve these goals. For each of these selections, in turn, I decided what subobjectives and what examples were needed to achieve the objectives assigned to that part of the book. I thus proceeded in a top-down fashion to a detailed outline of the book. Modifications to the original plan were found to be necessary, of course. However, before each modification was made, its impact upon the whole book was evaluated; necessary changes to the outline and to already written sections were made. I hope the final product has the hierarchical structure that was intended.

Cited References

1. D. E. Knuth, *The Art of Computer Programming, Vol. 1, Fundamental Algorithms*, 2nd ed. Addison Wesley, Reading, Massachusetts, 1973.
2. D. E. Knuth, *The Art of Computer Programming, Vol. 2, Seminumerical Algorithms*. Addison Wesley, Reading, Massachusetts, 1969.
3. D. E. Knuth, *The Art of Computer Programming, Vol. 3, Sorting and Searching*. Addison Wesley, Reading, Massachusetts, 1973.
4. E. W. Dijkstra, Notes on Structured Programming, in *Structured Programming* by O. J. Dahl, E. W. Dijkstra, and C. A. R. Hoare. Academic Press, New York, 1972.

ACKNOWLEDGMENTS

Much of the material in this book has been class tested at the Los Angeles IBM Systems Science Institute, first in a ten-day class "Performance Analysis of Communication Systems" and then in its successor "Performance Evaluation and Capacity Planning." I am grateful for the constructive suggestions of the students; some of their humorous comments have been included in the book as "Student Sayings."

It is my pleasure to acknowledge the help of several individuals. Professor David Cantor of UCLA and Gerald K. McAuliffe of IBM provided helpful advice. My colleague John Cunneen at the IBM Systems Science Institute provided encouragement and constructive suggestions. Mr. Lucian Bifano of IBM Los Angeles provided access to an IBM 5100 computer when I needed it most. Mr. John Hesse, also of IBM Los Angeles, provided a great deal of help in using the IBM 5100 and some useful programs for editing and listing the APL programs in Appendix C. Ms. Vi Ma of the IBM Research Library in San Jose, California, was most helpful in obtaining copies of important references. My manager Mr. J. Perry Free was most supportive in allowing me vacation time to finish the book. My wife Betty and my son John were valuable proof readers. Finally, I want to thank Elaine Barth for the outstanding way she converted my inscrutable scrawling into finished typing.

I hope the completed work shows the joy I felt in writing it. (Only the wisdom of the publisher kept me from calling the book "The Joy of Statistics.") I would appreciate hearing from readers. My address is: Dr. Arnold O. Allen, IBM Systems Science Institute, 3550 Wilshire Boulevard, Los Angeles, California 90010.

PROBABILITY, STATISTICS, AND QUEUEING THEORY

With Computer Science Applications

The theory of probabilities is at bottom nothing but common sense reduced to calculus.   Pierre Simon de Laplace

Chapter One

INTRODUCTION

This chapter is a preview of what the book is all about. As the title suggests, it is concerned with the application of probability, statistics, and queueing theory to computer science problems. It was written for the computer science (data processing) specialist or for one preparing for a career in this field. However, it should be of interest to many who are not computer science oriented, but who merely use a computer occasionally in their daily activities.

The book is divided into three parts: Probability, Queueing Theory, and Statistical Inference.

There are three chapters in Part I. In the first of these, Chapter 2, we take up the basic concept of probability and how to deal with it. In most areas of computer science we deal not with deterministic phenomena, but rather with probabilistic phenomena. The time it takes to write and check out a computer program, the time it takes to run a program (measured from the time it is submitted to a batch system or invoked via an on-line system), the time it takes to retrieve information from a storage device, as well as the number of jobs awaiting execution on a computer system, are all examples of probabilistic or random variables. This means that we cannot predict, in advance, what these values are. However, by using the concepts of probability, probability distributions, and random variables, we can make probability estimates (give the fraction of the time) that the values will fall into certain ranges, exceed certain limits, etc. These subjects are all covered in Chapter 2. We also discuss parameters of random variables (such as the response time, X, at an on-line terminal); these parameters include the mean or average value, the standard deviation (a measure of the deviation of values from the mean), as well as higher moments. We show how to use these parameters to make probability calculations. In the final part of Chapter 2 we discuss some powerful tools to use in dealing with random variables; these include conditional expectation, transform methods, and inequalities. Transform methods are important in studying random variables. We define and illustrate the use of the moment generating function, the generating function or z-transform, and the Laplace–Stieltjes transform. However, we do not make extensive use of transforms in this book, for to do so would take us too far from our primary goals. We do not want to "raise the Laplacian curtain" in the words of the eminent British statistician David Kendall.

In Chapter 3 we study the probability distributions most commonly used in applied probability, particularly for computer science applications. We give examples of the use of all of these except those used only in statistical inference, the subject of Part III of the book. A summary of the properties of the random variables discussed in Chapter 3 is given in Tables 1 and 2 of Appendix A.

In Chapter 4 the important concept of a stochastic process is defined, discussed, and illustrated with a number of examples. The chapter was written primarily as a support chapter for Part II, queueing theory. We examine the Poisson process because it is extremely important for queueing theory; similarly for the birth-and-death process. We finish the chapter by discussing Markov processes and chains—subjects that are important not only for queueing theory but also for other widely used models in computer science.

Part II of the book is the area that is likely to be most foreign to the reader, the discipline of queueing theory. Queueing theory is an applied branch of probability theory, itself, but some expressions and symbols are used differently in queueing theory than they are in other areas of probability and in statistics.

Figure 1.1, which also appears as Fig. 5.1.1 in Chapter 5, shows the elements of a queueing system. There is a "customer" population; a customer may be an inquiry to be processed by an on-line inquiry system, a job to be processed by a batch computer system, a message to be transmitted over a communication line, a request for service by a computer input/output channel, etc. Customers arrive in accordance with an "arrival process" of

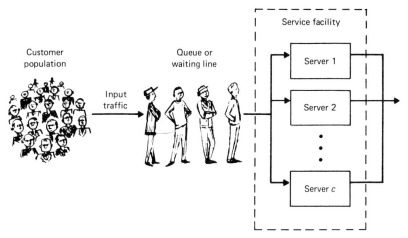

Fig. 1.1   Elements of a queueing system.

some type (a "Poisson arrival process" is one of the most common). Customers are provided service by a service facility which has one or more servers, each capable of providing service to a customer. Thus a server could be a program which processes an inquiry, a batch computer system, a communication line, a computer channel, a central processing unit, etc. If all the servers in the service facility are busy when a customer arrives at the queueing system, that customer must queue for service; that is, the customer must join a queue (waiting line) until a server is available. In Chapter 5 we study the standard (one might say "canonical") queueing systems and see how they can be applied in studying computer science problems. We have gathered most of the queueing theory formulas from Chapters 5 and 6 and put them in Appendix C. You will find this appendix to be very useful as a reference section once you have mastered Chapters 5 and 6. The APL programs to calculate the values given by the formulas in Appendix C have been collected in Appendix B. In Chapter 6 we discuss more sophisticated queueing system models that have been developed to study computer systems, particularly on-line computer systems where customers are typically represented by remote computer terminals. As in Chapter 5, a number of examples of "real world" use of the models are presented.

The subject matter of Part III, statistical inference, is rather standard statistical fare but we have attempted to give it a computer science orientation. Statistical inference could perhaps be defined as "the science of drawing conclusions about a population on the basis of a random sample from that population." For example, we may want to estimate the mean arrival rate of inquiries to an on-line inquiry system. We may also want to deduce what type of arrival process is involved. We can approach these tasks on the basis of a sample of the arrival times of inquiries during n randomly selected time periods. The first task is one of estimation. We want to estimate the mean arrival rate on the basis of the observed arrival rate during n time intervals. This is the subject of Chapter 7. In this chapter we learn not only how to make estimates but also how to make probability judgements concerning the accuracy of the estimates.

One of the important topics of Chapter 8 (on hypothesis testing) is goodness-of-fit tests, the kind of test needed in the second task of the previous paragraph. In that particular case we might want to test the hypothesis that the arrival process has a Poisson pattern (recall we said earlier that this is one of the most popular kinds of arrival patterns because of its desirable mathematical properties). We discuss the two most widely used goodness-of-fit tests, the chi-square and the Kolmogorov–Smirnov. We give several examples of the application of these tests. Chapter 8 also contains a number of other hypothesis tests of great utility. These tests are summarized in Table 8.5.1 at the end of the chapter. You will probably find this to be a useful reference, too, when you have mastered Chapter 8.

This completes the summary of the book. We hope you will find the study of this book entertaining as well as educational. We have tried to avoid being too solemn. For example, we chose whimsical names for mythical companies in our examples. We have attempted to make the examples as practical as possible within the constraints of a reasonably short description. We welcome your comments, suggestions, and observations.

PART ONE

PROBABILITY

Lest men suspect your tale untrue, Keep probability in view.
John Gay

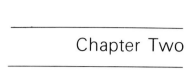

Chapter Two

PROBABILITY AND RANDOM VARIABLES

INTRODUCTION

One of the most noticeable aspects of many computer science related phenomena is the lack of certainty. When a job is submitted to a batch oriented computer system, the exact time the job will be completed is uncertain. The number of jobs that will be submitted tomorrow is probably not known, either. Similarly, the exact response time for an on-line inquiry system cannot be predicted. If the terminals attached to a communication line are polled until one is found which is ready to transmit, the required number of polls is not known in advance. Even the time it takes to retrieve a record from a disk storage device cannot be predicted exactly. Each of these phenomena has an underlying probabilistic mechanism. In order to work constructively with such observed, uncertain processes we need to put them into a mathematical framework. That is the purpose of this chapter.

2.1 SAMPLE SPACES AND EVENTS

To apply probability theory to the process under study we view it as a *random experiment*, that is, as an experiment whose outcome is not known in advance but for which the set of all possible individual outcomes is known. For example, if the ten terminals on a communication line are polled in a specified order until either (a) all are polled or (b) one is found with a message ready for transmission, then the number of polls taken describes the outcome of the polling experiment and can only be an integer between one and ten. The *sample space* of a random experiment is the set of all possible simple outcomes of the experiment. These individual outcomes are also called *sample points* or *elementary events*. A sample space is a set and thus is defined by specifying what objects are in it. One way to do this, if the set is small, is to list them all, such as, $\Omega = \{1, 2, 3\}$. When the set is large or infinite its elements are often specified by writing $\Omega = \{x : P(x)\}$, where $P(x)$ is a condition that x must satisfy to be an element of Ω. Thus $\Omega = \{x : P(x)\}$ means, "Ω is the set of all x such that $P(x)$ is true." The set of all nonnegative integers could be specified by writing $\{n : n$ is an integer and $n \geq 0\}$. Some examples of sample spaces follow.

Example 2.1.1   If the random experiment consists of tossing a die, then $\Omega = \{1, 2, 3, 4, 5, 6\}$ where the sample point n indicates that the die came to rest with n spots showing on the uppermost side.

Example 2.1.2   If the random experiment consists of tossing two fair dice,[1] then one possible sample space consists of the 36 sample points (1, 1),(1, 2), (1, 3), ..., (1, 6), ..., (6, 6), where the outcome (i, j) means that the first die showed i spots uppermost and the second one showed j.

Example 2.1.3   If the random experiment consists of polling the terminals on a communication line in sequence until either (a) one of the seven terminals on the line is found to be ready to transmit or (b) all the terminals have been polled, then the sample space could be represented by the sample points 1, 2, 3, 4, 5, 6, 7, 8, where an 8 signifies that no terminal had a message ready, while an integer n between 1 and 7 means that the nth terminal polled was the first in sequence found in the ready state.

Example 2.1.4   If the random experiment consists of tossing a fair coin[1] again and again until the first head appears, then the sample space can be represented by the sequence H, TH, TTH, TTTH, ..., where the 1st sample point corresponds to a head on the first toss, the second sample point to a head on the second toss, etc.

Example 2.1.5   The random experiment consists of measuring the elapsed time from the instant the last character of an inquiry is typed on an

[1]By a fair coin or a fair die we mean, of course, one for which each outcome is equally likely (whatever that means).

on-line terminal until the last character of the response from the computer system has been received and displayed at the terminal. This time is often called the "response time" although there are other useful definitions of response time. If it takes a minimum of 10 seconds for an inquiry to be transmitted to the central computer system, processed, a reply prepared, and the reply returned and displayed at the terminal, then $\Omega = \{\text{real } t : t \geq 10\}$.

Thus sample spaces can be *finite* as in Examples 2.1.1–2.1.3 or *infinite* as in Examples 2.1.4 and 2.1.5. Sample spaces are also classified as *discrete* if the number of sample points is finite or countably infinite (can be put into one-to-one correspondence with the positive integers). The sample space of Example 2.1.4 is countably infinite since each sample point can be associated uniquely with the positive integer giving the number of tosses represented by the sample point. For example, the sample point TTTH represents four tosses. A sample space is *continuous* if its sample points consist of all the numbers on some finite or infinite interval of the real line. Thus the sample space of Example 2.1.5 is continuous.

We use the notation (a, b) for the open interval $\{x : a < x < b\}$, $[a, b]$ for the closed interval $\{x : a \leq x \leq b\}$, $(a, b]$ for the half-open interval $\{x : a < x \leq b\}$, and $[a, b)$ for the half-open interval $\{x : a \leq x < b\}$, where all intervals are subsets of the real line.

An *event* is a subset of a sample space satisfying certain axioms (Axiom Set 2.2.1 described in Section 2.2). An event A is said to *occur* if the random experiment is performed and the observed outcome is an element of the set A.

Example 2.1.6 In Example 2.1.1, if $A = \{2, 3, 5\}$, then A is the event of rolling a prime number while the event $B = \{1, 3, 5\}$ is the event of rolling an odd number.

Example 2.1.7 In Example 2.1.2, if $A = \{(1, 6), (2, 5), (3, 4), (4, 3), (5, 2), (6, 1)\}$, then A is the event of rolling a seven. The event $B = \{(5, 6), (6, 5)\}$ corresponds to rolling an eleven.

Example 2.1.8 In Example 2.1.3, if $A = \{1, 2, 3, 4, 5\}$, then A is the event of requiring 5 polls or less, while $B = \{6, 7, 8\}$ is the event of taking more than 5 polls.

Example 2.1.9 In Example 2.1.4, $A = \{\text{TTH, TTTH}\}$ is the event that three or four tosses are required, while $B = \{\text{H, TH, TTH}\}$ is the event that not more than 3 tosses are needed.

Example 2.1.10 In Example 2.1.5 $A = \{t : 20 \leq t \leq 30\}$ is the event that the response time is between 20 and 30 seconds.

Since a sample space Ω is a set while an event A is a subset of Ω, we can form new events by using the usual operations of set theory. For some of these operations a somewhat different terminology is used in probability theory than in set theory—a terminology more indicative of the intuitive meaning of the operations in probability theory. Some of the set operations and corresponding probability statements are shown in Table 2.1.1. We will use probability statements and set theory statements, interchangeably, in this book.

TABLE 2.1.1

Some Set Operations and Corresponding Probability Statements

Set operation	Probability statement
$A \cup B$	At least one of A or B occurs
$A \cap B$	Both A and B occur
$\bar{A}$	A does not occur
$\varnothing$	The impossible event
$A \cap B = \varnothing$	A and B are mutually exclusive
$A \cap \bar{B}$	A occurs and B does not occur
$A \subset B$	A implies B

We indicate that the outcome ω is a sample point of event A by writing $\omega \in A$. We write $A = \varnothing$ to indicate that the event A contains no sample points. Here $\varnothing$ is the empty set, called the *impossible event* in probability theory. $\varnothing$ is considered to be an event just as Ω itself is. The reader should note that $\varnothing$ is *not* the Greek letter phi but rather a Danish letter pronounced " ugh," the sound one makes upon receiving an unexpected blow to the solar plexus. It has been rumored that the prevalence of $\varnothing$ in Danish words has been a leading cause of emphysema. Professor Richard Arens of UCLA has recommended the symbol $\bar{\varnothing}$ for the empty set, pronounced " uh uh," of course.

To every event A there corresponds the event $\bar{A}$, called the *complement of A*, consisting of all sample points of Ω which are not in A. Thus $\bar{A}$ is defined by the condition " A does not occur." As particular cases of the complement, $\bar{\Omega} = \varnothing$ and $\bar{\varnothing} = \Omega$. The concept of complement is illustrated by the *Venn diagram* of Fig. 2.1.1. In each Venn diagram that follows, the large rectangle will represent the sample space Ω, while simple geometric figures will be used to represent other events. A point thus represents an elementary event or outcome and the inside of a figure represents a collection of them (an event).

With each two events A and B are associated two new events which correspond to the intuitive ideas " either A or B occurs " and " both A and B

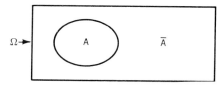

Fig. 2.1.1 An event A and its complement $\bar{A}$.

occur." The first of these events, $A \cup B$ (read: "A or B"), is the ordinary set union consisting of all sample points which are either in A or in B or (possibly) in both A and B. The second event $A \cap B$ (read: "A and B") is the ordinary set intersection, that is, all sample points which belong both to A and to B. If A and B have no sample points in common, that is, if $A \cap B = \varnothing$, we say that A and B are *mutually exclusive events*. Clearly, if A and B are mutually exclusive, then the occurrence of one of them precludes the occurrence of the other. These concepts are illustrated in the Venn diagrams of Figs. 2.1.2–2.1.4. In Fig. 2.1.2 $A \cup B$ is represented by the shaded area. $A \cap B$ is shaded in Fig. 2.1.3. The events A and B of Fig. 2.1.4 are mutually exclusive.

The concepts of union and intersection can be extended in a similar way to any finite collection of events such as $A \cup B \cup C$ or $A \cap B \cap C \cap D$. For a countable collection of events $A_1, A_2, A_3, \ldots$, the union $\bigcup_{n=1}^{\infty} A_n$ of

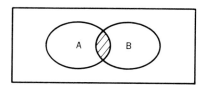

Fig. 2.1.2 The event $A \cup B$.

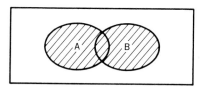

Fig. 2.1.3 The event $A \cap B$.

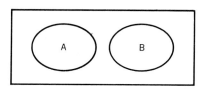

Fig. 2.1.4 Mutually exclusive events.

the events is defined to be the event consisting of all sample points which belong to at least one of the sets A_n, $n = 1, 2, \ldots$, while the intersection $\bigcap_{n=1}^{\infty} A_n$ of the events is the event consisting of all sample points which belong to each of the events A_n, $n = 1, 2, \ldots$.

If every sample point of event A is also a sample point of event B, so that A is a subset of B, we write $A \subset B$ and say that "event A implies event B." In this case, $B - A$ is defined to be the set of all sample points in B which are not in A. Thus, $\bar{A} = \Omega - A$ for every event A.

Example 2.1.11   Consider the sample space of Example 2.1.3. Let A be the event that at least five polls are required and B the event that not more than four polls are required $(A = \{5, 6, 7, 8\}$, $B = \{1, 2, 3, 4\})$. Then $A \cup B = \Omega$ while $A \cap B = \emptyset$ so A and B are mutually exclusive. They are also complements $(\bar{A} = B$ and $\bar{B} = A)$, although mutually exclusive events are not necessarily complementary.

2.2 PROBABILITY MEASURES

In the early or classical days of probability there was much concern with games of chance. Early workers in the field such as Cardano and Pascal were occupied with questions about the likelihood of winning in various games and in how to divide the purse if the game was discontinued before completion (called the "division problem" or the "problem of points"). The sample spaces were constructed in such a way that each elementary event or outcome was equally likely. In Example 2.1.2, if the two dice are perfectly formed, each of the 36 elementary outcomes is equally likely to occur on any given trial of the experiment so a probability of 1/36 is assigned to each sample point. Thus for finite sample spaces with n equiprobable sample points each event A was assigned the *probability* $P[A] = n_A/n$ where n_A is the number of sample points in A.

Example 2.2.1   Consider the two-die experiment of Example 2.1.2. We can draw the graph of Fig. 2.2.1 to help in calculating probabilities. Thus, if A is the event of rolling an 11, we can see from Fig. 2.2.1 that A consists of the sample points $(5, 6)$ and $(6, 5)$ so $P[A] = n_A/36 = 2/36 = 1/18$. Likewise, if B is the event of rolling a 7 or an 11, $P[B] = n_B/36 = (6 + 2)/36 = 2/9$. Other probabilities for this experiment can be calculated in a similar manner.

The classical definition of probability worked well for the kind of problem for which it was designed. However, the classical theory would not suffice to assign probabilities to the events of Example 2.1.3 because the elementary events are not equiprobable. Likewise, it would not help in

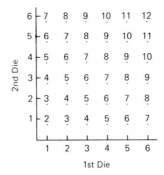

Fig. 2.2.1 Two-die experiment–sum rolled.

Examples 2.1.4 or 2.1.5, because these sample spaces are infinite. The classical definition has been generalized into a set of axioms that every probability measure should satisfy in assigning probabilities to events. However, some additional conditions must be imposed on the collection of events of a sample space before we can assign probabilities to them.

The family $\mathscr{F}$ of events of a sample space Ω is assumed to satisfy the following axioms (and thus form a σ-algebra):

Axiom Set 2.2.1 *(Axioms of a σ-Algebra)*

A1 $\varnothing$ and Ω are elements of $\mathscr{F}$.[1]
A2 If $A \in \mathscr{F}$, then $\bar{A} \in \mathscr{F}$.
A3 If $A_1, A_2, A_3, \ldots$ are elements of $\mathscr{F}$, so is $\bigcup_{n=1}^{\infty} A_n$.

It can be shown that these axioms also imply that, if each of the events $A_1, A_2, \ldots$ belongs to $\mathscr{F}$, then $\bigcap_{n=1}^{\infty} A_n$ is an element of $\mathscr{F}$, and similarly for finite intersections (see Exercise 8). Also, $A_1 \cup \cdots \cup A_n$ is in $\mathscr{F}$ if each A_i is. Likewise, if A, B are in $\mathscr{F}$, then $B - A = B \cap \bar{A}$ is in $\mathscr{F}$.

A probability measure $P[\cdot]$, regarded as a function on the family $\mathscr{F}$ of events of a sample space Ω, is assumed to satisfy the following axioms.

Axiom Set 2.2.2 *(Axioms of a Probability Measure)*

P1 $0 \leq P[A]$ for every event A.
P2 $P[\Omega] = 1$.
P3 $P[A \cup B] = P[A] + P[B]$ if the events A and B are mutually exclusive.
P4 If the events $A_1, A_2, A_3, \ldots$ are mutually exclusive (that is, $A_i \cap A_j = \varnothing$ if $i \neq j$), then

$$P\left[\bigcup_{n=1}^{\infty} A_n\right] = \sum_{n=1}^{\infty} P[A_n].$$

[1]Here we use the symbol $\in$ in the usual set theoretic sense; that is, it means "is an element of" or "belongs to."

It is immediate from **P3** by mathematical induction that for any finite collection $A_1, A_2, \ldots, A_n$ of mutually exclusive events

$$P[A_1 \cup A_2 \cup \cdots \cup A_n] = P[A_1] + \cdots + P[A_n].$$

Although there is not general agreement among statisticians and philosophers as to exactly what probability is, there *is* general agreement that a probability measure $P[\cdot]$ should satisfy the above axioms. The axioms are satisfied for the classical theory defined above. These axioms lead immediately to some consequences which are useful in computing probabilities. Some of them are listed in the following theorem.

THEOREM 2.2.1   Let $P[\cdot]$ be a probability measure defined on the family $\mathscr{F}$ of events of a sample space Ω. Then

(a) $P[\varnothing] = 0$;
(b) $P[A] = 1 - P[\bar{A}]$ for every event A;
(c) $P[A \cup B] = P[A] + P[B] - P[A \cap B]$ for any events A, B;
(d) $A \subset B$ implies $P[A] \leq P[B]$ for any events A, B.

Proof   (a) $A \cup \varnothing = A$. A and $\varnothing$ are   mutually   exclusive $(A \cap \varnothing = \varnothing)$ so, by Axiom **P3**, $P[A] = P[A \cup \varnothing] = P[A] + P[\varnothing]$. Hence, $P[\varnothing] = 0$.

(b) A and $\bar{A}$ are mutually exclusive by the definition of $\bar{A}$. Hence, by Axioms **P2** and **P3**,

$$1 = P[\Omega] = P[A \cup \bar{A}] = P[A] + P[\bar{A}].$$

Hence,

$$P[A] = 1 - P[\bar{A}].$$

(c) $A \cup B$ is the union of the mutually exclusive events $A \cap B, \bar{A} \cap B$, and $A \cap \bar{B}$, that is,

$$A \cup B = (A \cap B) \cup (\bar{A} \cap B) \cup (A \cap \bar{B}).$$

Therefore,

$$P[A \cup B] = P[A \cap B] + P[\bar{A} \cap B] + P[A \cap \bar{B}]. \qquad (2.2.1)$$

In addition, $A \cap B$ and $A \cap \bar{B}$ are disjoint events whose union is A. Hence,

$$P[A] = P[A \cap B] + P[A \cap \bar{B}]. \qquad (2.2.2)$$

Similarly,

$$P[B] = P[A \cap B] + P[\bar{A} \cap B]. \qquad (2.2.3)$$

Adding (2.2.2) to (2.2.3) yields

$$P[A] + P[B] = 2P[A \cap B] + P[A \cap \bar{B}] + P[\bar{A} \cap B]. \qquad (2.2.4)$$

Substituting (2.2.1) into (2.2.4) yields

$$P[A] + P[B] = P[A \cup B] + P[A \cap B] \qquad (2.2.5)$$

or

$$P[A \cup B] = P[A] + P[B] - P[A \cap B]. \qquad (2.2.6)$$

(d) Since $A \subset B$, B is the union of the disjoint events A and $B - A$. Thus,

$$P[B] = P[A] + P[B - A]. \qquad (2.2.7)$$

Since $P[B - A] \geq 0$, this means that $P[A] \leq P[B]$.

This completes the proof of Theorem 2.2.1.

Example 2.2.2 A collection of 100 computer programs was examined for various types of errors (bugs). It was found that 20 of them had syntax errors, 10 had input/output (I/O) errors that were not syntactical, five had other types of errors, six programs had both syntax errors and I/O errors, three had both syntax errors and other errors, two had both I/O errors and other errors, while one had all three types of error. A program is selected at random from this collection, that is, selected in such a way that each program was equally likely to be chosen. Let S be the event that the selected program has errors in syntax, I be the event it has I/O errors, and O the event that it has other errors. Since the program was randomly selected, Table 2.2.1 gives the probabilities associated with some of the events.

The probability that the program will have a syntax error or an I/O error or both is

$$P[S \cup I] = P[S] + P[I] - P[S \cap I]$$

$$= \frac{20}{100} + \frac{10}{100} - \frac{6}{100} = \frac{24}{100} = \frac{6}{25}, \qquad \text{by Theorem 2.2.1c.}$$

The probability that it will have some type of error is given by

$$P[S \cup I \cup O] = P[S] + P[I] + P[O] - P[S \cap I]$$

$$- P[S \cap O] - P[I \cap O] + P[S \cap I \cap O]$$

$$= \frac{20}{100} + \frac{10}{100} + \frac{5}{100} - \frac{6}{100} - \frac{3}{100} - \frac{2}{100} + \frac{1}{100}$$

$$= \frac{25}{100} = \frac{1}{4}.$$

In making the last calculation we used the formula

$$P[A \cup B \cup C] = P[A] + P[B] + P[C] - P[A \cap B]$$

$$- P[A \cap C] - P[B \cap C] + P[A \cap B \cap C].$$

TABLE 2.2.1
Probabilities for Example 2.2.2

Event	S	I	O	$S \cap I$	$S \cap O$	$I \cap O$	$S \cap I \cap O$
Probability	20/100	10/100	5/100	6/100	3/100	2/100	1/100

This formula follows from Theorem 2.2.1c and the *distributive law* $(A \cup B) \cap C = (A \cap C) \cup (B \cap C)$ (see Exercises 3 and 4).

For probability calculations involving finite sample spaces we need some results from combinatorial analysis.

2.3 COMBINATORIAL ANALYSIS

Combinatorial analysis is the science of counting—the number of elements in prescribed sets, the number of ways a particular selection can be made, etc.

One activity that is frequently employed in probability and statistics is drawing a few elements or items (a sample) from a collection or source (a population). Such a selection can be made with or without replacement. For example, if two cards are to be drawn from a 52-card bridge deck without replacement, one card is removed and then another without putting the first card back. Drawing with replacement requires that a card be drawn, recorded, and returned to the deck before the second card is drawn, so that the two cards drawn may be identical. We assume in all drawing, with or without replacement, that the collection Ω from which a drawing is made consists of n distinct objects $O_1, O_2, \ldots, O_n$. A *permutation of order k* is an ordered selection of k elements from Ω, where $0 \le k \le n$. A *combination of order k* is an unordered selection of k elements from Ω. The selections for both permutations and combinations can be made with or without replacement but are assumed to be made without replacement, unless otherwise stated.

Example 2.3.1 Suppose $\Omega = \{x, y, z\}$ and we draw two letters from $\Omega (k = 2)$. There are nine permutations of order 2 with replacement:

$$xx, \quad xy, \quad xz, \quad yx, \quad yy, \quad yz, \quad zx, \quad zy, \quad zz.$$

There are six permutations made without replacement: xy, xz, yx, yz, zx, zy. There are six combinations made with replacement: xx, xy, xz, yy, yz, zz (the permutations xy and yx, for example, are not distinguished, because combinations are unordered). There are three combinations without repetitions: xy, xz, yz.

One of the fundamental tools in combinatorics is the *multiplication principle* which we state formally as a theorem.

THEOREM 2.3.1 (*Multiplication Principle*) If a task A can be done in m different ways and, after it is completed in any of these ways, task B can be completed in n different ways, then A and B, together, can be performed in $m \times n$ ways.

COROLLARY Suppose k tasks $A_1, A_2, \ldots, A_k$ are to be done and that A_1 can be completed in n_1 ways, A_2 in n_2 ways after A_1 is completed, A_3 in n_3 ways after A_1 and A_2 are completed, ..., A_k in n_k ways after $A_1, A_2, \ldots,$ A_{k-1} are completed. Then the total task, $A_1, A_2, \ldots, A_k$ in succession, can be performed in $n_1 \times n_2 \times \cdots \times n_k$ ways.

The corollary follows immediately from the theorem by mathematical induction.

Hereafter we will refer to the "multiplication principle," even when, strictly speaking, we use the corollary to it.

We define $n!$ (pronounced "n factorial") for each nonnegative integer n by $0! = 1$, $n! = n(n-1)!$ for $n > 0$. Thus $1! = 1, 2! = 2, 3! = 6, 4! = 24$, etc., and we can write

$$n! = n \times (n-1) \times (n-2) \times \cdots \times 2 \times 1.$$

THEOREM 2.3.2 The number of permutations of n elements, taken k at a time, without repetitions, is

$$P(n, k) = \frac{n!}{(n-k)!} = n(n-1)(n-2) \cdots (n-k+1).$$

If repetitions are allowed the number of permutations is n^k.

Proof If repetitions are not allowed then the first element in the permutation can be selected in n different ways from the n elements in Ω. After the first selection is made there are $n-1$ elements left in Ω from which to make the second selection. After the second selection there are $n-2$ elements left in Ω from which to make the third selection, etc. Hence, by the multiplication principle,

$$P(n, k) = n(n-1)(n-2) \cdots (n-k+1) = \frac{n!}{(n-k)!}.$$

If repetitions are allowed there are n choices for each selection so $P(n, k) = n^k$.

This completes the proof.

Both the symbols $C(n, k)$ and $\binom{n}{k}$ are used to designate the number of combinations of k objects selected from a set of n elements.

THEOREM 2.3.3 There are $C(n, k) = \binom{n}{k} = n!/[k! \, (n-k)!]$ combinations of n objects, taken k at a time.

COROLLARY $\binom{n}{k}$ is the coefficient of $x^k y^{n-k}$ in the expansion of $(x + y)^n$, that is,

$$(x + y)^n = \sum_{k=0}^{n} \binom{n}{k} x^k y^{n-k}.$$

(This is why $\binom{n}{k}$ is often called a binomial coefficient.)

Proof If repetitions are not allowed, each combination of k elements forms $k!$ permutations of order k. Hence,

$$(k!)\binom{n}{k} = P(n, k) \quad \text{or} \quad \binom{n}{k} = \frac{P(n, k)}{k!} = \frac{n!}{k! \, (n - k)!}.$$

Proof of Corollary $(x + y)^n$ can be written as

$$(x + y)(x + y) \cdots (x + y) \quad (n \text{ factors}),$$

and the coefficient of $x^k y^{n-k}$ in the expansion is the number of ways we can choose x from k of these factors and y from the remaining $n - k$ factors. This is precisely $\binom{n}{k}$.

Example 2.3.2 Suppose 5 terminals are connected to an on-line computer system by attachment to one communication line. When the line is polled to find a terminal ready to transmit, there may be 0, 1, 2, 3, 4, or 5 terminals in the ready state. One possible sample space to describe the system state is $\Omega = \{(x_1, x_2, x_3, x_4, x_5): \text{each } x_i \text{ is either 0 or 1}\}$, where $x_i = 1$ means "terminal i is ready" and $x_i = 0$ means "terminal i is not ready." The sample point $(0, 1, 1, 0, 0)$ corresponds to terminals 2, 3 ready to transmit but terminals 1, 4, 5 not ready. By the multiplication principle, the number of sample points is $2^5 = 32$, since each x_i of $(x_1, x_2, x_3, x_4, x_5)$ can be selected in 2 ways. However, if we assume that exactly 3 terminals are in the ready state, then

$$\Omega = \{(x_1, x_2, x_3, x_4, x_5): \text{exactly 3 of the } x_i\text{'s are 1 and 2 are 0}\}.$$

In this case the number of sample points of Ω is the number of ways that the three terminals which are ready can be chosen from the five available, that is,

$$\binom{5}{3} = \frac{5!}{3! \, (5 - 3)!} = \frac{5!}{3! \, 2!} = \frac{5 \times 4}{2} = 10.$$

If the terminals are polled sequentially until a ready terminal is found, the number of polls required can be 1, 2, or 3. Let A_1, A_2, A_3 be the events that the required number of polls is 1, 2, 3, respectively. A_1 can occur only if $x_1 = 1$ and the other two 1's occur in the remaining 4 positions. Hence, the number of sample points favorable to A_1, n_1 is calculated as

$$n_1 = \binom{4}{2} = \frac{4!}{2!\,2!} = \frac{4 \times 3}{2} = 6 \quad \text{and} \quad P[A_1] = \frac{n_1}{n} = \frac{6}{10}.$$

A_2 can occur only if $x_1 = 0$, $x_2 = 1$, and the remaining two 1's are distributed in positions 3 through 5. Hence,

$$P[A_2] = \frac{\binom{3}{2}}{10} = \frac{3}{10}.$$

Similarly,

$$P[A_3] = \frac{\binom{2}{2}}{10} = \frac{1}{10}.$$

We have assumed, of course, that each terminal is equally likely to be in the ready condition.

2.4 CONDITIONAL PROBABILITY AND INDEPENDENCE

It is often useful to calculate the probability that an event A occurs when it is known that an event B has occurred, where B has positive probability. The symbol for this probability is $P[A\,|\,B]$ and reads "the conditional probability of A, given B."

Example 2.4.1 If, in Example 2.2.2, it is known that the program which was drawn has an error in syntax, what is the probability that it has an I/O error, also?

Solution Twenty programs have errors in syntax, and six of these also have I/O errors. Hence, the required probability is $6/20 = 3/10$. The knowledge that the selected program had a syntactical error effectively reduced the size of the sample space from 100 to 20.

In general, to calculate the probability that A occurs, given that B has occurred means reevaluating the probability of A in the light of the information that B has occurred. Thus B becomes our new sample space and we are only interested in that part of A which occurs with B, that is, $A \cap B$. Thus we must have the formula

$$P[A\,|\,B] = \frac{P[A \cap B]}{P[B]}, \tag{2.4.1}$$

if $P[B] > 0$. The conditional probability of A given B is not defined if $P[B] = 0$. In (2.4.1), $P[A \cap B]$ was divided by $P[B]$ so that $P[B|B] = 1$, making $P[\cdot|B]$ a probability measure. The event B in (2.4.1) is often called the *conditioning event*.

Formula (2.4.1) can be used to make the probability calculation in Example 2.4.1, with B the event that the program has at least one error in syntax and A the event that the program has at least one I/O error. Then,

$$P[A|B] = \frac{P[A \cap B]}{P[B]} = \frac{6/100}{20/100} = \frac{6}{20} = \frac{3}{10},$$

as before.

Formula (2.4.1) can be rewritten in a form called the *multiplication rule*.

THEOREM 2.4.1   (*Multiplication Rule*)   For events A and B

$$P[A \cap B] = P[A]P[B|A], \tag{2.4.2}$$

if $P[A] \neq 0$, and

$$P[A \cap B] = P[B]P[A|B], \tag{2.4.3}$$

if $P[B] \neq 0$. (If either $P[A] = 0$ or $P[B] = 0$ then $P[A \cap B] = 0$ by Theorem 2.2.1d.)

COROLLARY   (*General Multiplication Rule*)   For events $A_1, A_2, \ldots, A_n$,

$$P[A_1 \cap A_2 \cap \cdots \cap A_n] = P[A_1]P[A_2|A_1]P[A_3|A_1 \cap A_2] \cdots$$
$$\times P[A_n|A_1 \cap \cdots \cap A_{n-1}] \tag{2.4.4}$$

provided all the probabilities on the right are defined. A sufficient condition for this is that $P[A_1 \cap A_2 \cap \cdots \cap A_{n-1}] > 0$ since $P[A_1] \geq P[A_1 \cap A_2] \geq \cdots \geq P[A_1 \cap A_2 \cap \cdots \cap A_{n-1}]$.

Proof   Equations (2.4.2) and (2.4.3) are true by the definition of conditional probability, (2.4.1). The corollary follows by mathematical induction on n as follows. For $n = 2$ the result is the theorem and thus is true. Now suppose the corollary is true for $n = k \geq 2$ and $A_1, A_2, \ldots, A_k, A_{k+1}$ are events. Let $A = A_1 \cap A_2 \cap \cdots \cap A_k$ and $B = A_{k+1}$.

Then, by (2.4.2),

$$P[A_1 \cap A_2 \cap \cdots \cap A_k \cap A_{k+1}] = P[A \cap B] = P[A]P[B|A]$$
$$= P[A_1]P[A_2|A_1]P[A_3|A_1 \cap A_2] \cdots$$
$$\times P[A_k|A_1 \cap \cdots \cap A_{k-1}]$$
$$\times P[A_{k+1}|A_1 \cap \cdots \cap A_k],$$

where the last equality follows from the inductive assumption that

$$P[A_1 \cap A_2 \cap \cdots \cap A_k] = P[A_1]P[A_2|A_1]P[A_3|A_1 \cap A_2] \cdots$$
$$\times P[A_k|A_1 \cap A_2 \cap \cdots \cap A_{k-1}].$$

This completes the proof.

Example 2.4.2 Suppose a survey of 100 computer installations in a certain city shows that 75 of them have at least one brand x computer. If three of these installations are chosen at random, without replacement, what is the probability that each of them has at least one brand x machine?

Solution Let A_1, A_2, A_3 be the event that the first, second, third, selection, respectively, has a brand x computer. The required probability is

$$P[A_1 \cap A_2 \cap A_3] = P[A_1]P[A_2|A_1]P[A_3|A_1 \cap A_2]$$

by the general multiplication rule.
This value is

$$\frac{75}{100} \times \frac{74}{99} \times \frac{73}{98} = 0.418.$$

which is somewhat lower than intuition might lead one to believe.

One of the main uses of conditional probability is to assist in the calculation of unconditional probability by the use of the following theorem.

THEOREM 2.4.2 (*Law of Total Probability*) Let A_1, A_2, ..., A_n be events such that

(a) $A_i \cap A_j = \varnothing$ if $i \neq j$ (mutually exclusive events),
(b) $P[A_i] > 0$, $i = 1, 2, ..., n$,
(c) $A_1 \cup A_2 \cup \cdots \cup A_n = \Omega$.

(A family of events satisfying (a)–(c) is called a *partition of* Ω.)
Then, for any event A,

$$P[A] = P[A_1]P[A|A_1] + P[A_2]P[A|A_2] + \cdots + P[A_n]P[A|A_n]. \qquad (2.4.5)$$

Proof Let $B_i = A \cap A_i$, $i = 1, 2, 3, ..., n$. Then $B_i \cap B_j = \varnothing$ if $i \neq j$ because the events A_1, A_2, ..., A_n are mutually exclusive, and

$$A = B_1 \cup B_2 \cup \cdots \cup B_n \qquad (2.4.6)$$

because each element of A is in exactly one B_j.
Hence,

$$P[A] = P[B_1] + P[B_2] + \cdots + P[B_n]. \qquad (2.4.7)$$

But

$$P[B_i] = P[A \cap A_i] = P[A_i]P[A \mid A_i], \qquad i = 1, 2, \ldots, n. \qquad (2.4.8)$$

Substituting (2.4.8) into (2.4.7) yields (2.4.5) and completes the proof.

Example 2.4.3   Inquiries to an on-line computer system arrive on five communication lines. The percentage of messages received from lines 1, 2, 3, 4, 5, are 20, 30, 10, 15, and 25, respectively. The corresponding probabilities that the length of an inquiry will exceed 100 characters is 0.4, 0.6, 0.2, 0.8, and 0.9. What is the probability that a randomly selected inquiry will be longer than 100 characters?

Solution   Let A be the event that the selected message has more than 100 characters and A_i the event that it was received on line i, $i = 1, 2, 3, 4, 5$. Then, by the law of total probability,

$$P[A] = P[A_1]P[A \mid A_1] + \cdots + P[A_5]P[A \mid A_5]$$

$$= 0.2 \times 0.4 + 0.3 \times 0.6 + 0.1 \times 0.2$$

$$+ 0.15 \times 0.8 + 0.25 \times 0.9 = 0.625. \qquad (2.4.9)$$

Two events A and B are said to be *independent* if $P[A \cap B] = P[A]P[B]$. This does imply the usual meaning of independence; namely, that neither event influences the occurrence of the other. For, if A and B are independent (and both have nonzero probability), then

$$P[A \mid B] = \frac{P[A \cap B]}{P[B]} = \frac{P[A]P[B]}{P[B]} = P[A] \qquad (2.4.10)$$

and

$$P[B \mid A] = \frac{P[A \cap B]}{P[A]} = \frac{P[A]P[B]}{P[A]} = P[B]. \qquad (2.4.11)$$

The concept of two events A and B being independent should not be confused with the concept of their being mutually exclusive. In fact, if A and B are mutually exclusive,

$$0 = P[\varnothing] = P[A \cap B],$$

and thus $P[A \cap B]$ can be equal to $P[A]P[B]$ only if at least one of them has probability zero. Hence, mutually exclusive events are *not* independent except in the trivial case that at least one of them has zero probability.

The next example ties together some of the concepts discussed so far.

Example 2.4.4   Suppose that a certain department has 3 unbuffered terminals which can be connected to a computer via 2 communication lines. Terminal 1 has its own leased line while terminals 2 and 3 share a leased line so that at most one of the two can be in use at any particular time. During the working day terminal 1 is in use 30 minutes of each hour, terminal 2, 10 minutes of each hour, and terminal 3 is used 5 minutes of each hour—all times being average times. Assuming the communication lines operate independently, what is the probability that at least one terminal is in operation at a random time during the working day? If the operation of the two lines is not independent with the conditional probability that terminal 2 is in use given that terminal 1 is in operation equal to 1/3, and the corresponding conditional probability that terminal 3 is in use equal to 1/12; what is the probability that at least one line is in use?

Solution   *Case 1*   The lines operate independently. Let A, B, C be the events that terminals 1, 2, 3, respectively, are in use. The event that the first line is in use is A while the event that the second line is in use is $B \cup C$, and these events are independent. The event U that at least one terminal is in use is $A \cup (B \cup C)$.

By Theorem 2.2.1c

$$P[U] = P[A \cup (B \cup C)] = P[A] + P[B \cup C] - P[A \cap (B \cup C)]. \quad (2.4.12)$$

By the independence of A and $B \cup C$,

$$P[A \cap (B \cup C)] = P[A]P[B \cup C]. \quad (2.4.13)$$

Substituting (2.4.13) into (2.4.12) yields

$$P[U] = P[A] + P[B \cup C] - P[A]P[B \cup C]. \quad (2.4.14)$$

Since B and C are mutually exclusive, Axiom **P3** yields

$$P[B \cup C] = P[B] + P[C] = \tfrac{10}{60} + \tfrac{5}{60} = \tfrac{1}{4}. \quad (2.4.15)$$

Since $P[A] = 0.5$, substitution into (2.4.14) gives

$$P[U] = 0.5 + 0.25 - 0.5 \times 0.25 = 0.625. \quad (2.4.16)$$

Case 2   The communication lines are not independent. In this case (2.4.12) still applies but $P[A \cap (B \cup C)]$ has the formula,

$$P[A \cap (B \cup C)] = P[(A \cap B) \cup (A \cap C)], \quad (2.4.17)$$

by the distributive law for events (see Exercise 3).

The events $A \cap B$ and $A \cap C$ are mutually exclusive, since B and C are. Hence,

$$P[A \cap (B \cup C)] = P[(A \cap B) \cup (A \cap C)] = P[A \cap B] + P[A \cap C]$$
$$= P[A]P[B|A] + P[A]P[C|A] = \tfrac{1}{2} \times \tfrac{1}{3} + \tfrac{1}{2} \times \tfrac{1}{12} = \tfrac{5}{24},$$

$$(2.4.18)$$

by the multiplication rule. Substituting (2.4.18) into (2.4.12) gives

$$P[U] = \tfrac{1}{2} + \tfrac{1}{4} - \tfrac{5}{24} = \tfrac{13}{24} = 0.542. \tag{2.4.19}$$

Example 2.4.5 Suppose that, at a computer installation 80% of all new source programs are written in COBOL and 20% in PL/I; and that 20% of the COBOL programs as well as 60% of the PL/I programs compile on the first run. (By "compile on the first run" we mean that the compiler found no errors serious enough to prevent compilation of the source code.) If a randomly selected program has compiled on the first run, what is the probability that it is a COBOL program?

Solution Let E be the event that the selected program did compile, C the event that the program selected is written in COBOL, and PL the event that it is a PL/I program.
Then

$$P[C\,|\,E] = \frac{P[C \cap E]}{P[E]} = \frac{P[C]P[E\,|\,C]}{P[C]P[E\,|\,C] + P[PL]P[E\,|\,PL]}$$

$$= \frac{0.8 \times 0.2}{0.8 \times 0.2 + 0.2 \times 0.6} = \frac{0.16}{0.28} = 0.571, \tag{2.4.20}$$

where we have used the law of total probability to compute $P[E]$.

In the solution of Example 2.4.5 we have used a simple version of Bayes' theorem, which is stated below.

THEOREM 2.4.3 (*Bayes' Theorem*) Suppose the events $A_1, A_2, \ldots,$ A_n form a partition of Ω (for the definition of a partition see Theorem 2.4.2). Then, for any event A with $P[A] > 0$,

$$P[A_i\,|\,A] = \frac{P[A_i]P[A\,|\,A_i]}{P[A_1]P[A\,|\,A_1] + P[A_2]P[A\,|\,A_2] + \cdots + P[A_n]\,P[A\,|\,A_n]},$$

$$i = 1, 2, \ldots, n. \tag{2.4.21}$$

Proof For each i,

$$P[A_i\,|\,A] = \frac{P[A_i \cap A]}{P[A]} = \frac{P[A_i]P[A\,|\,A_i]}{P[A]}. \tag{2.4.22}$$

Equation (2.4.21) now follows from (2.4.22) by applying the law of total probability to calculate $P[A]$.

The $P[A_i]$, $i = 1, 2, \ldots, n$, are called *prior* (or *a priori*) probabilities and the $P[A_i\,|\,A]$, $i = 1, 2, \ldots, n$, are called *posterior* (or *a posteriori*) probabilities. To calculate the posterior probabilities using Bayes' theorem we must know both the prior probabilities $P[A_1]$, $P[A_2]$, ..., $P[A_n]$ and the conditional probabilities

$$P[A|A_1], ..., P[A|A_n].$$

Example 2.4.6 At the completion of a programming project five programmers from a team submit a collection of subroutines to an acceptance testing group. Each programmer has certified that each subroutine has been carefully tested and is free from errors. Table 2.4.1 shows how many routines each programmer submitted and the probability that a program certified by programmer i will pass the acceptance tests, $i = 1, 2, 3, 4, 5$, based on historical data. After the acceptance testing is completed, one of the subroutines is selected at random and it is found to have failed the tests. Assuming the probabilities in the third column of Table 2.4.1 have not changed, what is the probability that the subroutine was written by the first programmer? By the fifth?

TABLE 2.4.1
Statistics on Programmer Subroutines

Programmer (i)	Number of subroutines submitted	Probability that subroutine of programmer i will pass tests
1	12	0.98
2	4	0.75
3	10	0.95
4	5	0.85
5	2	0.55

Solution Let A_i be the event that the selected subroutine was written by programmer i, $i = 1, 2, 3, 4, 5$, and A the event that the selected program failed the tests. The data from Table 2.4.1 yield the probabilities shown in Table 2.4.2. By Bayes' theorem the probability that the program was written by the first programmer is given by

$$P[A_1|A] = \frac{P[A_1]P[A|A_1]}{P[A_1]P[A|A_1] + \cdots + P[A_5]P[A|A_5]}$$

$$= \frac{\frac{12}{33} \times 0.02}{\frac{12}{33} \times 0.02 + \frac{4}{33} \times 0.25 + \frac{10}{33} \times 0.05 + \frac{5}{33} \times 0.15 + \frac{2}{33} \times 0.45}$$

$$= \frac{0.24}{0.24 + 1 + 0.5 + 0.75 + 0.9} = \frac{0.24}{3.39} = 0.0708. \qquad (2.4.23)$$

TABLE 2.4.2

Probabilities from Table 2.4.1

i	$P[A_i]$	$P[A \mid A_i]$
1	12/33	0.02
2	4/33	0.25
3	10/33	0.05
4	5/33	0.15
5	2/33	0.45

Similarly, the probability that the program was written by the fifth programmer is

$$P[A_5 \mid A] = \frac{P[A_5]P[A \mid A_5]}{P[A]} = \frac{\frac{2}{33} \times 0.45}{3.39/33} = \frac{0.9}{3.39} = 0.2655. \quad (2.4.24)$$

2.5   RANDOM VARIABLES

In many random experiments we are interested in some number associated with the experiment rather than the actual outcome. Thus, in Example 2.1.2, we may be interested in the sum of the numbers shown on the dice. In Example 2.3.2 we may be interested in the number of polls taken to find the first ready terminal. We are thus interested in a function which associates a number with the outcome of an experiment—such a function is called a *random variable*. Formally, a *random variable* X is a real-valued function defined on a sample space Ω. Some examples of random variables of interest to computer science follow.

Example 2.5.1   (a)   Let X be the number of jobs processed by a computer center in one day. The sample space Ω might consist of collections of job numbers—an outcome is the set of job numbers of jobs run during the day. (We assume each job number is unique.) Thus, if $\omega = \{x_1, x_2, \ldots, x_n\}$ is a sample point consisting of the set of job numbers of jobs run during the day, then $X(\omega) = n$.

(b)   Let X be the number of communication lines in operation in an on-line computer system of n lines. The sample space Ω could be the collection of n-tuples $(x_1, x_2, \ldots, x_n)$ where each x_i is 1 if line i is in operation and otherwise 0.

To avoid cumbersome notation we will use abbreviations to denote some special events. If X is a random variable and x a real number we write

$$X = x$$

for the event

$$\{\omega : \omega \in \Omega \text{ and } X(\omega) = x\}.$$

Similarly, we write

$$X \leq x$$

for the event

$$\{\omega : \omega \in \Omega \text{ and } X(\omega) \leq x\}$$

and

$$y < X \leq x$$

for the event

$$\{\omega : \omega \in \Omega \text{ and } y < X(\omega) \leq x\}.$$

Another property required of a random variable is that the set $X \leq x$ be an event for each real x, that is $X \leq x$ be an element of $\mathcal{F}$ for each real x. This is necessary so that probability calculations can be made. A function having this property is said to be a *measurable function* or *measurable in the Borel sense* (see Cramér [2, page 37]).

For each random variable X we define its *distribution function* F for each real x by

$$F(x) = P[X \leq x].$$

Some intuitively clear properties of a distribution function are stated in the following proposition.

PROPOSITION 2.5.1 *(Properties of a Distribution Function)*

D1 F is a nondecreasing function, that is, $x < y$ implies $F(x) \leq F(y)$.
D2 $\lim_{x \to +\infty} F(x) = 1$.
D3 $\lim_{x \to -\infty} F(x) = 0$.

For a proof of Proposition 2.5.1 see Apostol [1].

The distribution function can be used to make certain probability calculations. For example, if $x < y$, then

$$P[x < X \leq y] = F(y) - F(x).$$

This is true, because the events $x < X \leq y$ and $X \leq x$ are disjoint; their union is $X \leq y$; and thus

$$F(y) = P[x < X \leq y] + P[X \leq x] = P[x < X \leq y] + F(x). \quad (2.5.1)$$

With each random variable X we associate another function $p(\cdot)$, called the *probability mass function* of X (abbreviated as pmf), defined for all real x by

$$p(x) = P[X = x].$$

Thus, if x is a value that X cannot assume, then $p(x) = 0$. The set T of all x such $p(x) > 0$ is either finite or countably infinite. For a proof see Cramér [2, p. 52]. (A set is countably infinite or denumerable if it can be put into one-to-one correspondence with the positive integers and thus enumerated $x_1, x_2, x_3, \ldots$.) The random variable X is said to be *discrete* if

$$\sum_{x \in T} p(x) = 1$$

where $T = \{x : p(x) > 0\}$. Thus X is discrete if T consists of (a) either a finite set, say, $x_1, x_2, \ldots, x_n$, or (b) an infinite set, say, $x_1, x_2, x_3, \ldots$ and in addition,

$$\sum_{x_i} p(x_i) = 1.$$

Thus a real-valued function $p(\cdot)$ defined on the whole real line is the probability mass function of a discrete random variable if and only if the following three conditions hold:

(i) $p(x) \geq 0$ for all real x.
(ii) $T = \{x \mid p(x) > 0\}$ is finite or countably infinite, that is, $T = \{x_1, x_2, \ldots\}$.
(iii) $\sum_{x_i \in T} p(x_i) = 1$.

If X is a discrete random variable the elements of T are called the *mass points of X*, and we say, "X assumes the values $x_1, x_2, x_3, \ldots$."

The random variables of Example 2.5.1 are discrete.

Example 2.5.2   In Example 2.3.2 we implicitly define a random variable X which counts the number of polls until a ready terminal is found. X is a discrete random variable which assumes (with positive probability) only the values 1, 2, 3. The probability mass function is defined by $p(1) = 0.6$, $p(2) = 0.3$, and $p(3) = 0.1$. The pmf $p(\cdot)$ of X is shown graphically in Fig. 2.5.1; the distribution function F is shown in Fig. 2.5.2. Thus the probability that two or fewer polls are required is $F(2) = p(1) + p(2) = 0.9$, which can be read from Fig. 2.5.2 or calculated from the probability mass function.

A random variable X is *continuous* if $p(x) = 0$ for all real x. The reason for the terminology is that the distribution function for a continuous random variable is a continuous function in the usual sense. By contrast, the distribution function for a discrete random variable has a discontinuity at each point of positive probability (mass point). We will be concerned only with those continuous random variables X which have a *density function* f with the following properties:

(a) $f(x) \geq 0$ for all real x.
(b) f is integrable and $P[a \leq X \leq b] = \int_a^b f(x)\, dx$ if $a < b$.

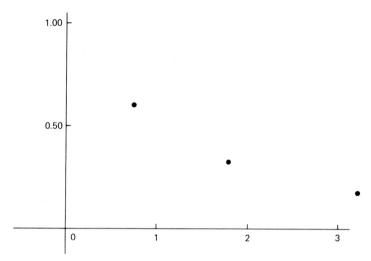

Fig. 2.5.1   Probability mass function for Example 2.5.2.

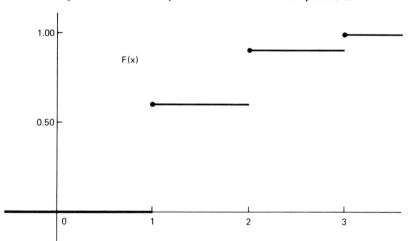

Fig. 2.5.2   Distribution function for Example 2.5.2.

(This means that the required probability is the area under the curve $y = f(x)$ between a and b.)

   (c) $\int_{-\infty}^{\infty} f(x)\, dx = 1.$

   (d) $F(x) = \int_{-\infty}^{x} f(t)\, dt$ for each real x.

By the fundamental theorem of calculus, at each point x where f is continuous

$$\frac{dF}{dx} = f(x).$$

Example 2.5.3   Let $\lambda > 0$. The random variable X is said to be an *exponential random variable with parameter λ* or to *have an exponential distribution with parameter λ* if it has the distribution function

$$F(x) = \begin{cases} 1 - e^{-\lambda x} & \text{for} \quad x > 0 \\ 0 & \text{for} \quad x \leq 0. \end{cases} \tag{2.5.2}$$

The density function $f = dF/dx$ is given by

$$f(x) = \begin{cases} \lambda e^{-\lambda x} & \text{for} \quad x > 0 \\ 0 & \text{for} \quad x \leq 0. \end{cases} \tag{2.5.3}$$

Suppose, for example, that $\lambda = 2$ and we wish to calculate the probability that X assumes a value between 1 and 3. This probability is the area under the curve $y = 2e^{-2x}$ (shown in Fig. 2.5.3) between $x = 1$ and $x = 3$. The probability may also be computed using the distribution function $F(x) = 1 - e^{-2x}$ shown in Fig. 2.5.4. We calculate

$$P[1 \leq X \leq 3] = F(3) - F(1) = (1 - e^{-6}) - (1 - e^{-2}) = e^{-2} - e^{-6}$$

$$= 0.135335 - 0.002479 = 0.132856. \tag{2.5.4}$$

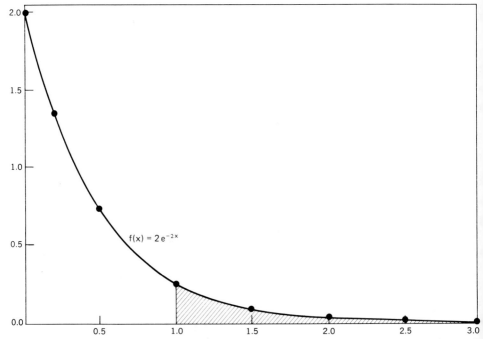

Fig. 2.5.3   Density function for exponential random variable with parameter 2. The shaded area represents the probability that X assumes a value between 1 and 3.

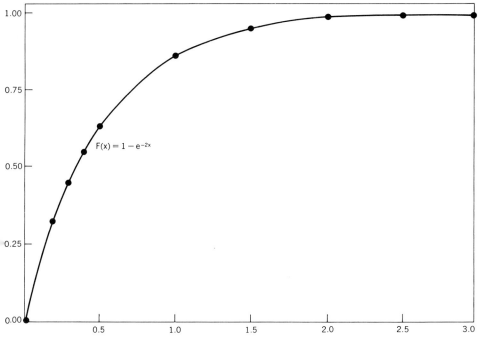

Fig. 2.5.4   Distribution function for exponential random variable with parameter 2.

In making this calculation we have used the fact that, for a continuous random variable X with $a < b$,

$$P[a \leq X \leq b] = P[a < X \leq b] = P[a \leq X < b] = P[a < X < b],$$
(2.5.5)

since

$$P[X = a] = P[X = b] = 0.$$

(This is true because, by definition, a continuous random variable X has the property that $P[X = x] = 0$ for all real x.)

Most random variables are either discrete or continuous but, occasionally, we shall encounter a random variable of mixed type; that is, a random variable which is continuous for some range of values and discrete for others. Usually a discrete random variable comes about when something is counted, such as number of jobs, inquiries, messages, etc. A continuous random variable often occurs when something is measured, such as the time between the arrival of two consecutive inquiries, the response time at a terminal, the time it takes to process a job, etc. Unless otherwise noted, we will assume that all random variables under consideration are either discrete or continuous and not of mixed type.

The distribution function F of a random variable X describes how the "probability mass" of X is "distributed" along the real line. From this point of view a random variable X determines how the probability mass of one unit is apportioned or spread out over the real numbers. A discrete random variable allocates the mass in nuggets or mass points while a continuous random variable diffuses the probability mass out in a continuous manner.

2.6   PARAMETERS OF RANDOM VARIABLES

All possible probability calculations involving a random variable X can be made from its pmf $p(\cdot)$, if it is discrete; from its density function f, if it is continuous; or from its distribution function F in either case. However, there are some parameters of a random variable which are important in summarizing its properties in a way that is easy to comprehend and to use for making probability estimates.

Let $h(X)$ be a function of a random variable such as $2X$, X^2, e^X, or log X, that is, a new random variable defined in terms of the original random variable. For example, e^X is the random variable defined on Ω by $e^X(\omega) = e^{X(\omega)}$. Then the *expected value of* $h(X)$, denoted by $E[h(X)]$, is defined by

$$E[h(X)] = (2.6.2) \text{ or } (2.6.3), \quad \text{provided} \quad E[|h(X)|] \quad \text{exists.} \qquad (2.6.1)$$

Thus, if X is discrete,

$$E[h(X)] = h(x_1)p(x_1) + h(x_2)p(x_2) + \cdots, \qquad (2.6.2)$$

if

$$|h(x_1)|p(x_1) + |h(x_2)|p(x_2) + \cdots < \infty,$$

while, if X is continuous,

$$E[h(X)] = \int_{-\infty}^{\infty} h(x)f(x)\, dx, \qquad (2.6.3)$$

provided

$$\int_{-\infty}^{\infty} |h(x)|\, f(x)\, dx < \infty.$$

The two most important parameters used to describe or summarize the properties of a random variable X are the *mean* or *expected value* $\mu = E[X]$ and the *standard deviation* σ, where σ^2 is the *variance of* X defined by $\sigma^2 = \mathrm{Var}[X] = E[(X - E[X])^2]$. Thus, if X is discrete,

$$\mu = E[X] = \sum_{x_i} x_i p(x_i) = x_1 p(x_1) + x_2 p(x_2) + \cdots, \qquad (2.6.4)$$

while, if X is continuous,

$$\mu = E[X] = \int_{-\infty}^{\infty} xf(x)\, dx. \tag{2.6.5}$$

For a discrete random variable X, $\mathrm{Var}[X]$ is given by

$$\sigma^2 = \mathrm{Var}[X] = \sum_{x_i} (x_i - E[X])^2 p(x_i)$$

$$= (x_1 - E[X])^2 p(x_1) + (x_2 - E[X])^2 p(x_2) + \cdots, \tag{2.6.6}$$

while, if X is continuous,

$$\sigma^2 = \mathrm{Var}[X] = \int_{-\infty}^{\infty} (x - E[X])^2 f(x)\, dx. \tag{2.6.7}$$

The reason that the mean or expected value is important is intuitively clear to almost everyone. For example, if X is a discrete random variable which describes how much one is to win for each outcome in a sample space Ω, then

$$\mu = E[X] = \sum_i x_i p(x_i)$$

is the weighted average of what one is to win. Thus calling μ the mean or expected value seems proper. The mean acts as a kind of summary of what we expect of the random variable. If we have only one number to use to describe a random variable, the mean would seem to be the proper one. However, the standard deviation, σ, does not have such an intuitive meaning to many of us, at first. However, as we will see in Chebychev's inequality (Theorem 2.10.2) and the one-sided inequality (Theorem 2.10.3), the standard deviation is the natural unit to measure the deviation of a random variable from its mean. Thus, if you were told that "Rockefeller and I, together, had an average income of over a million dollars last year," it would not give much information about my contribution. But if the statement was "The average income last year of Rockefeller and I was \$1,010,000, with a standard deviation of \$990,000," then it would be clear that our individual incomes were significantly different.

It is best not to worry as to exactly what σ is. We will show that it is a most useful quantity for many applied probability applications. We must ask the reader to "Have faith that truth will be revealed."

Example 2.6.1 Referring to Example 2.5.2, we see that

$$\mu = E[X] = \sum_{i=1}^{3} x_i p(x_i) = 1 \times 0.6 + 2 \times 0.3 + 3 \times 0.1 = 1.5, \tag{2.6.8}$$

$$\sigma^2 = \mathrm{Var}[X] = \sum_{i=1}^{3} (x_i - 1.5)^2 p(x_i)$$

$$= (1 - 1.5)^2 \times 0.6 + (2 - 1.5)^2 \times 0.3 + (3 - 1.5)^2 \times 0.1 = 0.45, \tag{2.6.9}$$

and thus

$$\sigma = (\text{Var}[X])^{1/2} = (0.45)^{1/2} = 0.6708. \tag{2.6.10}$$

This example can be generalized. Consider a communication line with m terminals attached of which n are ready to transmit. Let X be the number of polls required to find the first terminal that is ready. Then X can assume only the values $i = 1, 2, \ldots, m - n + 1$ with

$$p(i) = P[X = i] = \frac{\binom{m-i}{n-1}}{\binom{m}{n}} \qquad \text{(see Exercise 6).} \tag{2.6.11}$$

$E[X]$ and $\text{Var}[X]$ can then be calculated by the formulas

$$\mu = E[X] = \sum_{i=1}^{m-n+1} ip(i), \tag{2.6.12}$$

and

$$\sigma^2 = \text{Var}[X] = \sum_{i=1}^{m-n+1} (i - \mu)^2 p(i). \tag{2.6.13}$$

The APL function POLL (shown in Appendix B) with parameters m and n can be used in the general case to compute the probabilities that X assumes the values $1, 2, 3, \ldots, m - n + 1$. It also computes the expected value and standard deviation of X.

The APL function PARAM (see Appendix B) can be used to calculate the mean and standard deviation of a discrete random variable from the set of its possible values and the corresponding set of probabilities.

Example 2.6.2 Suppose X has an exponential distribution with parameter $\lambda > 0$. Then

$$\mu = E[X] = \int_0^\infty x\lambda e^{-\lambda x}\, dx. \tag{2.6.14}$$

Using integration by parts (the formula $\int u\, dv = uv - \int v\, du$) with $u = x$ and $dv = \lambda e^{-\lambda x}\, dx$, together with one application of the formula

$$\lim_{x \to \infty} xe^{-\lambda x} = 0, \tag{2.6.15}$$

brings (2.6.14) to the form

$$\mu = E[X] = \int_0^\infty e^{-\lambda x}\, dx = \left.\frac{-e^{\lambda x}}{\lambda}\right|_0^\infty = -\lim_{x \to \infty} \frac{e^{-\lambda x}}{\lambda} + \frac{1}{\lambda} = \frac{1}{\lambda}. \tag{2.6.16}$$

Similarly,

$$\sigma^2 = \text{Var}[X] = \int_0^\infty \left(x - \frac{1}{\lambda}\right)^2 \lambda e^{-\lambda x}\, dx = \frac{1}{\lambda^2}, \tag{2.6.17}$$

so that $\sigma = 1/\lambda = \mu$. Thus, for example, the exponential random variable with $\lambda = 2$ has $\mu = 0.5$ and $\sigma = 0.5$.

The sequence of *moments* of X defined by $E[X^k]$, $k = 1, 2, 3, \ldots$, is sometimes of interest. The first moment coincides with the mean or expected value. From the definition, we see that if X is discrete, then for each $k = 1, 2, 3, \ldots$,

$$E[X^k] = \sum_i x_i^k p(x_i) = x_1^k p(x_1) + x_2^k p(x_2) + \cdots, \qquad (2.6.18)$$

while, if X is continuous, then

$$E[X^k] = \int_{-\infty}^{\infty} x^k f(x)\, dx, \qquad k = 1, 2, 3, \ldots. \qquad (2.6.19)$$

It can be shown that under very general conditions (see Feller [3, pp. 227–228]), if all the moments for X exist, they uniquely determine the distribution of X, that is, if X and Y have the same sequence of moments, then $F_X = F_Y$. The following examples show that a random variable may not have *any* moments—not even a mean value.

Example 2.6.3 Let X assume the values $2, 2^2, 2^3, \ldots, 2^k, \ldots$ with the probability mass function $p(\cdot)$, defined by

$$p(x_k) = p(2^k) = \frac{1}{2^k}, \qquad k = 1, 2, 3, \ldots. \qquad (2.6.20)$$

Then $p(\cdot)$ is a true probability mass function since

$$\sum_{k=1}^{\infty} p(x_k) = \sum_{k=1}^{\infty} \frac{1}{2^k} = \frac{1}{2} \sum_{k=0}^{\infty} \frac{1}{2^k} = \frac{1}{2} \times \frac{1}{1 - \frac{1}{2}} = \frac{1}{2} \times 2 = 1. \quad (2.6.21)$$

However,

$$\sum_{i=1}^{\infty} x_i p(x_i) = 1 + 1 + 1 + \cdots \qquad (2.6.22)$$

diverges so that even the first moment fails to exist.

Example 2.6.4 Let X be the continuous random variable with density function f defined by

$$f(x) = \begin{cases} 0 & \text{for } x < 1 \\ \dfrac{1}{x^2} & \text{for } x \geq 1. \end{cases} \qquad (2.6.23)$$

Then

$$\int_{-\infty}^{\infty} f(x)\, dx = \int_{1}^{\infty} \frac{dx}{x^2} = -\frac{1}{x}\Big|_{1}^{\infty} = -\lim_{x \to \infty} \frac{1}{x} + 1 = 1, \qquad (2.6.24)$$

so f is a density function. However,

$$\int_{-\infty}^{\infty} xf(x)\,dx = \int_{1}^{\infty} \frac{dx}{x} = \ln x \Big|_{1}^{\infty} = +\infty, \qquad (2.6.25)$$

so that the first moment of X fails to exist. Clearly no higher order moments exist, either.

Although Examples 2.6.3 and 2.6.4 show that not all random variables have moments, most useful theoretical and empirically derived random variables do have moments.

2.7 JOINTLY DISTRIBUTED RANDOM VARIABLES

Sometimes it is of interest to investigate two or more random variables, simultaneously. Thus, if X and Y are two random variables defined on the same sample space Ω, we define the *joint distribution function F of X and Y* for all real x and y by

$$F(x, y) = P[X \le x, Y \le y] = P[(X \le x) \cap (Y \le y)]. \qquad (2.7.1)$$

Sometimes we write $F_{X,Y}$ for the joint distribution function of X and Y to emphasize that it is a joint distribution. Given $F_{X,Y}$, the individual distribution functions F_X and F_Y can be computed as follows:

$$F_X(x) = \lim_{y \to \infty} F_{X,Y}(x, y) \qquad \text{for each real} \quad x, \qquad (2.7.2)$$

and

$$F_Y(y) = \lim_{x \to \infty} F_{X,Y}(x, y) \qquad \text{for each real} \quad y. \qquad (2.7.3)$$

F_X is called the *marginal distribution function of X* corresponding to the joint distribution $F_{X,Y}$ and similarly for F_Y. Then the *joint probability mass function* $p(\cdot, \cdot)$ *of X and Y* is defined by

$$p(x, y) = P[X = x, Y = y]. \qquad (2.7.4)$$

Let $T = \{(x, y) : p(x, y) > 0\}$. Then, if $\sum_{(x,y) \in T} p(x, y) = 1$, we say that X and Y *are jointly discrete.* (It can be shown that T is either finite or countable.) If X and Y are jointly discrete then X and Y are each discrete and the probability mass functions p_X and p_Y of X and Y can be calculated as

$$p_X(x) = \sum_{\substack{y \text{ such that} \\ p(x,y) > 0}} p(x, y), \qquad (2.7.5)$$

$$p_Y(y) = \sum_{\substack{x \text{ such that} \\ p(x,y) > 0}} p(x, y). \qquad (2.7.6)$$

In this case p_X and p_Y are called *marginal probability mass functions.*

Example 2.7.1   Suppose a communication line is to be polled when it is known that two of the four terminals on the line are ready to transmit so, using the notation of Example 2.3.2, Ω consists of the six sample points $(1, 1, 0, 0), (1, 0, 1, 0), (1, 0, 0, 1), (0, 1, 1, 0), (0, 1, 0, 1), (0, 0, 1, 1)$. Let X be the number of polls until the first ready terminal is found and Y the number until the second ready terminal is found. The joint probability mass function and the marginal probability mass functions are shown in Table 2.7.1. It shows that $p(1, 2) = P[X = 1, Y = 2] = 1/6$, $p(2, 2) = p(3, 2) = p(3, 3) = 0$, etc. Also $p_X(1) = p(1, 2) + p(1, 3) + p(1, 4) = 1/2$, $p_X(2) = p(2, 3) + p(2, 4) = 1/3$, etc. (Obviously Y assumes only the values 2, 3, 4.)

TABLE 2.7.1

Joint Probability Mass Function
$p(\cdot, \cdot)$ **of** X **and** Y

X \ Y	2	3	4	p_X
1	1/6	1/6	1/6	1/2
2	0	1/6	1/6	1/3
3	0	0	1/6	1/6
p_Y	1/6	1/3	1/2	

It is also true, for two jointly discrete random variables, that

$$F(x, y) = \sum_{x_i \le x} \sum_{y_j \le y} p(x_i, y_j). \tag{2.7.7}$$

Thus, from Table 2.7.1, we see that

$$F(2, 3) = p(1, 2) + p(1, 3) + p(2, 2) + p(2, 3) = 0.5, \tag{2.7.8}$$

and

$$F(2, 4) = F(2, 3) + p(1, 4) + p(2, 4) = 5/6. \tag{2.7.9}$$

X and Y are *jointly continuous* if their joint distribution function F is continuous on the whole plane. We are interested only in jointly continuous random variables which have a *joint density function* f such that, if A is a set of real numbers as is B, then

$$P[X \in A, Y \in B] = \int_B \int_A f(x, y) \, dx \, dy. \tag{2.7.10}$$

In this case, for any real u and v,

$$F(u, v) = \int_{-\infty}^{v} \int_{-\infty}^{u} f(x, y) \, dx \, dy. \tag{2.7.11}$$

If the random variables X and Y are jointly continuous, then each is a continuous random variable and for each real x and y:

$$f_X(x) = \int_{-\infty}^{\infty} f(x, y)\, dy. \tag{2.7.12}$$

$$f_Y(y) = \int_{-\infty}^{\infty} f(x, y)\, dx. \tag{2.7.13}$$

The formulas (2.7.12) and (2.7.13) follow from (2.7.2), (2.7.3), and (2.7.11). For example,

$$F_X(u) = \lim_{v \to \infty} F(u, v) = \lim_{v \to \infty} \int_{-\infty}^{v} \int_{-\infty}^{u} f(x, y)\, dx\, dy$$

$$= \lim_{v \to \infty} \int_{-\infty}^{u} dx \int_{-\infty}^{v} f(x, y)\, dy = \int_{-\infty}^{u} dx \int_{-\infty}^{\infty} f(x, y)\, dy.$$

Hence $f_X(x) = \int_{-\infty}^{\infty} f(x, y)\, dy$ since

$$f_X(x) = \frac{dF_X}{dx}(x).$$

Example 2.7.2 Suppose the random variables X and Y have the joint density function

$$f(x, y) = \begin{cases} xy \exp\left[-\tfrac{1}{2}(x^2 + y^2)\right] & \text{for } x > 0 \text{ and } y > 0 \\ 0 & \text{otherwise.} \end{cases}$$

$$\tag{2.7.14}$$

Find $f_X(x)$, $f_Y(y)$, and $F(1, 1) = P[X \le 1,\ Y \le 1]$. This example is due to Parzen [4, p. 291].

Solution By (2.7.12)

$$f_X(x) = \int_0^{\infty} f(x, y)\, dy = \int_0^{\infty} xy \exp\left[-\tfrac{1}{2}(x^2 + y^2)\right] dy$$

$$= x \exp\left[-\tfrac{1}{2}x^2\right] \int_0^{\infty} y \exp\left[-\tfrac{1}{2}y^2\right] dy = x \exp\left[-\tfrac{1}{2}x^2\right](-\exp\left[-\tfrac{1}{2}y^2\right]\big|_0^{\infty})$$

$$= x \exp\left[-\tfrac{1}{2}x^2\right].$$

Similarly,

$$f_Y(y) = \int_0^{\infty} xy \exp\left[-\tfrac{1}{2}(x^2 + y^2)\right] dx = y \exp\left[-\tfrac{1}{2}y^2\right].$$

By (2.7.11)

$$F(1, 1) = \int_0^1 \int_0^1 f(x, y)\, dx\, dy = \int_0^1 x \exp\left[-\tfrac{1}{2}x^2\right] dx \int_0^1 y \exp\left[-\tfrac{1}{2}y^2\right] dy$$

$$= (-\exp\left[-\tfrac{1}{2}x^2\right]|_0^1)(-\exp\left[-\tfrac{1}{2}y^2\right]|_0^1)$$

$$= (1 - \exp\left[-0.5\right])(1 - \exp\left[-0.5\right])$$

$$= (1 - 0.606531)(1 - 0.606531) = 0.154818.$$

Suppose g is a function of two variables. Then, if X and Y are jointly distributed random variables, the mathematical expectation of $g(X, Y)$, $E[g(X, Y)]$, is defined as

$$E[g(X, Y)] = \sum_{\substack{\text{over all } (x,y) \\ \text{such that } p(x,y) > 0}} g(x, y)p(x, y), \qquad (2.7.15)$$

if X and Y are jointly discrete and $p(\cdot, \cdot)$ is the joint probability mass function.

$$E[g(X, Y)] = \int_{-\infty}^{\infty} \int_{-\infty}^{\infty} g(x, y)f(x, y)\, dx\, dy, \qquad (2.7.16)$$

if X and Y are jointly continuous with joint density function f.

Two random variables X and Y are said to be *independent* if any of the following relations hold:

(a)   Their joint distribution function can be expressed as a product

$$F(x, y) = F_X(x)F_Y(y) \qquad \text{for all real } x \text{ and } y. \qquad (2.7.17)$$

(b)   They are jointly discrete and their joint probability mass function can be written

$$p(x, y) = p_X(x)p_Y(y) \qquad \text{for all real } x \text{ and } y. \qquad (2.7.18)$$

(c)   They are jointly continuous and their joint density function can be written

$$f(x, y) = f_X(x)f_Y(y) \qquad \text{for all real } x \text{ and } y. \qquad (2.7.19)$$

Two random variables X and Y not independent by one of the above criteria are said to be *dependent*.

The next theorem gives some of the properties of the expected value operator $E[\cdot]$.

THEOREM 2.7.1   Suppose X and Y are random variables; c a constant; g and h arbitrary measurable functions. Then

(a) $E[c] = c$. (The expected value of a constant random variable is the constant.)

(b) $E[cX] = cE[X]$.
(c) $E[X + Y] = E[X] + E[Y]$. (X and Y need not be independent.)
(d) $E[g(X)h(Y)] = E[g(X)]E[h(X)]$ if X and Y are independent and the
expectations on the right exist.

Proof (a) Suppose X is the constant random variable c, that is,
$P[X = c] = 1$. X is discrete so, by definition, $E[X] = E[c] = cP[X = c] = c$.
(b) If X is discrete with pmf $p(\cdot)$, then

$$E[cX] = \sum_{x_i} cx_i p(x_i) = c \sum_{x_i} x_i p(x_i) = cE[X].$$

If X is continuous, then

$$E[cX] = \int_{-\infty}^{\infty} cxf(x)\,dx = c \int_{-\infty}^{\infty} xf(x)\,dx = cE[X].$$

(c) We give the proof for the case that X and Y are jointly continuous.
If they are jointly discrete the proof is similar.

$$E[X + Y] = \int_{-\infty}^{\infty} \int_{-\infty}^{\infty} (x + y)f(x, y)\,dx\,dy$$

$$= \int_{-\infty}^{\infty} x\,dx \int_{-\infty}^{\infty} f(x, y)\,dy + \int_{-\infty}^{\infty} y\,dy \int_{-\infty}^{\infty} f(x, y)\,dx$$

$$= \int_{-\infty}^{\infty} xf_X(x)\,dx + \int_{-\infty}^{\infty} yf_Y(y)\,dy = E[X] + E[Y].$$

The next to last equality follows from (2.7.12) and (2.7.13).
(d) We prove the theorems for the case that X and Y are jointly continuous. The jointly discrete case is similar.

$$E[g(X)h(Y)] = \int_{-\infty}^{\infty} \int_{-\infty}^{\infty} g(x)h(y)f(x, y)\,dx\,dy$$

$$= \int_{-\infty}^{\infty} \int_{-\infty}^{\infty} g(x)h(y)f_X(x)f_Y(y)\,dy\,dx$$

$$= \int_{-\infty}^{\infty} g(x)f_X(x)\,dx \int_{-\infty}^{\infty} h(y)f_Y(y)\,dy = E[g(X)]E[h(Y)].$$

This completes the proof of Theorem 2.7.1.

The *covariance of X and Y*, written $\text{Cov}[X, Y]$, is defined by

$$\text{Cov}[X, Y] = E[(X - E[X])(Y - E[Y])]$$

$$= E[XY - XE[Y] - YE[X] + E[X]E[Y]]$$

$$= E[XY] - E[X]E[Y] - E[Y]E[X] + E[X]E[Y]$$

$$= E[XY] - E[X]E[Y]. \tag{2.7.20}$$

If $\text{Cov}[X, Y] = 0$, X and Y are said to be *uncorrelated*. Theorem 2.7.1d implies that any two independent random variables X and Y are uncorrelated. However, not all uncorrelated random variables are independent.

The next theorem gives some useful properties of the variance operator $\text{Var}[\cdot]$.

THEOREM 2.7.2 Suppose X and Y are random variables; c a constant; and all the variances in the formulas below exist. Then

(a) $\text{Var}[c] = 0$.
(b) $\text{Var}[cX] = c^2\,\text{Var}[X]$.
(c) $\text{Var}[X + Y] = \text{Var}[X] + \text{Var}[Y] + 2\,\text{Cov}[X, Y]$.
(d) $\text{Var}[X] = E[X^2] - (E[X])^2$.

Proof (a) If c is a constant random variable then, by Theorem 2.7.1a, $E[c] = c$. Hence, $\text{Var}[c] = E[(c - c)^2] = E[0] = 0$.

(b) $\text{Var}[cX] = E[(cX - E[cX])^2] = E[(c(X - E[X]))^2]$

$\qquad = c^2 E[(X - E[X])^2] = c^2\,\text{Var}[X]$.

(c) $\text{Var}[X + Y] = E[\{(X + Y) - (E[X] + E[Y])\}^2]$

$\qquad = E[\{(X - E[X]) + (Y - E[Y])\}^2]$

$\qquad = E[(X - E[X])^2 + (Y - E[Y])^2$

$\qquad\qquad + 2(X - E[X])(Y - E[Y])]$

$\qquad = E[(X - E[X])^2] + E[(Y - E[Y])^2$

$\qquad\qquad + 2E[(X - E[X])(Y - E[Y])]$

$\qquad = \text{Var}[X] + \text{Var}[Y] + 2\,\text{Cov}[X, Y]$.

(d) $\text{Var}[X] = E[(X - E[X])^2] = E[X^2 - 2XE[X] + (E[X])^2]$

$\qquad = E[X^2] - 2E[X]E[X] + (E[X])^2 = E[X^2] - (E[X])^2$.

We have used Theorem 2.7.1 freely in the above equalities. This completes the proof.

It should be noted that, if X and Y are independent and thus uncorrelated, then

$$\text{Var}[X + Y] = \text{Var}[X] + \text{Var}[Y]. \qquad (2.7.21)$$

An application of mathematical induction shows that

$$\text{Var}[X_1 + X_2 + \cdots + X_n] = \text{Var}[X_1] + \text{Var}[X_2] + \cdots + \text{Var}[X_n]$$

$$(2.7.22)$$

for any finite collection of mutually independent random variables.

Although we defined the joint distribution function F for only two random variables in (2.7.1), the concept can be extended, in a natural way, to any finite number of random variables. Thus, if $X_1, X_2, X_3, \ldots, X_n$ are random variables, their *joint distribution function* F (or $F_{X_1, X_2, \ldots, X_n}$ if it is desired to make names of the random variables explicit) is defined by

$$F(x_1, x_2, \ldots, x_n) = P[X_1 \le x_1, X_2 \le x_2, \ldots, X_n \le x_n],$$

$$\text{for all real} \quad x_1, x_2, \ldots, x_n. \quad (2.7.23)$$

All the other concepts we have discussed are then defined just as they were for two random variables. Thus, for example, one condition that means the random variables $X_1, X_2, \ldots, X_n$ are independent is that

$$F(x_1, x_2, \ldots, x_n) = F_{X_1}(x_1) F_{X_2}(x_2) \cdots F_{X_n}(x_n)$$

$$\text{for all real} \quad x_1, x_2, \ldots, x_n. \quad (2.7.24)$$

Some particular functions of several random variables are important for applications. The properties of one such function are given in the next theorem.

THEOREM 2.7.3   Let $X_1, X_2, \ldots, X_n$ be n independent random variables with distribution functions $F_{X_1}, F_{X_2}, \ldots, F_{X_n}$. Let $Y = g(X_1, \ldots, X_n)$ be the random variable defined by

$$Y(\omega) = \max\{X_1(\omega), X_2(\omega), \ldots, X_n(\omega)\} \quad \text{for each} \quad \omega \in \Omega. \quad (2.7.25)$$

Then the distribution function F_Y is given by

$$F_Y(y) = F_{X_1}(y) F_{X_2}(y) \cdots F_{X_n}(y) \quad \text{for each real} \quad y. \quad (2.7.26)$$

Proof   From the definition of Y, $Y \le y$ if and only if $X_1 \le y$, $X_2 \le y, \ldots, X_n \le y$. Hence,

$$\begin{aligned} F_Y(y) &= P[Y \le y] = P[X_1 \le y, \ldots, X_n \le y] \\ &= P[X_1 \le y] P[X_2 \le y] \cdots P[X_n \le y] \\ &= F_{X_1}(y) F_{X_2}(y) \cdots F_{X_n}(y). \end{aligned} \quad (2.7.27)$$

In the next-to-last equality in (2.7.27) we used the independence of the random variables.

Example 2.7.3   An on-line airline reservation system uses two identical duplexed computer systems, each of which has an exponential time to failure with a mean of 2000 hours. Each computer system has built in redundancy so failures are rare. The system fails only if both computers fail. What is the probability that the system will not fail during one week (168 hours) of continuous operation? 30 days?

Solution Theorem 2.7.3 applies. The distribution function of the time to failure X, in one system, is

$$F(t) = P[X \le t] = 1 - e^{-t/2000}, \qquad t \quad \text{in hours.} \qquad (2.7.28)$$

$F(t)$ in (2.7.28) is the probability that a failure will occur *before* time t in one of the systems. Hence, the probability of a system failure (both computer systems down) within a week, is, by Theorem 2.7.3,

$$(1 - e^{-168/2000})^2 = (0.080569)^2 = 0.006491.$$

Thus, the probability of no system failure for at least a week is

$$1 - 0.006491 = 0.993509.$$

The corresponding probability for 30 days is

$$1 - (1 - e^{-720/2000})^2 = 1 - (0.302324)^2 = 1 - 0.0914 = 0.9086.$$

If the system was not duplexed, that is, consisted of only one computer system, the probability of no failure within 30 days is 0.69768. Thus, if it is desired that the probability of failure-free operation for at least a week is to exceed 0.95, a duplex system is required.

The next theorem is similar to Theorem 2.7.3 but describes the distribution of the minimum of several random variables.

THEOREM 2.7.4 Let $X_1, X_2, \ldots, X_n$ be independent random variables. Let $Y = g(X_1, \ldots, X_n)$ be the random variable defined by

$$Y(\omega) = \min\{X_1(\omega), X_2(\omega), \ldots, X_n(\omega)\} \qquad \text{for each} \quad \omega \in \Omega. \qquad (2.7.29)$$

Then the distribution function F_Y is given by

$$F_Y(y) = 1 - (1 - F_{X_1}(y))(1 - F_{X_2}(y)) \cdots (1 - F_{X_n}(y))$$

$$\text{for each real} \quad y. \qquad (2.7.30)$$

Proof For each real y, $Y > y$ if and only if $X_1 > y, X_2 > y, \ldots, X_n > y$. Hence

$$\begin{aligned} P[Y > y] &= P[X_1 > y, X_2 > y, \ldots, X_n > y] \\ &= P[X_1 > y]P[X_2 > y] \cdots P[X_n > y] \\ &= (1 - F_{X_1}(y))(1 - F_{X_2}(y)) \cdots (1 - F_{X_n}(y)). \end{aligned}$$

Therefore,

$$F_Y(y) = 1 - P[Y > y] = 1 - (1 - F_{X_1}(y))(1 - F_{X_2}(y)) \cdots (1 - F_{X_n}(y)).$$

Example 2.7.4 A computer system consists of n subsystems, each of which has the same exponential distribution of time to failure. Each subsystem is independent but the whole computer system fails if any of the subsystems do. Find the distribution function F for system time to failure. If the mean time to failure of each subsystem is 2000 hours, and there are four subsystems, find the mean time to system failure and the probability that the time to failure exceeds 100 hours.

Solution   By Theorem 2.7.4, if there were n subsystems, then

$$F(t) = 1 - (e^{-\mu t})^n = 1 - e^{-n\mu t}, \qquad (2.7.31)$$

where $1/\mu$ is the average time to failure, since the distribution function for time to failure is

$$F_X(t) = 1 - e^{-\mu t}, \qquad (2.7.32)$$

for each subsystem. Thus, the system time to failure has an exponential distribution with mean value $1/(n\mu) = (1/\mu)/n$. If $1/\mu = 2000$ and $n = 4$, then the system mean time to failure is $2000/4 = 500$ hours. Hence, the distribution function for system time to failure is

$$F(t) = 1 - e^{-t/500}, \qquad (2.7.33)$$

and thus the probability it exceeds 100 hours is $e^{-100/500} = e^{-0.2} = 0.8187$.

The individual subsystems need not have the same mean time to failure in order for the overall system time to failure to have an exponential distribution. If each subsystem has an exponentially distributed time to failure with mean values $1/\mu_i$, $i = 1, 2, \ldots, n$, then the distribution function F for the system time to failure is given by

$$F(t) = 1 - \exp\left[-t(\mu_1 + \mu_2 + \cdots + \mu_n)\right], \qquad (2.7.34)$$

that is, the time to failure has an exponential distribution with mean value

$$1/(\mu_1 + \mu_2 + \cdots + \mu_n).$$

In the above example with $n = 4$, if the mean time to failure had been 1000 hours, 2000 hours, 3000 hours, 4000 hours, respectively, then the average time to failure would be

$$1 / \left(\frac{1}{1000} + \frac{1}{2000} + \frac{1}{3000} + \frac{1}{4000} \right) = \frac{1}{(12 + 6 + 4 + 3)/12,000}$$

$$= \frac{12,000}{25} = 480 \quad \text{hours.}$$

By (2.7.34) the probability that the system time to failure exceeds 100 hours is

$$e^{-100/480} = 0.81194.$$

In the next theorem the convolution method is given for calculating the pmf $p(\cdot)$ or the density function $f(\cdot)$ for the sum of two independent random variables. We will see, later, that transform methods make it easier to find $p(\cdot)$ or $f(\cdot)$ than the method of Theorem 2.7.5.

THEOREM 2.7.5 (*Convolution Theorem*) Let X and Y be jointly distributed random variables with $Z = X + Y$. Then the following hold.

(a) If X and Y are independent discrete random variables with each taking on the values 0, 1, 2, 3, 4, ..., then Z takes on the values $k = i + j$ $(i, j = 0, 1, 2, 3, ...)$ and

$$P[Z = k] = \sum_{i+j=k} p_X(i)p_Y(j) = \sum_{i=0}^{k} p_X(i)p_Y(k - i). \qquad (2.7.35)$$

(b) If X and Y are independent continuous random variables then

$$P[Z \le z] = \int_{-\infty}^{\infty} f_X(x)F_Y(z - x)\,dx = \int_{-\infty}^{z} f_Z(x)\,dx \qquad (2.7.36)$$

where the density function of Z is given by

$$f_Z(z) = \int_{-\infty}^{\infty} f_X(x)f_Y(z - x)\,dx = \int_{-\infty}^{\infty} f_X(z - y)f_Y(y)\,dy. \qquad (2.7.37)$$

Proof (a) The event $Z = k$ can be represented as

$$[Z = k] = [(X = 0) \cap (Y = k)] \cup [(X = 1) \cap (Y = k - 1)]$$
$$\cup [(X = 2) \cap (Y = k - 2)] \cup \cdots \cup [(X = k) \cap (Y = 0)].$$

Hence,

$$P[Z = k] = p(0, k) + p(1, k - 1) + \cdots + p(k, 0) = \sum_{i=0}^{k} p(i, k - i) \qquad (2.7.38)$$

where $p(\cdot, \cdot)$ is the joint density function of X and Y. But X and Y are independent so (2.7.38) yields (2.7.35) because $p(i, j) = p_X(i)p_Y(j)$ for each i and j.

(b)

$$P[Z \le z] = \iint_{x+y\le z} f_{X,Y}(x, y)\,dx\,dy$$

$$= \int_{-\infty}^{\infty} dx \int_{-\infty}^{z-x} f_{X,Y}(x, y)\,dy. \qquad (2.7.39)$$

This equation is valid for all jointly continuous random variables, independent or not. If we assume X and Y independent so that $f_{X,Y}(x, y) =$

$f_X(x)f_Y(y)$, then (2.7.39) yields

$$P[Z \le z] = \int_{-\infty}^{\infty} dx \int_{-\infty}^{z-x} f_X(x)f_Y(y)\, dy$$

$$= \int_{-\infty}^{\infty} f_X(x)\, dx \int_{-\infty}^{z-x} f_Y(y)\, dy$$

$$= \int_{-\infty}^{\infty} f_X(x)F_Y(z-x)\, dx. \tag{2.7.40}$$

Differentiation of (2.7.40) gives (2.7.37).
This completes the proof.

The sum $\sum_{i=0}^{k} p_X(i)p_Y(k-i)$ is called the *convolution* of p_X and p_Y and designated $p_X * p_Y(k)$. Thus Theorem 2.7.5 asserts that the pmf of the sum of two independent discrete random variables each of which assumes only nonnegative integer values, is the convolution of the individual pmf's. Similarly, the integral $\int_{-\infty}^{\infty} f_X(x)f_Y(z-x)\, dx$ is called the *convolution* of f_X and f_Y. By symmetry it can also be calculated as $\int_{-\infty}^{\infty} f_X(z-y)f_Y(y)\, dy$. Theorem 2.7.5 shows that the density of the sum of two independent continuous random variables is the convolution of the individual densities.

Example 2.7.5 Let X and Y be independent random variables each having an exponential distribution with parameter λ. Find the density function of $Z = X + Y$.

Solution The density function is the same for each random variable:

$$f_X(x) = f_Y(x) = \begin{cases} \lambda e^{-\lambda x} & \text{for} \quad x > 0 \\ 0 & \text{for} \quad x \le 0. \end{cases}$$

Thus $f_{X+Y}(z) = 0$ for $z \le 0$ and by (2.7.37), for $z > 0$

$$f_{X+Y}(z) = \int_{0}^{z} f_X(x)f_Y(z-x)\, dx.$$

(We have used the fact that $f_Y(z) = 0$ for $z < 0$.) Hence,

$$f_{X+Y}(z) = \int_{0}^{z} \lambda e^{-\lambda x} \lambda e^{-\lambda(z-x)}\, dx = \lambda^2 e^{-\lambda z} \int_{0}^{z} dx = \lambda^2 z e^{-\lambda z}.$$

Thus $Z = X + Y$ has an Erlang-2 density function. We will study the Erlang family of random variables in Chapter 3.

Example 2.7.6 If $\lambda > 0$, the discrete random variable which assumes the values 0, 1, 2, 3, ... and which has the pmf $p(\cdot)$ defined by

$$p(k) = e^{-\lambda}\frac{\lambda^k}{k!}, \qquad k = 0, 1, 2, \ldots, \tag{2.7.41}$$

is called a *Poisson random variable with parameter* λ or is said to have a *Poisson distribution*. We will investigate Poisson random variables in more detail in Chapter 3. Assuming that (2.7.41) does define a pmf, let us calculate the pmf $p(\cdot)$ for the sum of two independent Poisson distributed random variables, one with parameter λ and one with parameter μ. If X is the first random variable and Y the second, then by (2.7.35) of Theorem 2.7.5 and (2.7.41),

$$p(k) = \sum_{i=0}^{k} p_X(i)p_Y(k-i) = \sum_{i=0}^{k} e^{-\lambda}\frac{\lambda^i}{i!}e^{-\mu}\frac{\mu^{k-i}}{(k-i)!}$$

$$= \frac{e^{-(\lambda+\mu)}}{k!}\sum_{i=0}^{k}\frac{k!}{i!\,(k-i)!}\lambda^i\mu^{k-i}$$

$$= \frac{e^{-(\lambda+\mu)}}{k!}\sum_{i=0}^{k}\binom{k}{i}\lambda^i\mu^{k-i} = \frac{e^{-(\lambda+\mu)}}{k!}(\lambda+\mu)^k. \qquad (2.7.42)$$

The last equality is true by the corollary to Theorem 2.3.3. Thus the sum of two independent Poisson random variables is another Poisson random variable whose parameter is the sum of the original parameters, that is, $X + Y$ is Poisson with parameter $\lambda + \mu$. For this reason Poisson random variables are said to have the *reproductive property*. Exponential random variables do *not* have this property as we saw in Example 2.7.5.

2.8 CONDITIONAL EXPECTATION

In Example 2.6.1 we assumed that a communication line had five terminals attached, three of which were ready to transmit, and we calculated the mean and standard deviation of X, the number of polls required to find the first ready terminal. X depends upon the number of terminals in the ready state. Thus, if we let Y be the random variable giving the number of ready terminals, we are interested in the average value of X given that Y assumes one of the values 0, 1, 2, 3, 4, or 5. This is the *conditional expectation of X given Y*, which we now define, formally.

Suppose X and Y are discrete random variables assuming the values x_1, $x_2, \ldots$, and $y_1, y_2, \ldots$, respectively. Then for each y_j such that $p_Y(y_j) > 0$ we define the *conditional probability mass function of X given that $Y = y_j$* by

$$p_{X|Y}(x_i|y_j) = \frac{p(x_i, y_j)}{p_Y(y_j)}, \qquad i = 1, 2, 3, \ldots, \qquad (2.8.1)$$

where $p(\cdot, \cdot)$ is the joint probability mass function of X and Y. We then define the *conditional expectation of X given that $Y = y_j$* for all y_j such that

$p_Y(y_j) > 0$ by

$$E[X \mid Y = y_j] = \sum_{x_i} x_i p_{X \mid Y}(x_i \mid y_j). \tag{2.8.2}$$

Similarly, we define the *conditional kth moment of X given that* $Y = y_j$ by

$$E[X^k \mid Y = y_j] = \sum_{x_i} x_i^k p_{X \mid Y}(x_i \mid y_j), \qquad k = 1, 2, \ldots. \tag{2.8.3}$$

The motivation for (2.8.1) is that

$$P[X = x_i \mid Y = y_j] = P[X = x_i, Y = y_j]/P[Y = y_j] = \frac{p(x_i, y_j)}{p_Y(y_j)}.$$

In reality the pmf $p(\cdot)$ we defined in Example 2.6.1 is $p_{X \mid Y}(\cdot, 3)$, and we calculated the expected value of X given that $Y = 3$. We give the other conditional expectation values for Example 2.6.1 in Table 2.8.1. We use formula (2.6.11), which is a general formula for the case of m terminals on a line with n ready to transmit where $1 \leq n \leq m$. Formula (2.6.11) is *not* valid when $n = 0$, that is, no terminals are ready to transmit; we assume that it takes m polls to discover this fact, so that

$$E[X \mid Y = 0] = m.$$

The values shown in Table 2.8.1 for $y \geq 1$ can easily be calculated using the APL function POLL in Appendix B.

TABLE 2.8.1

Conditional Expectation of Number of Polls for Example 2.6.1

y:	0	1	2	3	4	5
$E[X \mid Y = y]$:	5.0	3.0	2.0	1.5	1.2	1.0

Suppose X and Y are jointly continuous with the joint density function f. Then the *conditional probability density function of X given that* $Y = y$, is defined for all values of y such that $f_Y(y) > 0$, by

$$f_{X \mid Y}(x \mid y) = \frac{f(x, y)}{f_Y(y)}. \tag{2.8.4}$$

The *conditional expectation of X, given that* $Y = y$, is defined for all values of y such that $f_Y(y) > 0$, by

$$E[X \mid Y = y] = \int_{-\infty}^{\infty} x f_{X \mid Y}(x \mid y) \, dx. \tag{2.8.5}$$

The *conditional* kth *moment of* X, *given that* $Y = y$, is defined for all values of y such that $f_Y(y) > 0$, by

$$E[X^k \mid Y = y] = \int_{-\infty}^{\infty} x^k f_{X|Y}(x \mid y) \, dx, \qquad k = 1, 2, 3, \ldots. \tag{2.8.6}$$

Thus the 1st conditional moment is the conditional expectation.

Example 2.8.1 The jointly continuous random variables X and Y of Example 2.7.2 have the joint density function

$$f(x, y) = \begin{cases} xy \exp[-\tfrac{1}{2}(x^2 + y^2)] & \text{for } x > 0 \text{ and } y > 0 \\ 0 & \text{otherwise.} \end{cases} \tag{2.8.7}$$

and

$$f_Y(y) = \begin{cases} y \exp[-\tfrac{1}{2}y^2] & \text{for } y > 0 \\ 0 & \text{otherwise.} \end{cases} \tag{2.8.8}$$

Hence, if $y > 0$,

$$f_{X|Y}(x \mid y) = \frac{f(x, y)}{f_Y(y)} = x \exp[-\tfrac{1}{2}x^2] = f_X(x) \qquad \text{for all } y > 0. \tag{2.8.9}$$

Thus, $f_{X|Y}(x, y)$ is independent of the particular value of y. This is to be expected since X and Y are independent. (They are independent since $f(x, y) = f_X(x) f_Y(y)$.) Hence, for each $y > 0$

$$E[X \mid Y = y] = E[X] = \int_0^{\infty} x^2 \exp[-\tfrac{1}{2}x^2] \, dx.$$

This integral is difficult to evaluate but its value is $\sqrt{2\pi}/2$.

Given jointly distributed random variables X and Y, $E[X \mid Y = y]$ is a function of the random variable Y, say $h(Y) = E[X \mid Y]$. Thus $h(Y)$ is a random variable having an expected value $E[h(Y)] = E[E[X \mid Y]]$. If Y is discrete and assumes the values $y_1, y_2, \ldots$, then

$$E[E[X \mid Y]] = \sum_{y_j} E[X \mid Y = y_j] P[Y = y_j], \tag{2.8.10}$$

while if Y is continuous with density function f_Y, then

$$E[E[X \mid Y]] = \int_{-\infty}^{\infty} E[X \mid Y = y] f_Y(y) \, dy. \tag{2.8.11}$$

Formulas (2.8.10) and (2.8.11) can be formally represented by the equation

$$E[E[X \mid Y]] = \int_{-\infty}^{\infty} E[X \mid Y = y] \, dF_Y(y), \tag{2.8.12}$$

where the integral in question is a *Stieltjes* integral, which, of course, is calculated by (2.8.10), when Y is discrete, and by (2.8.11), when Y is continuous. The Stieltjes integral can also be used to evaluate $E[h(Y)]$ when Y is neither discrete nor continuous, but this is beyond the scope of the book. The interested reader can consult Parzen [4, pp. 233–235] or Apostol [5, Chapter 7].

The next theorem shows how to evaluate $E[X]$, $E[X^k]$ in terms of $E[X \mid Y]$ and $E[X^k \mid Y]$.

THEOREM 2.8.1   Let X and Y be jointly distributed random variables. Then

$$E[E[X \mid Y]] = \int_{-\infty}^{\infty} E[X \mid Y = y]\, dF_Y(y) = E[X] \qquad (2.8.13)$$

and

$$E[E[X^k \mid Y]] = \int_{-\infty}^{\infty} E[X^k \mid Y = y]\, dF_Y(y) = E[X^k], \qquad k = 1, 2, 3, \ldots .$$
$$(2.8.14)$$

(Equation (2.8.13) is known as the *law of total expectation* and (2.8.14) as the *law of total moments*.)

Proof   We prove (2.8.14) for the cases

(a) X and Y are discrete, and
(b) X and Y are continuous.

Equation (2.8.13) is a special case of (2.8.14). We omit the proof for the cases when one of X, Y is continuous and the other is discrete.

Case (a)   Suppose that X and Y are discrete.

$$E[E[X^k \mid Y]] = \sum_{y_j} E[X^k \mid Y = y_j] p_Y(y_j) = \sum_{y_j} \sum_{x_i} x_i^k P[X = x_i \mid Y = y_j] p_Y(y_j)$$

$$= \sum_{y_j} \sum_{x_i} x_i^k \frac{p(x_i, y_j)}{p_Y(y_j)} p_Y(y_j) = \sum_{y_j} \sum_{x_i} x_i^k p(x_i, y_j)$$

$$= \sum_{x_i} x_i^k \sum_{y_j} p(x_i, y_j) = \sum_{x_i} x_i^k p_X(x_i) = E[X^k].$$

This proves that $E[X^k] = E[E[X^k \mid Y]]$ when X and Y are both discrete.

Case (*b*) Suppose that X and Y are continuous. Then

$$E[E[X^k \mid Y]] = \int_{-\infty}^{\infty} E[X^k \mid Y] f_Y(y)\, dy = \int_{-\infty}^{\infty} \left[\int_{-\infty}^{\infty} x^k f_{X|Y}(x \mid y)\, dx \right] f_Y(y)\, dy$$

$$= \int_{-\infty}^{\infty} \int_{-\infty}^{\infty} x^k \frac{f(x,y)}{f_Y(y)} f_Y(y)\, dx\, dy = \int_{-\infty}^{\infty} \int_{-\infty}^{\infty} x^k f(x,y)\, dx\, dy$$

$$= \int_{-\infty}^{\infty} x^k\, dx \int_{-\infty}^{\infty} f(x,y)\, dy = \int_{-\infty}^{\infty} x^k f_X(x)\, dx = E[X^k].$$

This completes the proof of Case (b).

Example 2.8.2 Consider Example 2.6.1 in which five terminals were connected to one communication line. Let the value of X be the number of polls until the first ready terminal is found when there is a ready terminal and five, otherwise; let Y be the number of terminals ready. We can find $E[X]$ by Theorem 2.8.1 if we know the pmf for Y. Let us assume that $p_Y(\cdot)$ is as shown in Table 2.8.2. This pmf was calculated assuming that each terminal was independent of the others and had probability 0.5 of being ready to transmit. Y has a *binomial distribution*, which is discussed in Example 2.9.5 and in Chapter 3. Using the data in the table, we see that

$$E[X] = (5 \times 1 + 3 \times 5 + 2 \times 10 + 1.5 \times 10 + 1.2 \times 5 + 1)/32 = 1.9375$$

and

$$E[X^2] = (25 + 11 \times 5 + 5 \times 10 + 2.7 \times 10 + 1.6 \times 5 + 1)/32 = 5.1875.$$

TABLE 2.8.2

Data for Example 2.8.2

k	$p_Y(k)$	$E[X \mid Y = k]$	$E[X^2 \mid Y = k]$
0	1/32	5.0	25
1	5/32	3.0	11
2	10/32	2.0	5.0
3	10/32	1.5	2.7
4	5/32	1.2	1.6
5	1/32	1.0	1.0

The APL function MPOLL calculates the expected number of polls, and second moment of number of polls, as well as $E[X \mid Y = y]$ and $E[X^2 \mid Y = y]$ for $y = 0, 1, \ldots, m$, given the pmf of Y, using the APL functions POLLM and POLL2M. (The pmf of X is given by (2.6.11) which the reader is asked to prove in Exercise 6.)

Example 2.8.3   An on-line computer system receives inquiry messages
of n different types. The message length distribution for each type i is a
random variable X_i, $i = 1, 2, \ldots, n$. The message type of the current message
is given by the discrete random variable Y which assumes the values 1
through n. The message length X of the current message is determined by X_i
if and only if $Y = i$. Thus $E[X \mid Y = i] = E[X_i]$ and $E[X^2 \mid Y = i] = E[X_i^2]$.
Therefore, by Theorem 2.8.1,

$$E[X] = \sum_{i=1}^{n} E[X \mid Y = i]p_Y(i) = \sum_{i=1}^{n} E[X_i]p_Y(i), \qquad (2.8.15)$$

and

$$E[X^2] = \sum_{i=1}^{n} E[X^2 \mid Y = i]p_Y(i) = \sum_{i=1}^{n} E[X_i^2]p_Y(i). \qquad (2.8.16)$$

From (2.8.15) and (2.8.16) we can calculate

$$\mathrm{Var}[X] = E[X^2] - (E[X])^2.$$

As an example, suppose ten different types of messages arrive at the central
computer system. The fraction of each type, as well as the mean and standard deviation of the message length (in characters) of each type, are shown
in Table 2.8.3. Find the mean and the standard deviation of the message
length of the mix of all messages which arrive at the central computer
system.

TABLE 2.8.3
Message Length Data for Example 2.8.3

Message type	$p_Y(i)$ (fraction of type)	$E[X_i]$ (mean length)	σ_{X_i} (standard deviation)
1	0.100	100	10
2	0.050	120	12
3	0.200	200	20
4	0.050	75	5
5	0.025	300	25
6	0.075	160	40
7	0.150	360	36
8	0.050	50	4
9	0.150	60	3
10	0.150	130	10

Solution   Theorem 2.8.1 applies as outlined above. The expected value
$E[X]$ is calculated by (2.8.15) to yield 164.25 characters. To apply (2.8.16) we
calculate $E[X_i^2]$ by the formula $E[X_i^2] = E[X_i]^2 + \mathrm{Var}[X_i]$ for $i = 1, 2, \ldots,$

10, to obtain the respective values 10,100; 14,544; 40,400; 5,650; 90,625; 27,200; 130,896; 2,516; 3,609; and 17,000. Then, by (2.8.16), we calculate

$$E[X^2] = \sum_{i=1}^{10} E[X_i^2]p_Y(i) = 37{,}256.875.$$

Thus

$$\text{Var}[X] = E[X^2] - E[X]^2 = 37{,}256.875 - 26{,}978.0625 = 10{,}278.8125$$

so that $\sigma_X = 101.38$ characters. The reader might be tempted to conclude that

$$\text{Var}[X] = \text{Var}[X_1]p_Y(1) + \text{Var}[X_2]p_Y(2) + \cdots + \text{Var}[X_{10}]p_Y(10),$$

but this formula is incorrect. In the present case it would yield $\text{Var}[X] = 445.625$ or $\sigma_X = 21.11$, although the correct value of σ_X is 101.38. The APL function CONDEXPECT can be used to make calculations such as in this example with little effort.

2.9 TRANSFORM METHODS

Calculating the mean, the variance, and the moments of a random variable can be a tedious process. In the case of a discrete random variable it is often true that complicated sums must be evaluated; for a continuous random variable the integrals involved may be difficult to evaluate. These difficulties can often be overcome by transform methods.

The *moment generating function* $\psi(\cdot)$ (or $\psi_X(\cdot)$) of a random variable X is defined by $\psi(\theta) = E[e^{\theta X}]$ for all real θ such that $E[e^{\theta X}]$ is finite. Thus,

$$\psi(\theta) = \begin{cases} \sum\limits_{x_i} e^{\theta x_i} p(x_i) & \text{if } X \text{ is discrete} \\ \int_{-\infty}^{\infty} e^{\theta x} f(x)\, dx & \text{if } X \text{ is continuous.} \end{cases} \tag{2.9.1}$$

We say that X *has a moment generating function* if there exists a $\delta > 0$ such that $\psi(\theta)$ is finite for all $|\theta| \le \delta$. There are random variables without moment generating functions, such as the random variable of Example 2.6.3 and that of Example 2.6.4. However, most random variables of concern to us have moment generating functions. In defining the moment generating function $\psi_X(\cdot)$ we have *transformed* the random variable X, which is defined on a sample space, into the function $\psi_X(\cdot)$ defined for some set of real numbers. The next theorem gives some important properties of the moment generating function.

THEOREM 2.9.1 (*Properties of the Moment Generating Function*) Let X and Y be random variables for which the moment generating functions $\psi_X(\cdot)$ and $\psi_Y(\cdot)$ exist. Then the following hold.

(a) $F_X = F_Y$ if and only if $\psi_X(\cdot) = \psi_Y(\cdot)$ (uniqueness).

(b) The coefficient of θ^n in the power series for $\psi_X(\theta)$,

$$\psi_X(\theta) = \sum_{n=0}^{\infty} \psi_X^{(n)}(0) \frac{\theta^n}{n!},\qquad (2.9.2)$$

is $E[X^n]/n!$, so that

$$E[X^n] = \frac{d^n \psi_X(\theta)}{d\theta^n}\bigg|_{\theta=0}.\qquad (2.9.3)$$

(c) $\psi_{X+Y}(\theta) = \psi_X(\theta)\psi_Y(\theta)$ for all θ, if X and Y are independent.

Proof (a) The proof of (a) is beyond the scope of this book. The proof is carried out by showing that an inverse transform exists which maps $\psi_X(\cdot)$ to $F_X(\cdot)$.

(b) The power series

$$e^{\theta x} = 1 + x\theta + x^2 \frac{\theta^2}{2!} + \cdots + x^n \frac{\theta^n}{n!} + \cdots \qquad (2.9.4)$$

converges uniformly in x for $|\theta| \le \delta_1 < \delta$.

Hence, we can calculate the expectation term by term to get

$$\psi_X(\theta) = E[e^{\theta x}] = 1 + E[X]\theta + \cdots + E[X^n]\frac{\theta^n}{n!} + \cdots. \qquad (2.9.5)$$

Since the infinite series representation of a function is unique, a comparison of the coefficients of θ^n in (2.9.5) with those of (2.9.2) shows that (2.9.3) is true.

(c) If X and Y are independent then,

$$\psi_{X+Y}(\theta) = E[e^{\theta(X+Y)}] = E[e^{\theta X}e^{\theta Y}] = E[e^{\theta X}]E[e^{\theta Y}] = \psi_X(\theta)\psi_Y(\theta). \qquad (2.9.6)$$

The third equality is true by Theorem 2.7.1. This completes the proof of Theorem 2.9.1.

It is immediate by mathematical induction that if $X_1, X_2, \ldots, X_n$ are independent random variables, then

$$\psi_{X_1+X_2+\cdots+X_n}(\theta) = \psi_{X_1}(\theta)\psi_{X_2}(\theta) \cdots \psi_{X_n}(\theta),$$

for all θ such that

$$\psi_{X_i}(\theta) \quad \text{is defined for} \quad i = 1, 2, \ldots, n.$$

We now give some examples of how Theorem 2.9.1 can be applied.

Example 2.9.1 Let X be an exponential random variable with parameter λ (see Examples 2.5.3 and 2.6.2). Then

$$\psi(\theta) = \int_0^\infty \lambda e^{\theta x} e^{-\lambda x} \, dx = \lambda \int_0^\infty e^{-x(\lambda - \theta)} \, dx.$$

If $\theta < \lambda$, then

$$\psi(\theta) = -\frac{\lambda}{\lambda - \theta} e^{-x(\lambda - \theta)} \Big|_0^\infty = \frac{\lambda}{\lambda - \theta}. \qquad (2.9.7)$$

Hence,

$$\frac{d\psi}{d\theta} = \frac{\lambda}{(\lambda - \theta)^2}$$

so that by Theorem 2.9.1b,

$$E[X] = \frac{d\psi}{d\theta}\Big|_{\theta = 0} = \frac{1}{\lambda}.$$

Also,

$$\frac{d^2\psi}{d\theta^2} = \frac{2\lambda}{(\lambda - \theta)^3}$$

so that again by Theorem 2.9.1b,

$$E[X^2] = \frac{d^2\psi}{d\theta^2}\Big|_{\theta = 0} = \frac{2}{\lambda^2}.$$

Thus,

$$\mathrm{Var}[X] = E[X^2] - (E[X])^2 = \frac{2}{\lambda^2} - \frac{1}{\lambda^2} = \frac{1}{\lambda^2}.$$

We can use (2.9.7) to generate a simple formula for all the moments of X. If $\theta < \lambda$, then

$$\psi(\theta) = \frac{\lambda}{\lambda - \theta} = \frac{1}{1 - (\theta/\lambda)} = 1 + \frac{\theta}{\lambda} + \left(\frac{\theta}{\lambda}\right)^2 + \cdots + \left(\frac{\theta}{\lambda}\right)^n + \cdots. \quad (2.9.8)$$

Here, we have used the fact that, if $|x| < 1$ then

$$\frac{1}{1 - x} = 1 + x + x^2 + \cdots + x^n + \cdots \qquad \text{(the geometric series)}.$$

Equating the coefficients of θ^n in (2.9.5) and (2.9.8) yields

$$\frac{E[X^n]}{n!} = \frac{1}{\lambda^n} \qquad \text{or} \qquad E[X^n] = \frac{n!}{\lambda^n} = n! \, E[X]^n, \qquad n = 1, 2, 3, \ldots.$$

Thus we have found all the moments of the exponential distribution with very little effort.

Example 2.9.2 Let X be a Poisson random variable with parameter λ (see Example 2.7.6). Then

$$\psi(\theta) = \sum_{k=0}^{\infty} e^{\theta k} p(k) = \sum_{k=0}^{\infty} e^{k\theta} e^{-\lambda} \frac{\lambda^k}{k!} = e^{-\lambda} \sum_{k=0}^{\infty} \frac{(\lambda e^{\theta})^k}{k!} = e^{-\lambda} e^{\lambda e^{\theta}} = e^{\lambda(e^{\theta}-1)}.$$

Thus

$$E[X] = \frac{d\psi}{d\theta}\bigg|_{\theta=0} = \lambda e^{\theta} e^{\lambda(e^{\theta}-1)}\bigg|_{\theta=0} = \lambda,$$

and

$$E[X^2] = \frac{d^2\psi}{d\theta^2}\bigg|_{\theta=0} = \lambda e^{\theta} e^{\lambda(e^{\theta}-1)}(1 + \lambda e^{\theta})\bigg|_{\theta=0} = \lambda(1 + \lambda) = \lambda + \lambda^2.$$

Hence,

$$\text{Var}[X] = E[X^2] - (E[X])^2 = \lambda^2 + \lambda - \lambda^2 = \lambda.$$

Thus both the mean and variance of a Poisson random variable with parameter λ are equal to λ.

Example 2.9.3 Let X and Y be independent Poisson random variables with parameters λ and μ, respectively. Using moment generating functions show that the random variable $X + Y$ is also a Poisson random variable with parameter $\lambda + \mu$.

Solution (We have already proven the result in Example 2.7.6 using the method of convolutions.) By Theorem 2.9.1 and Example 2.9.2,

$$\psi_{X+Y}(\theta) = \psi_X(\theta)\psi_Y(\theta) = e^{\lambda(e^{\theta}-1)} e^{\mu(e^{\theta}-1)} = e^{(\lambda+\mu)(e^{\theta}-1)}. \qquad (2.9.9)$$

Since (2.9.9) is the moment generating function of a Poisson random variable with parameter $\lambda + \mu$, we conclude, by the uniqueness of the moment generating function, that $X + Y$ has a Poisson distribution with parameter $\lambda + \mu$. That is,

$$P[X + Y = k] = e^{-(\lambda+\mu)} \frac{(\lambda + \mu)^k}{k!}, \qquad k = 0, 1, 2, \ldots. \qquad (2.9.10)$$

The uniqueness of the moment generating function guarantees that no random variable which does *not* have a Poisson distribution can have the *same* moment generating function as a Poisson random variable has.

Let X be a discrete random variable assuming only nonnegative integer values and let $p(j) = P[X = j] = p_j, j = 0, 1, 2, \ldots$. Then the function $g(z) =$

$g_X(z)$ defined by

$$g(z) = \sum_{j=0}^{\infty} p_j z^j = p_0 + p_1 z + p_2 z^2 + \cdots \qquad (2.9.11)$$

is called the *generating function* of X or the *z-transform of X*. Since $g(1) = p_0 + p_1 + p_2 + \cdots = 1$, $g(z)$ converges at least for $|z| \le 1$. The next theorem states some of the useful properties of the generating function.

THEOREM 2.9.2 (*Properties of the Generating Function or z-transform*) Let X and Y be discrete random variables assuming only non-negative integer values. Then the following hold.

(a) X and Y have the same distribution, if and only if, $g_X(z) = g_Y(z)$ (uniqueness).

(b) $\displaystyle p_n = \frac{1}{n!} \left. \frac{d^n g_X(z)}{dz^n} \right|_{z=0}, \qquad n = 0, 1, 2, \ldots.$

(c) $E[X] = g_X'(1)$ and $\mathrm{Var}[X] = g_X''(1) + g_X'(1) - (g_X'(1))^2$.

(d) $g_{X+Y}(z) = g_X(z)g_Y(z)$, if X and Y are independent.

Proof (a) Since

$$g_X(z) = p_0 + p_1 z + p_2 z^2 + \cdots$$

is a convergent power series for $|z| \le 1$, $g_X(z)$ is unique by the uniqueness of power series.

(b) Also true by uniqueness of power series.

(c) $\displaystyle g_X'(z) = \sum_{j=1}^{\infty} j p_j z^{j-1} = p_1 + 2p_2 z + 3p_3 z^2 + \cdots.$

Hence,

$$g_X'(1) = \sum_{j=1}^{\infty} j p_j = E[X].$$

$$g_X''(z) = 2p_2 + 3 \times 2p_3 z + 4 \times 3p_4 z^2 + 5 \times 4p_5 z^3 + \cdots$$

$$= \sum_{j=2}^{\infty} j(j-1) p_j z^{j-2}.$$

Thus

$$g_X''(1) = \sum_{j=1}^{\infty} j(j-1) p_j = E[X(X-1)] = E[X^2] - E[X].$$

Therefore,

$$\mathrm{Var}[X] = E[X^2] - (E[X])^2 = g_X''(1) + g_X'(1) - (g_X'(1))^2.$$

(d)   Let $a_k = P[X = k]$, $b_k = P[Y = k]$, $c_k = P[X + Y = k]$, $k = 0, 1,$ $2, \ldots$. Then we know by Theorem 2.7.5 that the sequence $\{c_k\}$ is the convolution of the sequences $\{a_k\}$ and $\{b_k\}$, that is,

$$c_k = \sum_{i=0}^{k} a_i b_{k-i}, \qquad k = 0, 1, 2, \ldots. \tag{2.9.12}$$

Moreover, if we formally multiply together the power series for $g_X(z)$ and $g_Y(z)$, we get

$$g_X(z)g_Y(z) = \left(\sum_{k=0}^{\infty} a_k z^k \right) \left(\sum_{k=0}^{\infty} b_k z^k \right) = \sum_{k=0}^{\infty} \left(\sum_{i=0}^{k} a_i b_{k-i} \right) z^k = g_{X+Y}(z).$$

This completes the proof of Theorem 2.9.2.

The reader should note that a discrete random variable which takes on only nonnegative integer values has the moment generating function

$$\psi(\theta) = \sum_{k=0}^{\infty} e^{k\theta} p_k = \sum_{k=0}^{\infty} (e^{\theta})^k p_k = g(e^{\theta}), \tag{2.9.13}$$

that is, the moment generating function is obtained from the generating function (z-transform) by a simple change of variable. Similarly, $g(z) = \psi(\ln z)$, if $z > 0$.

Example 2.9.4   A random variable X is called a Bernoulli random variable (has a Bernoulli distribution) if it can assume only two values, usually taken to be 1 and 0; the first with probability p and the second with probability $q = 1 - p$. Find the mean and variance of a Bernoulli random variable X, using its generating function.

Solution   The generating function of X is

$$g(z) = q + pz. \tag{2.9.14}$$

Hence, $g'(z) = p$ and $g''(z) = 0$. Therefore, by Theorem 2.9.2,

$$E[X] = g'(1) = p$$

and

$$\text{Var}[X] = g''(1) + g'(1) - (g'(1))^2 = 0 + p - p^2 = p(1 - p) = pq.$$

Example 2.9.5   The Bernoulli random variable X, discussed in Example 2.9.4, is often used to describe a random experiment with but two outcomes, "success" or "failure." X is 1 for a success and 0 for a failure. Such an experiment is called a *Bernoulli trial*. A sequence of n such trials is called a *Bernoulli sequence of trials* if the probability of success does not change from trial to trial. An example is tossing a coin repeatedly, with a head considered a success. Let Y be the random variable which counts the number of successes in a Bernoulli sequence of n trials, where $n \geq 1$. Then we can write

$$Y = X_1 + X_2 + \cdots + X_n, \tag{2.9.15}$$

where $X_1, X_2, \ldots, X_n$ is a collection of identical independent Bernoulli random variables. Hence by Theorem 2.9.2,

$$g_Y(z) = (q + pz)^n = \sum_{k=0}^{n} P[Y = k]z^k. \tag{2.9.16}$$

But, by the binomial theorem (the corollary to Theorem 2.3.3)

$$(pz + q)^n = \sum_{k=0}^{n} \binom{n}{k} (pz)^k q^{n-k}. \tag{2.9.17}$$

Equating coefficients of z^k in (2.9.17) and (2.9.16) gives

$$p(k) = P[Y = k] = \binom{n}{k} p^k q^{n-k}, \qquad k = 0, 1, 2, \ldots, n. \tag{2.9.18}$$

Y is called a *binomial random variable* or said to have a *binomial distribution*. (The binomial distribution will be discussed more completely in Chapter 3.)

Since $g_Y(z) = (q + pz)^n$, $g_Y'(z) = n(q + pz)^{n-1}p$ and $E[Y] = g_Y'(1) = np$. Also, $g_Y''(z) = n(n-1)(q + pz)^{n-2}p^2$. Hence, by Theorem 2.9.2,

$$\mathrm{Var}[X] = g_Y''(1) + g_Y'(1) - (g_Y'(1))^2 = n(n-1)p^2 + np - n^2p^2$$

$$= np(1-p) = npq.$$

Let X be a random variable such that $P[X < 0] = 0$. Then the *Laplace–Stieltjes transform of X* is defined for $\theta \geq 0$, by

$$X^*(\theta) = E[e^{-\theta X}] = \begin{cases} \displaystyle\int_0^{\infty} e^{-\theta x} f(x)\, dx & \text{if } X \text{ is continuous} \\ \displaystyle\sum_{x_i} e^{-\theta x_i} p(x_i) & \text{if } X \text{ is discrete.} \end{cases} \tag{2.9.19}$$

Sometimes $X^*(\theta)$ is called the Laplace–Stieltjes transform of F. The integral $\int_0^{\infty} e^{-\theta x} f(x)\, dx$ is called the *Laplace transform of f*. Many authors write

$$X^*(\theta) = \int_0^{\infty} e^{-\theta x}\, dF(x), \tag{2.9.20}$$

where the integral is called a *Stieltjes integral*. However, the integral is always evaluated as we have shown in (2.9.19), that is, as $\int_0^{\infty} e^{-\theta x} f(x)\, dx$ if X is continuous and as $\sum_{x_i} e^{-\theta x_i} p(x_i)$ if X is discrete.

THEOREM 2.9.3 (*Properties of the Laplace–Stieltjes Transform*)   Let X and Y be random variables with Laplace–Stieltjes transforms $X^*(\cdot)$ and $Y^*(\cdot)$. Then the following hold.

(a) $F_X = F_Y$ if and only if $X^*(\cdot) = Y^*(\cdot)$ (uniqueness).

(b)   For $\theta > 0$, $X^*(\theta)$ has derivatives of all orders given by

$$\frac{d^n X^*}{d\theta^n} = \begin{cases} (-1)^n \displaystyle\int_0^\infty e^{-\theta x} x^n f(x)\, dx & \text{if } X \text{ is continuous} \\[2mm] (-1)^n \displaystyle\sum_{x_i} e^{-\theta x_i} x_i^{\,n} p(x_i) & \text{if } X \text{ is discrete.} \end{cases} \qquad (2.9.21)$$

(c)   If $E[X^n]$ exists, then

$$E[X^n] = (-1)^n \frac{d^n X^*}{d\theta^n}\bigg|_{\theta=0}. \qquad (2.9.22)$$

In particular, if $E[X]$ and $E[X^2]$ exist, then

$$E[X] = -\frac{dX^*}{d\theta}(0), \qquad E[X^2] = \frac{d^2 X^*}{d\theta^2}(0). \qquad (2.9.23)$$

(d) $(X + Y)^*(\theta) = X^*(\theta)Y^*(\theta)$, if X and Y are independent.

The proof of Theorem 2.9.3 is beyond the scope of this book but may be found in Feller [3].

Example 2.9.6   Let X be an exponential random variable with parameter λ, that is,

$$f(x) = \begin{cases} \lambda e^{-\lambda x} & \text{if } 0 < x \\ 0 & \text{otherwise.} \end{cases} \qquad (2.9.24)$$

Then, if $\theta < \lambda$,

$$X^*(\theta) = \int_0^\infty \lambda e^{-\theta x} e^{-\lambda x}\, dx = \lambda \int_0^\infty e^{-(\theta + \lambda)x}\, dx = \frac{\lambda}{\lambda + \theta}. \qquad (2.9.25)$$

2.10   INEQUALITIES AND APPLICATIONS

In this section we consider some inequalities and their uses. One important application is the derivation of the law of large numbers.

THEOREM 2.10.1   (*Markov's Inequality*)   Let X be a random variable with expected value $E[X]$ and such that $P[X < 0] = 0$. Then, for each $t > 0$,

$$P[X \ge t] \le \frac{E[X]}{t}. \qquad (2.10.1)$$

Proof We give the proof for discrete X. The proof when X is continuous is similar.

$$E[X] = \sum_{x_i} x_i p(x_i) = \sum_{x_i < t} x_i p(x_i) + \sum_{t \leq x_i} x_i p(x_i) \geq \sum_{t \leq x_i} x_i p(x_i) \geq \sum_{t \leq x_i} t p(x_i)$$

$$= t P[X \geq t].$$

Hence, $P[X \geq t] \leq E[X]/t$.

Example 2.10.1 Suppose an on-line computer system is proposed for which it is estimated that the mean response time $E[T]$ is 10 seconds. Use Markov's inequality to estimate the probability that the response time T will be 20 seconds or more.

Solution By Markov's inequality,

$$P[T \geq 20] \leq E[T]/20 = \tfrac{10}{20} = \tfrac{1}{2}.$$

It should be noted that Markov's inequality implies that

$$E[X \geq kE[X]] \leq 1/k, \qquad k > 0.$$

This inequality usually gives rather crude estimates because only the value of $E[X]$ is assumed to be known. Chebychev's inequality, in which the standard deviation is also assumed to be known, gives better probability estimates.

THEOREM 2.10.2 *(Chebychev's Inequality)* Let X be a random variable with mean $E[X]$ and standard deviation σ. Then for every $t > 0$,

$$P[|X - E[X]| \geq t] \leq \sigma^2/t^2, \tag{2.10.2}$$

or

$$P[|X - E[X]| \geq t\sigma] \leq 1/t^2. \tag{2.10.3}$$

Proof Applying Markov's inequality to $(X - E[X])^2$, with t^2 in place of t, yields

$$P[(X - E[X])^2 \geq t^2] \leq E[(X - E[X])^2]/t^2 = \sigma^2/t^2. \tag{2.10.4}$$

However, $(X - E[X])^2 \geq t^2$, if and only if, $|X - E[X]| \geq t$. Substituting this relation into (2.10.4) yields (2.10.2). Equation (2.10.3) follows from (2.10.2) by using $t\sigma$ in place of t in (2.10.2).

Example 2.10.2 Suppose that, for the proposed on-line computer system of Example 2.10.1, it is estimated that the standard deviation of response time is 2 seconds. Use Chebychev's inequality to estimate the probability that the response time will be between 4 and 16 seconds.

Solution

$$P[(T \leq 4) \cup (T \geq 16)] = P[|T - 10| \geq 6] \leq 2^2/6^2 = (\tfrac{1}{3})^2 = \tfrac{1}{9}.$$

Hence,

$$P[4 < T < 16] = 1 - P[|T - 10| \geq 6] \geq 1 - \tfrac{1}{9} = \tfrac{8}{9} = 0.8889.$$

Chebychev's inequality often gives poor probability estimates. For example, if X has an exponential distribution with mean $E[X] = 2$, then

$$P[|X - E[X]| \geq 4] = P[|X - 2| \geq 4] = 1 - P[X \leq 6]$$
$$= 1 - (1 - e^{-6/2}) = e^{-3} = 0.0498,$$

although the Chebychev inequality shows only that this probability does not exceed 0.25. However, the next example shows that the Chebychev inequality cannot be improved without strengthening the hypotheses.

Example 2.10.3   Suppose a discrete random variable X can assume only the values $-2, 0, 2$ with $p(-2) = p(2) = \tfrac{1}{8}$ and $p(0) = \tfrac{1}{2}$. Then

$$E[X] = -2 \times \tfrac{1}{8} + 0 \times \tfrac{1}{2} + 2 \times \tfrac{1}{8} = 0, \qquad E[X^2] = 4 \times \tfrac{1}{8} + 0 + 4 \times \tfrac{1}{8} = 1,$$

$$\text{Var}[X] = E[X^2] - (E[X])^2 = 1 - 0 = 1, \quad \text{and}$$

$\sigma = 1$. Then, by Chebychev's inequality, $P[|X - E[X]| \geq 2] \leq \tfrac{1}{4}$. However,

$$P[|X - E[X]| \geq 2] = P[(X = 2) \cup (X = -2)] = \tfrac{1}{8} + \tfrac{1}{8} = \tfrac{1}{4}.$$

Hence, the value estimated by Chebychev's inequality is the exact value.

In many computer science applications we are more interested in calculating one tail of a probability distribution than in calculating both tails, that is, we want an estimate of the size of $P[X - E[X] > t]$ or $P[X - E[X] < t]$ rather than the estimate of $P[|X - E[X]| \geq t]$ provided by Chebychev's inequality. The one-sided inequality gives us this estimate.

THEOREM 2.10.3   (*One-Sided Inequality*)   Let X be a random variable with mean $E[X]$ and variance σ^2. Then,

$$P[X \leq t] \leq \frac{\sigma^2}{\sigma^2 + (t - E[X])^2} \qquad \text{if} \quad t < E[X], \qquad (2.10.5)$$

and

$$P[X > t] \leq \frac{\sigma^2}{\sigma^2 + (t - E[X])^2} \qquad \text{if} \quad t > E[X]. \qquad (2.10.6)$$

Proof   Cramér [2, p. 256], using advanced methods, shows that if X is a random variable with mean 0 and standard deviation σ, then

$$P[X \leq t] \leq \frac{\sigma^2}{\sigma^2 + t^2} \qquad \text{for} \quad t < 0, \qquad (2.10.7)$$

and

$$P[X > t] \leq \frac{\sigma^2}{\sigma^2 + t^2} \qquad \text{for} \quad t > 0. \qquad (2.10.8)$$

Now, if X is an arbitrary random variable with a mean and variance, then $E[X - E[X]] = E[X] - E[X] = 0$, and

$$\mathrm{Var}[X - E[X]] = E[(X - E[X])^2] = \mathrm{Var}[X].$$

Hence, if $t < E[X]$, then $t - E[X] < 0$ and by (2.10.7),

$$P[X \leq t] = P[X - E[X] \leq t - E[X]] \leq \frac{\sigma^2}{\sigma^2 + (t - E[X])^2}.$$

If $t > E[X]$, $t - E[X] > 0$ and (2.10.8) gives

$$P[X > t] = P[X - E[X] > t - E[X]] \leq \frac{\sigma^2}{\sigma^2 + (t - E[X])^2}.$$

This completes the proof of Theorem 2.10.3.

Example 2.10.4 A mathematical model of a proposed on-line computer system gives a mean time to retrieve a record from a direct access storage device of 400 milliseconds with a standard deviation of 116 milliseconds. One design criterion requires that 90% of all retrieval times must not exceed 750 milliseconds. Use the one-sided inequality to test the design criterion.

Solution Let T be the retrieval time. The design criterion is that $P[T \leq 750$ milliseconds$] \geq 0.90$.

By the one-sided inequality,

$$P[T > 750] \leq \frac{116^2}{116^2 + (750 - 400)^2} = \frac{1}{1 + (350/116)^2} = 0.09897.$$

Hence,

$$P[T \leq 750] \geq 1 - 0.09897 = 0.90103,$$

and the design criterion is met. The best estimate we could make with Chebychev's inequality is

$$P[X \geq 750] = P[X - 400 \geq 350] \leq P[|X - 400| \geq 350] \leq (116/350)^2$$
$$= 0.1098.$$

This does not indicate that the design criterion has been met.

Most of us have an intuitive feel for what the probability of an event A, such as rolling a 7 with a pair of dice, "really" is, which is close to the "relative frequency" school of thought about probability. We have the feeling that, if we perform the random experiment n times and let S_n be the number of times that event A occurs, then S_n/n is approximately $P[A]$, at least in the sense that $\lim_{n \to \infty} S_n/n = P[A]$. The *law of large numbers* makes this intuitive notion more precise (and shows that it is true).

Let A be an event which has probability $P[A]$ and suppose we perform a Bernoulli sequence of n trials as described in Example 2.9.5, where a "success" corresponds to the occurrence of event A. Let S_n be the number of successes in the n trials. As we saw in Example 2.9.5, S_n has a binomial distribution with $E[S_n] = nP[A]$ and $\mathrm{Var}[S_n] = nP[A](1 - P[A])$. We are interested in the ratio S_n/n. We calculate

$$E\left|\frac{S_n}{n}\right| = \frac{1}{n} E[S_n] = P[A]$$

and

$$\mathrm{Var}\left[\frac{S_n}{n}\right] = \frac{1}{n^2} \mathrm{Var}[S_n] = \frac{P[A](1 - P[A])}{n}.$$

Let $\varepsilon > 0$ be arbitrary. Then by (2.10.2) of Chebychev's inequality,

$$P\left[\left|\frac{S_n}{n} - P[A]\right| \geq \varepsilon\right] \leq \frac{P[A](1 - P[A])}{n\varepsilon^2}. \tag{2.10.9}$$

The expression on the right of (2.10.9) can be made as small as desired, for fixed values of ε and A, by choosing n sufficiently large. This proves the following theorem.

THEOREM 2.10.4　(*Weak Law of Large Numbers*)　Let A be an event and S_n the number of times that A occurs in a Bernoulli sequence of n trials. Then for each $\varepsilon > 0$,

$$\lim_{n \to \infty} P\left[\left|\frac{S_n}{n} - P[A]\right| \geq \varepsilon\right] = 0. \tag{2.10.10}$$

There is a stronger form of Theorem 2.10.4, called the *strong law of large numbers*, which uses a more restrictive definition of the intuitive idea that $\lim_{n \to \infty} S_n/n = P[A]$ (see Feller [6, pp. 202–204]). However, a more useful form of the law is immediate from the *central limit theorem*, which is discussed in Chapter 3.

The weak law of large numbers shows that $p = P[A]$ can be estimated by S_n/n and that this estimate converges to p. However, it does not give any information as to how large n should be to guarantee that the error is less than a given value for a certain probability level. The Chebychev inequality does provide crude estimates. For if $\delta > 0$ then, by Chebychev's inequality,

$$P\left[\left|\frac{S_n}{n} - p\right| \geq \delta\right] \leq \mathrm{Var}[S_n/n]/\delta^2 = p(1 - p)/n\delta^2. \tag{2.10.11}$$

It is easy to show that $p(1 - p)$ has its maximum value at $p = 1/2$ (see Exercise 2). Hence, no matter what value p actually has, we have

$$P\left[\left|\frac{S_n}{n} - p\right| \geq \delta\right] \leq \frac{1}{4n\delta^2}. \tag{2.10.12}$$

Suppose now that δ and $\varepsilon > 0$ are given and we want to find how many trials of the experiment we need to be sure that

$$P\left[\left|\frac{S_n}{n} - p\right| \geq \delta\right] \leq \varepsilon. \tag{2.10.13}$$

If we know approximately what the value of p is, we see that (2.10.13) will be satisfied if $p(1 - p)/n\delta^2 \leq \varepsilon$ or $n \geq p(1 - p)/\varepsilon\delta^2$. If we have no idea what the value of p is, we can use (2.10.12) to conclude that $n \geq 1/(4\varepsilon\delta^2)$ trials will suffice. Since Chebychev's inequality usually yields poor estimates, we would expect either of these estimates to yield conservative estimates for n. In Chapter 3 we show that the central limit theorem can be applied to give a better estimate.

Example 2.10.5 Assuming that each terminal in an on-line system has the same probability p of being in use during the peak period of the day (the load is evenly distributed over the terminals), how many observations n need be made so that

$$P\left[\left|\frac{S_n}{n} - p\right| \geq 0.1\right] \leq 0.05?$$

If the first 100 observations indicate that p is approximately 0.2, how many more trials are needed?

Solution The estimate, based on (2.10.12), is $n = 1/(4 \times 0.05 \times 0.01) = 500$. If p is approximately 0.2, then we can use (2.10.11) to conclude that we need a total of $n = 0.2 \times 0.8/(0.05 \times 0.01) = 320$ observations or an additional 220. In Example 3.3.3 we show that this estimate can be improved by using the central limit theorem.

2.11 SUMMARY

The purpose of this chapter was to set up the mathematical machinery necessary to come to grips with probabilistic phenomena, that is, to deal with "random" or "nondeterministic" variables. Most of the interesting variables in modern computer systems, such as the response time of an on-line inquiry system or the retrieval time of a record from a direct access storage device, are *random variables*. This means that it is impossible to predict in advance the exact value of such a variable but that, if we know its *probability distribution*, then we can predict the *probability* that it will assume a value within a certain range of values. Here we interpret " probability" to mean the fraction of the time that a certain result will occur. Thus, if we say that the 90th percentile response time for an on-line inquiry system is 2 seconds, we mean that 90 percent of all response times will be less than or equal to 2 seconds and only 10 percent of them will exceed 2 seconds.

We gave some formal definitions to explain the probability concept, as well as examples to illustrate the definitions. Then we defined "random variable" and explained how the probability distribution of a random variable is defined. We gave some examples of random variables of the two main types, *discrete* and *indiscrete* (usually called continuous). We showed how some parameters of a random variable X, such as expected value $E[X]$ and variance $\text{Var}[X]$ can be used to make interesting probability calculations and estimates. Finally, we showed how transform methods (the use of moment generating functions, the Laplace–Stieltjes transform, etc.) simplify many probability calculations.

Student Sayings

Our observation of Nancy's distribution has given us many fine moments.

Well, what did you expect from a conditional expectation?

Exercises

1. [C20]   The on-line order entry system of the WEWE Diaper Company can receive order messages from Los Angeles, San Diego, Bakersfield, and San Francisco. Ordering activity in each city is independent of that from the other cities. The probability that the system receives one or more orders during any one minute time interval (during the peak load period of the day) from Los Angeles, San Diego, Bakersfield, or San Francisco, respectively, is 0.8, 0.3, 0.05, 0.5.

(a)   What is the probability that ordering activity occurs from exactly one of the cities during any one minute period?

(b)   Exactly two cities?

(c)   Not more than two cities?

(d)   No city?

2. [HM15]   In discussing the weak law of large numbers we claimed that the function $pq = p(1 - p)$ has a unique maximum value of $\frac{1}{4}$ at $p = \frac{1}{2}$. Prove this claim.

3. [20]   Suppose A, B, and C are events in some sample space Ω, and thus are subsets of Ω. Prove the distributive law

$$(A \cup B) \cap C = (A \cap C) \cup (B \cap C).$$

4. [18] Prove that, if A, B, and C are events, then

$$P[A \cup B \cup C] = P[A] + P[B] + P[C] - P[A \cap B] - P[A \cap C]$$
$$- P[B \cap C] + P[A \cap B \cap C].$$

[*Hint*: Use Theorem 2.2.1c and the result of Exercise 3.]

5. [15] Assume that a single depth charge has a probability of $\frac{1}{3}$ of sinking a submarine, $\frac{1}{2}$ of damage, and $\frac{1}{6}$ of missing. Assume also that two damaging explosions sink the sub. What is the probability that four depth charges will sink the sub?

6. [C25] Seven terminals in an on-line system are attached to a communication line to the central computer. Exactly four of the seven terminals are ready to transmit a message. Assume that each terminal is equally likely to be in the ready state. Let X be the random variable whose value is the number of terminals polled until the first ready terminal is located.

(a) What values may X assume?

(b) What is the probability that X will assume each of these values? Assume that terminals are polled in a fixed sequence without repetition.

(c) Suppose the communication line has m terminals attached of which n are ready to transmit. Show that X can assume only the values $i = 1, 2, \ldots, m - n + 1$ with $p(i) = P[X = i] = \binom{m-i}{n-1}/\binom{m}{n}$.

7. [18] Assume $A_1, A_2, A_3, \ldots$ are subsets of some set Ω. Prove De Morgan's formulas:

(a) $\overline{A_1 \cup A_2 \cup \cdots \cup A_N} = \overline{A_1} \cap \overline{A_2} \cap \cdots \cap \overline{A_N}$.

(b) $\overline{A_1 \cap A_2 \cap \cdots \cap A_N} = \overline{A_1} \cup \overline{A_2} \cup \cdots \cup \overline{A_N}$.

(c) $\overline{\bigcup_{n=1}^{\infty} A_n} = \bigcap_{n=1}^{\infty} \overline{A_n}$.

(d) $\overline{\bigcap_{n=1}^{\infty} A_n} = \bigcup_{n=1}^{\infty} \overline{A_n}$.

8. [15] Let $A_1, A_2, \ldots$ be events in some sample space Ω. Use Axiom Set 2.2.1 and the results of Exercise 7 to prove that

(a) $A_1 \cap A_2 \cap \cdots \cap A_N$ is an event for each positive integer N.

(b) $\bigcap_{n=1}^{\infty} A_n$ is an event.

9. [10] An on-line computer system has four incoming communication lines with the properties described in Table E2.9. What is the probability that a randomly chosen message has been received without error?

TABLE E2.9

Line	Fraction of traffic	Fraction of messages without error
1	0.4	0.9998
2	0.3	0.9999
3	0.1	0.9997
4	0.2	0.9996

10. [C25] (This rating is [T30] if nothing more powerful than a nonprogrammable calculator is available and [15] if APL is available so that you can copy and use MPOLL.) Suppose seven terminals are connected to a communication line of an on-line computer system. Each terminal operates independently and has probability 0.2 of being ready to transmit. Thus, if Y is the random variable which counts the number of terminals ready to transmit, Y has a binomial distribution with parameters $n = 7$ and $p = 0.2$. Find the mean and standard deviation of the number of polls necessary to find the first ready terminal.

11. [5] A discrete random variable X is called a *truncated Poisson* random variable if its mass points are 0, 1, 2, 3, ..., N and its probability mass function $p(\cdot)$ is given by $p(k) = Ce^{-\lambda}\lambda^k/k!$, $k = 0, 1, 2, 3, ..., N$. What is the value of the constant C?

12. [18] The average length of messages received at a message switching center is 50 characters with a standard deviation of 10 characters. Use the one-sided inequality to estimate how many bytes (characters) of storage should be provided for each message buffer to ensure that 95% of all messages fit into one buffer.

13. [00] Twas Brillig, who can program equally well (poorly) in FORTRAN and ALGOL, needs two coding sheets to write a program. Twas asks his friend Slithy Toves for two coding sheets of either kind. If Slithy has FORTRAN and ALGOL coding sheets mixed in a drawer what is the smallest number he must select sight unseen to satisfy Twas's request?

14. [15] Big Bored Securities has two brands of tape drives in a computer room, brand y and brand z. If two drives are selected at random, the probability that both are brand y is $\frac{1}{2}$. What is the smallest number of tape drives that could be in the room?

15. [20] Swann Dive, a systems programmer at Poly Unsaturated, offered his friend Charlie Tuna, an application programmer, the following proposition. On each roll of three dice Swann would pay Charlie one dollar for each ace that showed; if no aces were turned Charlie would pay Swann one dollar. Charlie reasons that the probability of rolling an ace on the first die is $\frac{1}{6}$; similarly for the second and third die. Hence the probability is $3 \times \frac{1}{6} = \frac{1}{2}$ of getting at least one ace and he might get two or even three of them.

(a) Is Charlie right, that is, is it a good proposition for him?

(b) What is the probability that Charlie will roll one, two, or three aces, respectively?

(c) What is the average amount of money Charlie can expect to win each time the dice are rolled? (Swann didn't tell him, but this game is known as chuck-a-luck at carnivals.)

16. [30] A single disk storage device has N concentric tracks and one access arm. It has been loaded with data in such a way that successive movements of the access arm (called track seeks) are independent of one another. The probability that a randomly chosen seek will take the arm to track i is p_i. Let X represent the number of tracks passed between consecutive seeks assuming that no physical repositioning of the access arm takes place between successive seek operations. Show the following.

(a) X assumes the values 0, 1, ..., $N - 1$ and has the pmf $p(\cdot)$ defined by

$$p(j) = P[X = j] = \begin{cases} \sum_{i=1}^{N} p_i^2, & j = 0 \\ 2 \sum_{i=1}^{N-j} p_i p_{i+j}, & j = 1, 2, \ldots, N - 1. \end{cases} \tag{1}$$

(b) For the case that $p_i = 1/N$ for all i, it is true that

$$E[X] = (N^2 - 1)/3N \cong N/3 \tag{2}$$

$$E[X^2] = (N^2 - 1)/6 \cong N^2/6 \tag{3}$$

and

$$\text{Var}[X] \cong N^2/18. \tag{4}$$

(c) Suppose T, the seek time, is a linear function of X; that is,

$$T = AX + B,$$

where A and B are constants. (A is then given by

$$A = \frac{\text{maximum seek time} - \text{minimum seek time}}{N - 1}$$

and B is minimum seek time.)
Then it is true that

$$E[T] = [A(N^2 - 1)/3N] + B \cong (AN/3) + B \tag{5}$$

and

$$\text{Var}[T] = A^2 \, \text{Var}[X] \cong A^2 N^2/18. \tag{6}$$

17. [15] A certain access method, called method A, has been found to give a mean record retrieval time of 36 milliseconds with a standard deviation of 7 milliseconds, while method B has a mean retrieval time of 42 milliseconds with a standard deviation of 4 milliseconds.

(a) If a major design objective is to have 90% of all individual retrievals completed in 55 milliseconds or less, which method should be selected?

(b) Does the chosen method meet the objective?

References

1. T. M. Apostol, *Calculus*, Vol. II, 2nd ed. Ginn (Blaisdell), Boston, 1969.
2. H. Cramér, *Mathematical Methods of Statistics*. Princeton Univ. Press, Princeton, New Jersey, 1946.
3. W. Feller, *An Introduction to Probability Theory and Its Applications*, Vol. 2, 2nd ed. Wiley, New York, 1971.
4. E. Parzen, *Modern Probability Theory and Its Applications*. Wiley, New York, 1960.
5. T. M. Apostol, *Mathematical Analysis*, 2nd ed. Addison-Wesley, Reading, Massachusetts, 1974.
6. W. Feller, *An Introduction to Probability Theory and Its Applications*, Vol. 1, 3rd ed., rev. ptg. Wiley, New York, 1968.

Chapter Three

PROBABILITY DISTRIBUTIONS

INTRODUCTION

In Chapter 2 we defined a random variable X to be a real-valued function defined on a sample space. Thus to each outcome ω of a random experiment the random variable X assigns the value $X(\omega)$. The "randomness" in the name "random variable" comes about because of the uncertainty of the outcome of the experiment, before the experiment is performed; once the outcome of the experiment has been determined, so has the value of the random variable. Thus, if the random variable X counts the number of spots turned up when two dice are tossed, as soon as the dice are rolled, the value of X is known. The usefulness of the random variable concept depends upon the ability to determine the probability that the *values* of the random variable occur in a given set of real numbers. That is, the *probability distribution* of a random variable is its most important property. For this reason the two statements

(1) "X is a Poisson random variable," and
(2) "X has a Poisson distribution,"

are used interchangeably. The same is true for any other type of random variable, of course. If the probability distribution of a random variable is known, the actual underlying sample space is not important. Thus, if we know the distribution function $F(\cdot)$ of X, defined for all real x by $F(x) = P[X \leq x]$, we can calculate the probability $P[X \in A]$ where A is a set of real numbers satisfying very weak restrictions. In most practical examples we are interested in probabilities such as $P[a < X \leq b] = F(b) - F(a)$; the distribution function $F(\cdot)$ enables us to make this type of calculation easily. If X is a discrete random variable, its distribution function can be calculated from its probability mass function $p(\cdot)$, defined for all real x by $p(x) = P[X = x]$. If X is continuous, its distribution function can be calculated from its density function $f(\cdot)$, which, itself, is characterized by the properties:

(i) $f(x) \geq 0$ for all real x,
(ii) $\int_{-\infty}^{\infty} f(x)\, dx = 1$, and
(iii) $P[a \leq X \leq b] = \int_a^b f(x)\, dx$ for all real a, b with $a < b$.

Then $F(x) = \int_{-\infty}^{x} f(t)\, dt$.

In this chapter we will study some common random variables which are especially useful for computer science applications.

3.1   DISCRETE RANDOM VARIABLES

A random variable X is discrete if

$$\sum_{x \in T} p(x) = \sum_{x \in T} P[X = x] = 1,$$

where $T = \{\text{real } x : p(x) > 0\}$. The set T is either finite or countably infinite. (For a proof see Apostol [9, p. 511].) Each point of T is called a *mass point of* X. We sometimes indicate the mass points of X by writing " X *assumes the values* x_1, x_2, x_3, ..." Just as the distribution function of X can be calculated from the pmf $p(\cdot)$ using the formula

$$F(x) = \sum_{x_i \leq x} p(x_i),$$

the pmf $p(\cdot)$ can be calculated from $F(\cdot)$ at all mass points by $p(x_i) = \lim_{x \to x_i^-} \{F(x_i) - F(x)\}$. That is, the graph of $F(\cdot)$ is a step function with a jump at each mass point x_i, the jump having magnitude $p(x_i)$.

We summarize the properties of some useful discrete random variables in Table 1 of Appendix A.

3.1.1   Bernoulli Random Variables

Several important discrete random variables are derived from the concept of a Bernoulli sequence of trials. A *Bernoulli trial* is a random experi-

ment in which there are only two possible outcomes, usually called
"success" or "failure," with respective probabilities p and q, where
$p + q = 1$. We assume that $0 < p < 1$, for otherwise the results are trivial. A
sequence of such trials is a *Bernoulli sequence* if the trials are independent
and the probability of success (or of failure) is constant from trial to trial. A
Bernoulli random variable describes a Bernoulli trial and thus assumes only
two values: 1 (for success) with probability p and 0 (for failure) with probability q, where $q = 1 - p$.

An example of a Bernoulli trial can be constructed from any sample
space Ω which has an event A such that $0 < P[A] < 1$ by identifying the
occurrence of A with "success" and $\bar{A}$ with "failure." The corresponding
Bernoulli random variable X is defined to be 1 for every point of A and to be
0 at all points of $\bar{A}$.

A Bernoulli random variable X is completely determined by the value p
(or q) and thus is said to have one parameter, namely p. As we saw in
Example 2.9.4, such a random variable has the moment generating function

$$g(z) = q + pz;$$

also

$$E[X] = p \quad \text{and} \quad \text{Var}[X] = pq.$$

A Bernoulli random variable, like a bad penny, is not very useful, itself,
but is the basis of other important random variables, including the binomial
and geometric random variables; just as a lot of bad pennies add up to form
good dollars.

3.1.2 Binomial Random Variables

Consider a Bernoulli sequence of n trials where the probability of success
on each trial is p. The random variable X, which counts the number of
successes in the n trials is called a *binomial random variable* with parameters
n and p. Thus X can assume only the values $0, 1, 2, \ldots, n$ with positive
probability.

A sequence of n Bernoulli trials can be represented as a string $a_1 a_2 \cdots a_n$
where each a_i is either s for a success or f for a failure. Thus a sequence of 5
trials, in which 3 successes are followed by 2 failures would be represented as
sssff. If now the random variable X has parameters n and p; and k is an
integer between 0 and n (inclusive); then any string $a_1 a_2 \cdots a_n$ representing
k successes and $n - k$ failures has probability $p^k q^{n-k}$, since each trial is
independent. (The probability can be calculated by replacing each s in
$a_1 a_2 \cdots a_n$ by a p, each f by a q, and then multiplying the resulting numbers.)
The number of strings $a_1 a_2 \cdots a_n$ representing k successes and $n - k$ failures
is just the number of ways the k indices representing success can be chosen

from the n indices, that is, $\binom{n}{k}$. Hence, the pmf $b(\cdot\,; n, p)$ of a binomial random variable with parameters n and p is defined by

$$b(k; n, p) = \binom{n}{k} p^k q^{n-k}, \qquad k = 0, 1, 2, \ldots, n, \qquad (3.1.1)$$

where $q = 1 - p$.

X can be represented as

$$X = X_1 + X_2 + \cdots + X_n, \qquad (3.1.2)$$

where $X_1, X_2, \ldots, X_n$ are independent, identically distributed Bernoulli random variables. By Theorem 2.7.1, we have

$$E[X] = E[X_1] + E[X_2] + \cdots + E[X_n] = np, \qquad (3.1.3)$$

since $E[X_i] = p$ for each i.

By Theorem 2.7.2,

$$\mathrm{Var}[X] = \mathrm{Var}[X_1] + \mathrm{Var}[X_2] + \cdots + \mathrm{Var}[X_n] = npq, \qquad (3.1.4)$$

since $\mathrm{Var}[X_i] = pq$ for each i.

(In Example 2.9.5 we calculated the mean and variance of X by using generating functions.)

The APL function BINOMIAL can be used to calculate binomial probabilities if n is not too large. The APL function BINΔSUM can be used to sum binomial probabilities.

Example 3.1.1 A master file of 120,000 records is stored as a sequential file on a direct access storage device in blocks of six records. Each day the transaction file is run against the master file and approximately 5% of the master records are updated. The records to be updated are assumed to be distributed uniformly over the master file. An entire block of records must be updated if one or more records in the block need updating. What is the mean and standard deviation of the number of blocks that must be updated? Use Chebychev's inequality to estimate the probability that between 5000 and 5600 blocks must be updated.

Solution Let X be the random variable which counts the number of records in a block that must be updated. It is reasonable to assume that X is a binomial random variable with parameters $n = 6$ and $p = 0.05$. (A Bernoulli trial consists of checking a record to determine whether or not it must be updated, that is, to check whether or not it is listed in the transaction file.) A given block must be updated if $X \geq 1$, that is, with probability $P[X \geq 1] = 1 - P[X = 0]$. Hence, the probability that any given block must be updated is

$$1 - b(0; 6, 0.05) = 1 - (0.95)^6 = 1 - 0.735 = 0.265.$$

Let Y be the number of blocks that need to be updated; Y is a binomial random variable with parameters $n = 20{,}000$ and $p = 0.265$. Therefore, the average number of blocks to be updated is $E[Y] = 20{,}000 \times 0.265 = 5300$ with standard deviation

$$\sigma = (20{,}000 \times 0.265 \times 0.735)^{1/2} = (3895.5)^{1/2} = 62.414$$

blocks. By Chebychev's inequality the probability that Y is between 5000 and 5600 blocks is

$$P[\,|\,Y - 5300\,| \;\leq 300] \geq 1 - \left(\frac{62.414}{300}\right)^2 = 1 - 0.0433 = 0.9567.$$

Example 3.1.2   The on-line computer system at Gnu Glue has 20 communication lines. The lines operate independently. The probability that any particular line is in use is 0.6. What is the probability that 10 or more lines are in operation?

Solution   The number of lines in operation X has a binomial distribution with parameters $n = 20$ and $p = 0.6$. The required probability is

$$P[X \geq 10] = \sum_{k=10}^{20} \binom{20}{k} (0.6)^k (0.4)^{20-k} = 0.872479.$$

This is a tedious calculation to carry out manually, but it can be calculated by the APL statement

$$+/(K!\,20) \times (0.6*K) \times (0.4*20 - K \leftarrow 9 + \iota 11)$$
0.8724787539.

This probability can also be approximated by the *normal distribution* as we will show later in this chapter. The APL function BINΔSUM may also be used to make the above calculation.

3.1.3   Geometric Random Variables

Suppose a sequence of Bernoulli trials is continued until the first success occurs. Let X be the random variable which counts the number of trials *before* the trial at which the first success occurs. Then X can assume the values 0, 1, 2, 3, ... X assumes the value zero if and only if the first trial yields a success; hence with probability p. X assumes the value 1 if and only if the first trial yields a failure and the second a success; hence with probability qp, where $q = 1 - p$. Continuing in this way we see that the pmf of X is given by

$$p(k) = q^k p, \qquad k = 0, 1, 2, \dots. \tag{3.1.5}$$

The probability generating function of X is thus

$$g(z) = \sum_{k=0}^{\infty} q^k p z^k = p \sum_{k=0}^{\infty} (qz)^k = \frac{p}{1 - qz}. \tag{3.1.6}$$

In order for (3.1.6) to hold we must have $qz < 1$ or $z < 1/q$. Then

$$g'(z) = \frac{pq}{(1 - qz)^2} \quad \text{and} \quad g''(z) = \frac{2pq^2}{(1 - qz)^3}.$$

Hence, by Theorem 2.9.2,

$$E[X] = g'(1) = \frac{pq}{(1 - q)^2} = \frac{q}{p}, \tag{3.1.7}$$

and

$$\text{Var}[X] = g''(1) + g'(1) - (g'(1))^2 = 2\frac{q^2}{p^2} + \frac{q}{p} - \frac{q^2}{p^2}$$

$$= \frac{q^2}{p^2} + \frac{q}{p} = \frac{q(q + p)}{p^2} = \frac{q}{p^2}. \tag{3.1.8}$$

The geometric random variable is important in queueing theory and other areas of applied probability.

Example 3.1.3 Consider Example 3.1.1. Let X be the number of blocks of the master file that are read before the first block is found which must be updated. Then X is a geometric random variable with parameter $p = 0.265$. The expected value of X is $q/p = 0.735/0.265 = 2.774$ blocks with standard deviation $\sqrt{q}/p = 3.235$.

Suppose now that during each time interval of a fixed length, say h, that an event of some kind, called an *arrival*, may or may not occur. Suppose further that the occurrence or nonoccurrence of the arrival in each interval is determined by a Bernoulli random variable with a fixed probability p of success from one interval to the next, that is, by a Bernoulli sequence of trials. Then the *interarrival time T* is defined to be the number of trials (time intervals of length h) before the first success (arrival). Thus T has a geometric distribution with parameter p. Now suppose we are given that there were no arrivals during the first m intervals of length h and we wish to calculate the probability that there will be k more time intervals with no arrivals before the next arrival, that is, $P[T = k + m \mid T \geq m]$ for $k = 0, 1, 2, \ldots$. By the definition of conditional probability (see Section 2.4) we have

$$P[T = k + m \mid T \geq m] = \frac{P[(T = k + m) \cap (T \geq m)]}{P[T \geq m]}. \tag{3.1.9}$$

But

$$(T = k + m) \cap (T \ge m) = (T = k + m), \qquad (3.1.10)$$

and

$$P[T \ge m] = pq^m(1 + q + q^2 + \cdots = pq^m/(1 - q) = q^m. \qquad (3.1.11)$$

Hence

$$P[T = k + m \mid T \ge m] = pq^{m+k}/q^m = pq^k = P[T = k]. \qquad (3.1.12)$$

Equation (3.1.12) shows the Markov or memoryless property of the geometric distribution, that is, the presence or absence of an arrival at any point in time has no effect upon the interarrival time to the next arrival. The system simply does not "remember" when the last arrival (success) occurred. Thus, in Example 3.1.3, the average interarrival time between any two successive blocks which must be updated is 2.774; also at any arbitrary point in time, it is the average number of blocks to be read before the next block is found that must be updated.

3.1.4   Poisson Random Variables

We say that a random variable X is a Poisson random variable with parameter $\lambda > 0$ if X has the mass points 0, 1, 2, 3, ..., and if its probability mass function $p(\cdot\,; \lambda)$ is given by

$$p(k; \lambda) = P[X = k] = e^{-\lambda}(\lambda^k/k!), \qquad k = 0, 1, 2, \dots. \qquad (3.1.13)$$

This probability distribution was discovered by the French mathematician Siméon D. Poisson (1781–1840).

Equation (3.1.13) does define a pmf because $p(k; \lambda)$ is clearly nonnegative for all nonnegative integers k and

$$\sum_{k=0}^{\infty} p(k; \lambda) = e^{-\lambda} \sum_{k=0}^{\infty} \frac{\lambda^k}{k!} = e^{-\lambda}e^{\lambda} = e^0 = 1.$$

In Example 2.9.2 we showed that a Poisson random variable with parameter λ has the moment generating function

$$\psi(\theta) = \exp\left[\lambda(e^\theta - 1)\right], \qquad (3.1.14)$$

and furthermore, that

$$E[X] = \lambda \qquad \text{and} \qquad \text{Var}[X] = \lambda.$$

In Example 2.9.3 we showed that, if $Z = X + Y$, where X is Poisson distributed with parameter λ and Y is Poisson distributed with parameter μ;

then Z is Poisson distributed with parameter $\lambda + \mu$. Thus Poisson random variables have the reproductive property.

The Poisson random variable is one of the four or five most important random variables for applied probability and statistics. One reason for this importance is that a great many natural and man-made phenomena are described by Poisson random variables.

The following phenomena have Poisson distributions.

(a)   The number of alpha particles emitted from a radioactive substance per unit time (see Bateman [1], Rutherford and Geiger [2], and Lippman [3, pp. 76, 77]). (Geiger is the inventor of the celebrated Geiger counter which counts not geigers but rather radiation levels.)

(b)   The number of flying-bomb hits in the south of London during World War II (see Clarke [4] and Feller [5, pp. 160, 161]).

(c)   The number of vacancies per year in the United States Supreme Court (see Wallis [6] and Parzen [7, pp. 256, 257]). Other examples include misprints per page of a book, raisins per cubic inch of raisin bread, horse kicks per year in the Prussian army (this observation was made in the nineteenth century), and the number of chromosome interchanges in organic cells caused by X-ray radiation.

A number of random variables of interest to computer science have been found to have Poisson distributions. We shall discuss some of these in this book.

Another reason for the importance of the Poisson distribution is that its pmf, given by (3.1.13), is easy to calculate. Furthermore, a binomial random variable can often be approximated by a Poisson random variable—in fact this is the way Poisson originally conceived the probability distribution that bears his name.

THEOREM 3.1.1   Suppose X has a binomial distribution with parameters n and p. Then, if n is large and p is small with $\lambda = np$, $b(k; n, p)$ is approximately $p(k; \lambda)$ in the sense that

$$\lim_{n \to \infty} b\left(k; n, \frac{\lambda}{n}\right) = p(k; \lambda), \qquad k = 0, 1, 2, \dots.$$

Proof   Fix k with $0 \le k \le n$. Then

$$b(k; n, p) = \binom{n}{k}\left(\frac{\lambda}{n}\right)^k \left(1 - \frac{\lambda}{n}\right)^{n-k} = \frac{n!}{k!\,(n-k)!}\,\frac{\lambda^k}{n^k}\,\frac{(1 - (\lambda/n))^n}{(1 - (\lambda/n))^k}$$

$$= \frac{\lambda^k}{k!}\left(1 - \frac{\lambda}{n}\right)^n \left[\frac{n!}{n^k(n-k)!}\right]\left(1 - \frac{\lambda}{n}\right)^{-k}. \tag{3.1.15}$$

Consider the term in square brackets in (3.1.15). It can be written as

$$\frac{n!}{n^k(n-k)!} = \frac{n(n-1)(n-2)\cdots(n-k+1)}{n^k}$$

$$= \left(1 - \frac{1}{n}\right)\left(1 - \frac{2}{n}\right)\cdots\left(1 - \frac{k-1}{n}\right). \qquad (3.1.16)$$

Hence,

$$\lim_{n\to\infty} \frac{n!}{n^k(n-k)!} = 1. \qquad (3.1.17)$$

Also, since k is fixed,

$$\lim_{n\to\infty}\left(1 - \frac{\lambda}{n}\right)^{-k} = \lim_{n\to\infty}\frac{1}{(1-(\lambda/n))^k} = \frac{1}{\lim_{n\to\infty}(1-(\lambda/n))^k} = \frac{1}{1} = 1. \qquad (3.1.18)$$

By a well-known property of the exponential function

$$\lim_{n\to\infty}\left(1 - \frac{\lambda}{n}\right)^n = e^{-\lambda}. \qquad (3.1.19)$$

Combining (3.1.15)–(3.1.19) we see that

$$\lim_{n\to\infty} b\left(k; n, \frac{\lambda}{n}\right) = e^{-\lambda}\frac{\lambda^k}{k!}, \qquad k = 0, 1, 2, \ldots. \qquad (3.1.20)$$

The import of Theorem 3.1.1 is that, if n is large and p is small so that np is not close to either p or n, then the binomial random variable with parameters n and p can be approximated by a Poisson random variable with the parameter $\lambda = np$.

The APL functions POISSON and POISSONΔSUM can be used to make Poisson probability calculations.

Example 3.1.4 Suppose the Wildgoose Errcraft computer installation has a library of 100 subroutines and that each week, on the average, bugs are found (and corrected) in two of the subroutines. Assuming that the number of subroutines per week with newly discovered and corrected bugs has a binomial distribution, use the Poisson approximation to calculate the probability that errors will be found in not more than three subroutines next week.

Solution Using the APL functions POISSON and BINOMIAL; and rounding to five decimal places, we compute the values in Table 3.1.1. The

TABLE 3.1.1

Data From Example 3.1.4

k	$P[X = k]$	Poisson approximation $e^{-2}2^k/k!$
0	0.13262	0.13534
1	0.27065	0.27067
2	0.27341	0.27067
3	0.18228	0.18045
Total probability	0.85896	0.85713

true value of the required probability is 0.85896; the value given by the Poisson approximation is 0.85713. These values are close, although some individual probabilities are a little off, e.g., for $k = 2$ the error of the approximation is 0.00274.

The value of the Poisson distribution as a means of approximating the binomial distribution is minor compared to its value in describing random variables which occur in computer science and other sciences.

Example 3.1.5 Suppose it has been determined that the number of inquiries which arrive per second at the central computer installation of the Varoom Broom on-line computer system can be described by a Poisson random variable with an average rate of 10 messages per second. What is the probability that no inquiries arrive in a one-second period? What is the probability that 15 or fewer inquiries arrive in a one-second period?

Solution By hypothesis

$$P[X = k] = e^{-10}(10^k/k!), \qquad k = 0, 1, 2, \dots.$$

Hence the probability that no inquiry arrives in a one-second period is $e^{-10} = 4.54 \times 10^{-5}$. The answer to the second question is

$$e^{-10} \sum_{k=0}^{15} \frac{10^k}{k!} = 0.95126.$$

This is a laborious calculation to make without a computer but can be made easily with the APL function POISSONΔDIST.

In Chapter 4 we give the conditions which characterize a random phenomenon which is described by a Poisson random variable. It will be evident that these conditions are characteristic of many real-life situations.

3.1.5 Discrete Uniform Random Variables

A random variable X which assumes a finite number of values $x_1, x_2, \ldots,$ x_n, each with the same probability, $1/n$, is called a *discrete uniform* random variable. Often the values are taken to be multiples of some value L, such as $L, 2L, 3L, \ldots, nL$. The expected value is given by

$$E[X] = \frac{1}{n} \sum_{i=1}^{n} x_i. \tag{3.1.21}$$

The second moment $E[X^2]$ is given by

$$E[X^2] = \frac{1}{n} \sum x_i^2, \tag{3.1.22}$$

and the variance can be calculated by the formula

$$\text{Var}[X] = E[X^2] - (E[X])^2. \tag{3.1.23}$$

3.2 CONTINUOUS RANDOM VARIABLES

A continuous random variable X is characterized by the property that $P[X = x] = 0$ for all real x, that is, its probability mass function assumes only the value zero. In this book each continuous random variable we consider is described by a density function $f(\cdot)$, with properties defined in Section 2.5. It is *not* true, in general, that $f(x) = P[X = x]$; it *is* true that, for each real x and for small Δx, the probability that the value of X lies between x and $x + \Delta x$ is about $f(x)\,\Delta x$. Some of the properties of the continuous random variables we discuss in this section are summarized in Table 2 of Appendix A. (It is the author's belief that continuous random variables should be known as *indiscrete random variables*, since they clearly are not discrete; this would also add a little spice to a subject with a reputation for dullness.)

3.2.1 Continuous Uniform Random Variables

A continuous random variable X is said to be a *uniform random variable on the interval a to b* or to be *uniformly distributed on the interval a to b*, if its density function is given by

$$f(x) = \begin{cases} \dfrac{1}{b-a} & \text{for} \quad a < x < b \\ 0 & \text{otherwise.} \end{cases} \tag{3.2.1}$$

The corresponding distribution function is easily calculated by integration to give

$$F(x) = \begin{cases} 0 & \text{for} \quad x < a \\ \dfrac{x-a}{b-a} & \text{for} \quad a \leq x < b \\ 1 & \text{for} \quad x \geq b. \end{cases} \tag{3.2.2}$$

Figure 3.2.1 is a graph of the density function of a random variable which is uniformly distributed on the interval a to b, and Fig. 3.2.2 is the corresponding distribution function. Thus the probability that the values of X will lie in any subinterval of the interval from a to b is merely the ratio of the length of the subinterval to the length of the whole interval, that is, the probability that X will lie in any subinterval of length δ is $\delta/(b-a)$.

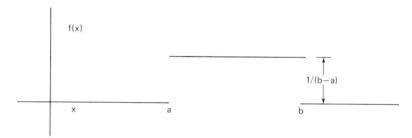

Fig. 3.2.1 Density function for uniform distribution on a to b.

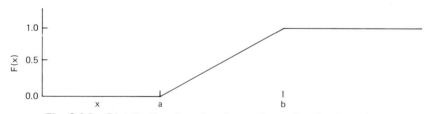

Fig. 3.2.2 Distribution function for uniform distribution on a to b.

It is an easy exercise to show (see Exercise 13) that

$$E[X] = \int_a^b x f(x)\, dx = \frac{a+b}{2}, \tag{3.2.3}$$

and that

$$\text{Var}[X] = \int_a^b (x - E[X])^2 f(x)\, dx = \frac{(b-a)^2}{12}. \tag{3.2.4}$$

Example 3.2.1   Suppose the disks in a disk memory device rotate once every 25 milliseconds. When a read/write head is positioned over a track to read a particular record from the track the record can be anywhere on the track. Hence the rotational delay, T, before the required record is in position to be read is uniformly distributed on the interval from 0 to 25 milliseconds. Thus $E[T] = 12.5$ milliseconds,

$$\text{Var}[T] = \frac{25^2}{12} = 52.0833,$$

and

$$\sigma_T = \sqrt{52.0833} = 7.2169 \quad \text{milliseconds.}$$

The probability that the rotational delay is between 5 and 15 milliseconds is $10/25 = 0.4$, the same as the probability that it is between 15 and 25 milliseconds.

3.2.2   Exponential Random Variables

A continuous random variable X has an *exponential distribution with parameter* $\lambda > 0$ if its density function f is defined by

$$f(x) = \begin{cases} \lambda e^{-\lambda x}, & x > 0 \\ 0, & x \le 0. \end{cases} \tag{3.2.5}$$

The distribution function F is then given by

$$F(x) = \begin{cases} 1 - e^{-\lambda x} = 1 - \exp(-x/E[X]), & x > 0 \\ 0, & x \le 0. \end{cases} \tag{3.2.6}$$

(Students who have learned the importance of the exponential distribution have been known to shout "Eureka!" upon seeing formula (3.2.6) appear. Therefore I call this formula the "Eureka formula." It should not be confused with the motto of the state of California.)

Figure 2.5.3 shows the density function for an exponential random variable with $\lambda = 2$ and Fig. 2.5.4 is the graph of the corresponding distribution function. As shown in Example 2.6.2, an exponential random variable with parameter λ has mean $E[X] = 1/\lambda$ and $\text{Var}[X] = 1/\lambda^2 = E[X]^2$.

In Example 2.9.1 we proved that the moments are given by

$$E[X^k] = k!/\lambda^k = k!\, E[X]^k, \quad k = 1, 2, 3, \ldots. \tag{3.2.7}$$

One reason for the importance of the exponential distribution to queueing theory and elsewhere is the *Markov property*, sometimes called the "memoryless" property, given by

$$P[X > t + h \mid X > t] = P[X > h], \quad t > 0, \quad h > 0. \tag{3.2.8}$$

One interpretation of (3.2.8) is that, if X is the waiting time until a particular event occurs and t units of time have produced no event, then the distribution of further waiting time is the same as it would be if no waiting time had passed—that is, the system does not "remember" that t time units have produced no "arrival." To prove (3.2.8) we note that, by (3.2.6), $P[X > x] = e^{-\lambda x}$ for all real positive x. Hence

$$P[X > t + h | X > t] = \frac{P[(X > t + h) \cap (X > t)]}{P[X > t]}$$

$$= \frac{P[X > t + h]}{P[X > t]} = \frac{e^{-\lambda(t+h)}}{e^{-\lambda t}} = \frac{e^{-\lambda t} e^{-\lambda h}}{e^{-\lambda t}}$$

$$= e^{-\lambda h} = P[X > h].$$

Figure 2.5.3 shows that the density function for the exponential distribution is not symmetrical about the mean but is highly skewed. In fact, for an exponential random variable X with parameter λ, it is true that

$$P[X \le E[X]] = 1 - e^{-E[X]/E[X]} = 1 - e^{-1} = 0.63212. \tag{3.2.9}$$

For any random variable, such as a uniform random variable, which is symmetrical about the mean, $P[X \le E[X]] = 0.5$. Thus, for an exponential random variable, values of X between 0 and $E[X]$ are more likely to occur than values between $E[X]$ and $2E[X]$, although each interval is one standard deviation long ($\text{Var}[X] = E[X]^2$ so that $\sigma_X = E[X]$).

For any random variable X, its rth *percentile value*, $\pi(r)$, is defined by $P[X \le \pi(r)] = r/100$. Thus the 90th percentile value of an exponential distributed random variable is defined by

$$P[X \le \pi(90)] = 0.9 \quad \text{or} \quad 1 - e^{-\lambda \pi(90)} = 0.9.$$

Hence

$$e^{-\lambda \pi(90)} = 0.1. \tag{3.2.10}$$

By taking the natural logarithm of both sides of (3.2.10) and solving for $\pi(90)$, we get

$$\pi(90) = -\frac{\ln(0.1)}{\lambda} = E[X] \ln 10 = 2.30259 E[X] \approx 2.3 E[X]. \tag{3.2.11}$$

(Here $\ln x$ means the logarithm of x to the base e.)
Similarly,

$$\pi(95) = E[X] \ln 20 = 2.99573 E[X] \approx 3 E[X], \tag{3.2.12}$$

and for $r > 0$

$$\pi(r) = E[X] \ln\left(\frac{100}{100 - r}\right). \tag{3.2.13}$$

In this book we will use the approximations

$$\pi(90) \approx E[X] + 1.3\sigma_X = 2.3E[X], \tag{3.2.14}$$

and

$$\pi(95) \approx E[X] + 2\sigma_X = 3E[X] \tag{3.2.15}$$

for the exponential distribution.

Example 3.2.2   Personnel of an engineering firm use an on-line terminal to make routine engineering calculations. If the time each engineer spends in a session at a terminal has an exponential distribution with an average value of 36 minutes, find

(a)   the probability that an engineer will spend less than 30 minutes at the terminal,

(b)   more than an hour.

(c)   If an engineer has already been at the terminal for 30 minutes, what is the probability that he or she will spend at least another hour at the terminal?

(d)   Ninety percent of the sessions end in less than R minutes. What is R?

Solution   Let T be the time an engineer spends using the terminal. By (3.2.6) the probability that T does not exceed t minutes is

$$1 - e^{-t/36}.$$

Hence, the probability that T is less than 30 minutes is

$$1 - e^{-30/36} = 1 - e^{-5/6} = 1 - 0.43460 = 0.5654.$$

By taking complements, (3.2.6) yields

$$P[T > t] = e^{-t/36}, \qquad t \quad \text{in minutes.}$$

Thus, the probability that over an hour is spent at the terminal in one session is

$$e^{-60/36} = e^{-5/3} = 0.1889,$$

or slightly less than 20% of the time. By the Markov property, the fact that an engineer has already been using the terminal for 30 minutes has no effect on the probability that he or she will use it for at least another hour. Hence, this probability is 0.1889. R is $\pi(90)$ or $2.3E[X] = 2.3 \times 36 = 82.8$ minutes.

We summarize the properties of the exponential distribution in the following theorem.

THEOREM 3.2.1 (*Properties of the Exponential Distribution*) Let X be an exponential random variable with parameter $\lambda > 0$. Then the following hold.

(a) If $\theta < \lambda$, then the moment generating function $\psi(\cdot)$ is given by

$$\psi(\theta) = \lambda/(\lambda - \theta), \tag{3.2.16}$$

and the Laplace–Stieltjes transform, $X^*(\theta)$, by

$$X^*(\theta) = \lambda/(\lambda + \theta). \tag{3.2.17}$$

(b) $E[X^k] = k!/\lambda^k = k!\ E[X]^k, k = 1, 2, 3, \dots.$
(c) $E[X] = 1/\lambda, \quad \text{Var}[X] = 1/\lambda^2 = E[X]^2.$
(d) X has the Markov property

$$P[X > t + h \mid X > t] = P[X > h], \qquad t > 0, \quad h > 0.$$

(e) The rth percentile value $\pi(r)$ defined by $P[X \le \pi(r)] = r/100$ is given by

$$\pi(r) = E[X] \ln\left(\frac{100}{100 - r}\right).$$

(f) Suppose the number of arrivals, Y, of some entity per unit of time is described by a Poisson random variable with parameter λ. Then the time T between any two successive arrivals (the interarrival time) is independent of the interarrival time of any other successive arrivals and has an exponential distribution with parameter λ. Thus, $E[T] = 1/\lambda$, and $P[T \le t] = 1 - e^{-\lambda t}$ for $t \ge 0$.

(g) Suppose the interarrival times of customers to a queueing system are independent, identically distributed, exponential random variables, each with mean $1/\lambda$. Then the number of arrivals, Y_t, in any interval of length $t > 0$, has a Poisson distribution with parameter λt, that is,

$$P[Y_t = k] = e^{-\lambda t}[(\lambda t)^k/k!], \qquad k = 0, 1, 2, \dots.$$

(h) Suppose $X_1, X_2, X_3, \dots, X_n$ are independent exponential random variables with parameters $\lambda_1, \lambda_2, \lambda_3, \dots, \lambda_n$, respectively, and $Y = \min\{X_1, X_2, \dots, X_n\}$. Then Y has an exponential distribution with parameter $\lambda = \lambda_1 + \lambda_2 + \dots + \lambda_n$. In particular, if each $\lambda_i = \lambda$, then Y is exponential with parameter $n\lambda$.

Proof (a)–(e) have been proven, above, except for (3.2.17), which was calculated in Example 2.9.6. The proof of (f) is given in Chapter 4 (Theorem 4.2.2), (g) follows from Theorems 4.2.3 and 4.2.1; (h) follows from Theorem 2.7.4.

3.2.3   Normal Random Variables

A continuous random variable X is said to be a *normal* random variable with parameters μ and $\sigma > 0$ if it has the density function

$$f(x) = \frac{1}{\sigma\sqrt{2\pi}} \exp\left[-\tfrac{1}{2}((x - \mu)/\sigma)^2\right], \qquad x \quad \text{real.} \qquad (3.2.18)$$

We indicate this fact by writing " X is $N(\mu, \sigma^2)$ ". A *standard normal* random variable is one with parameters $\mu = 0$ and $\sigma = 1$. Thus the standard normal density $\varphi(\cdot)$ is defined by

$$\varphi(x) = \frac{1}{\sqrt{2\pi}} \exp\left[-\tfrac{1}{2}x^2\right], \qquad x \quad \text{real.} \qquad (3.2.19)$$

The corresponding *standard normal distribution function* $\Phi(\cdot)$ is therefore defined by

$$\Phi(x) = \int_{-\infty}^{x} \varphi(t)\,dt = \int_{-\infty}^{x} \frac{\exp\left[-\tfrac{1}{2}t^2\right]}{\sqrt{2\pi}}\,dt. \qquad (3.2.20)$$

The standard normal distribution is important because every normal distribution can be calculated in terms of it. Thus, if X is normally distributed with parameters μ and σ (X is $N(\mu, \sigma^2)$), then

$$F_X(x)$$

$$= \int_{-\infty}^{x} \frac{\exp\left[-\tfrac{1}{2}((t - \mu)/\sigma)^2\right]}{\sigma\sqrt{2\pi}}\,dt = \int_{-\infty}^{(x-\mu)/\sigma} \frac{\exp\left[-\tfrac{1}{2}y^2\right]}{\sqrt{2\pi}}\,dy = \Phi\left(\frac{x - \mu}{\sigma}\right).$$

$$(3.2.21)$$

The second integral in (3.2.21) is the result of the change of variable $y = (t - \mu)/\sigma$. Unfortunately $\Phi(\cdot)$ cannot be calculated in closed form but must be approximated using numerical methods. The APL function NDIST displayed in Appendix B calculates values of the standard normal distribution using formula (26.2.17) of Abramowitz and Stegun [8]. It was used to create Table 3 of Appendix A, a table of values of the standard normal distribution function $\Phi(\cdot)$.

In order to prove a number of useful properties of the normal distribution we first need to prove that

$$\int_{-\infty}^{\infty} \varphi(x)\,dx = \int_{-\infty}^{\infty} \frac{\exp[-z^2/2]}{\sqrt{2\pi}}\,dz = 1. \qquad (3.2.22)$$

This, in particular, will show that (3.2.18) defines a density function, since $f(x) > 0$ for all real x because the exponential function assumes only positive values and

$$\int_{-\infty}^{\infty} \frac{\exp[-\frac{1}{2}((x-\mu)/\sigma)^2]}{\sigma\sqrt{2\pi}} dx = \int_{-\infty}^{\infty} \frac{\exp[-z^2/2]}{\sqrt{2\pi}} dz, \qquad (3.2.23)$$

under the change of variable $z = (x - \mu)/\sigma$.

To prove (3.2.22) we write

$$\left(\int_{-\infty}^{\infty} \varphi(x)\, dx\right)^2 = \left(\frac{1}{\sqrt{2\pi}} \int_{-\infty}^{\infty} \exp[-x^2/2]\, dx\right)$$

$$\times \left(\frac{1}{\sqrt{2\pi}} \int_{-\infty}^{\infty} \exp[-y^2/2]\, dy\right)$$

$$= \frac{1}{2\pi} \int_{-\infty}^{\infty} \int_{-\infty}^{\infty} \exp[-(x^2+y^2)/2]\, dx\, dy. \qquad (3.2.24)$$

Now we can transform to polar coordinates. Thus $r^2 = x^2 + y^2$ and $dx\, dy = r\, dr\, d\theta$. (See Apostol [9] for a discussion of how to convert from Cartesian to polar coordinates.) Making the polar coordinates substitution in (3.2.24) gives

$$\left|\int_{-\infty}^{\infty} \varphi(x)\, dx\right|^2 = \frac{1}{2\pi} \int_0^{2\pi} \int_0^{\infty} \exp[-r^2/2]\, r\, dr\, d\theta$$

$$= \frac{1}{2\pi} \int_0^{2\pi} [-\exp[-r^2/2]]\Big|_{r=0}^{\infty} d\theta$$

$$= \frac{1}{2\pi} \int_0^{2\pi} d\theta = 1. \qquad (3.2.25)$$

This proves (3.2.22).

If X is $N(\mu, \sigma^2)$, then the moment generating function of X (see Section 2.9) is given by

$$\psi(\theta) = E[\exp(\theta X)] = \frac{1}{\sigma\sqrt{2\pi}} \int_{-\infty}^{\infty} \exp(\theta x) \exp[-\frac{1}{2}((x-\mu)/\sigma)^2]\, dx.$$

$$(3.2.26)$$

Let $z = (x - \mu)/\sigma$. Then (3.2.26) yields

$$\psi(\theta) = \frac{\exp(\mu\theta)}{\sqrt{2\pi}} \int_{-\infty}^{\infty} \exp\left(\sigma\theta z - \frac{z^2}{2}\right) dz$$

$$= \frac{\exp(\mu\theta)}{\sqrt{2\pi}} \int_{-\infty}^{\infty} \exp[-\frac{1}{2}(z^2 - 2\sigma\theta z + \sigma^2\theta^2 - \sigma^2\theta^2)]\, dz$$

$$= \frac{\exp(\mu\theta)}{\sqrt{2\pi}} \exp[(\sigma\theta)^2/2] \int_{-\infty}^{\infty} \exp[-\frac{1}{2}(z - \sigma\theta)^2]\, dz$$

$$= \exp(\mu\theta) \exp[(\sigma\theta)^2/2] \int_{-\infty}^{\infty} \frac{\exp[-w^2/2]}{\sqrt{2\pi}}\, dw = \exp[\mu\theta + (\sigma^2\theta^2)/2].$$

$$(3.2.27)$$

In the next to last integral we substituted $w = z - \sigma\theta$, and used (3.2.22). Thus the moment generating function of a normal random variable X with parameters μ and σ is $\exp[\mu\theta + (\sigma^2\theta^2/2)]$. Hence,

$$\frac{d\psi}{d\theta} = (\mu + \sigma^2\theta)\exp[\mu\theta + (\sigma^2\theta^2/2)],$$

and

$$\frac{d^2\psi}{d\theta^2} = \sigma^2\exp[\mu\theta + (\sigma^2\theta^2/2)] + (\mu + \sigma^2\theta)^2\exp[\mu\theta + (\sigma^2\theta^2/2)].$$

Therefore, by Theorem 2.9.1,

$$E[X] = \frac{d\psi}{d\theta}\bigg|_{\theta=0} = \mu, \qquad E[X^2] = \frac{d^2\psi}{d\theta^2}\bigg|_{\theta=0} = \sigma^2 + \mu^2,$$

and

$$\text{Var}[X] = E[X^2] - (E[X])^2 = \sigma^2 + \mu^2 - \mu^2 = \sigma^2.$$

Thus the parameters μ and σ are, respectively, the mean and standard deviation of X.

We summarize what we have just shown in the following theorem.

THEOREM 3.2.2 (*Properties of a Normal Random Variable*) Suppose X is a normal random variable with parameters μ and σ (X is $N(\mu, \sigma^2)$). Then

$$E[X] = \mu, \qquad \text{Var}[X] = \sigma^2, \tag{3.2.28}$$

and

$$\psi(\theta) = E[e^{\theta X}] = \exp[\mu\theta + (\sigma^2\theta^2/2)]. \tag{3.2.29}$$

If $X_1, X_2, \ldots, X_n$ are n independent random variables having normal $N(\mu_1, \sigma_1{}^2)$, $N(\mu_2, \sigma_2{}^2)$, ..., $N(\mu_n, \sigma_n{}^2)$ distributions, respectively, the moment generating function of

$$Y = X_1 + X_2 + \cdots + X_n,$$

is, by Theorem 2.9.1,

$$\psi_Y(\theta) = \psi_{X_1}(\theta)\psi_{X_2}(\theta)\cdots\psi_{X_n}(\theta) = \exp\left[\theta\sum_{i=1}^{n}\mu_i + \frac{\theta^2}{2}\sum_{i=1}^{n}\sigma_i{}^2\right], \tag{3.2.30}$$

which is the moment generating function of a random variable which is

$$N\left(\sum_{i=1}^{n}\mu_i, \sum_{i=1}^{n}\sigma_i{}^2\right).$$

We have proven the following theorem.

THEOREM 3.2.3   Suppose $X_1, X_2, \ldots, X_n$ are n independent random variables such that X_1 is $N(\mu_1, \sigma_1^2)$, X_2 is $N(\mu_2, \sigma_2^2)$, $\ldots$, X_n is $N(\mu_n, \sigma_n^2)$. Then $Y = X_1 + X_2 + \cdots + X_n$ is normally distributed with mean $\mu_1 + \mu_2 + \cdots + \mu_n$ and variance $\sigma_1^2 + \sigma_2^2 + \cdots + \sigma_n^2$.

Figure 3.2.3 shows several normal density functions. The symmetry of the normal densities about the mean follows from (3.2.18) that is, $f(\mu + x) = f(\mu - x)$ for all real x.

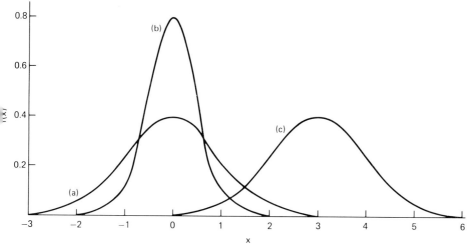

Fig. 3.2.3 The density function of the normal distribution with **(a)** $\mu = 0$, $\sigma = 1$; **(b)** $\mu = 0$, $\sigma = \frac{1}{2}$; **(c)** $\mu = 3$, $\sigma = 1$.

As we saw from formula (3.2.21), probability calculations for any normal distribution can be made from the standard normal distribution. If X is $N(\mu, \sigma^2)$ then the change of variable

$$Z = (X - \mu)/\sigma \qquad (3.2.31)$$

yields a normalized random variable which is $N(0, 1)$. The numerator in (3.2.31) is a shift of origin transformation which transforms the mean value to 0. Division by σ converts the value of $X - \mu$ into units of standard deviation, σ. The fact that Z is normally distributed follows from the uniqueness of the moment generating function as follows:

$$\psi(\theta) = E[e^{\theta((X-\mu)/\sigma)}] = e^{-\mu\theta/\sigma}E[e^{\theta X/\sigma}], \qquad (3.2.32)$$

by Theorem 2.7.1(b).

Now X/σ is a normal random variable with mean μ/σ and variance 1 by the properties of mean and variance (Theorems 2.7.1 and 2.7.2) and the fact that dividing a random variable by a constant does not change the nature of the random variable, but only the scale. Hence the moment generating function of X/σ is

$$E[\exp(\theta X/\sigma)] = \exp[(\mu\theta/\sigma) + \theta^2/2], \qquad (3.2.33)$$

by (3.2.29). Substituting (3.2.33) into (3.2.32) yields

$$\psi(\theta) = \exp(-\mu\theta/\sigma)\exp[(\mu\theta/\sigma) + \theta^2/2] = \exp(\theta^2/2). \qquad (3.3.34)$$

Since (3.3.34) is the moment generating function of a standard normal random variable, Z is $N(0, 1)$.

As illustrated in Fig. 3.2.4, the probability that a randomly selected value x of a normal random variable X will fall within one standard deviation of

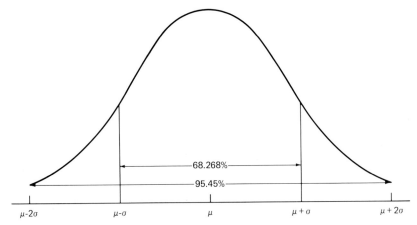

Fig. 3.2.4   Area under normal density curve.

the mean μ is 0.68268 or 68.268%. This can be calculated, using Table 3 of Appendix A, by noting that the probability that z is greater than 1 is $1 - 0.84134 = 0.15866$, so that, by the symmetry of the standard normal density

$$P[\mu - \sigma \le X \le \mu + \sigma] = P[-1 \le Z \le 1] = 1 - 2(0.15866) = 0.68268.$$

Similarly,

$$P[\mu - 1.96\sigma \le X \le \mu + 1.96\sigma] = 0.95.$$

Example 3.2.3   Suppose the number of message buffers in use in the Leevy Stress on-line inquiry system, X, has a normal distribution with a mean of 100 and a standard deviation of 10. Calculate the probability that the number of buffers in use does not exceed 120, lies between 80 and 120, exceeds 130, respectively.

Solution 120 is 2 standard deviations above the mean so the first probability requested is the probability that z does not exceed 2, which, by Table 3, is 0.97725. If x is between 80 and 120, it is not more than 2 standard deviations from the mean; hence, by Fig. 3.2.4, the second probability is 0.9545. Since 130 is 3 standard deviations above the mean,

$$P[X > 130] = P[Z > 3] = 1 - 0.99865 = 0.00135.$$

The normal distribution is the most important distribution in applied probability and statistics because many useful random variables have normal or nearly normal distributions and (more importantly) because of the central limit theorem, which is discussed in Section 3 of this chapter.

3.2.4 Gamma Random Variables

A continuous random variable X is said to have a *gamma distribution with parameters* $\alpha > 0$ and $\lambda > 0$ if its density function f is given by

$$f(x) = \begin{cases} \dfrac{\lambda(\lambda x)^{\alpha-1} e^{-\lambda x}}{\Gamma(\alpha)}, & x > 0 \\ 0, & x \le 0. \end{cases} \tag{3.2.35}$$

Here $\Gamma(\cdot)$ is the celebrated gamma function defined by

$$\Gamma(t) = \int_0^\infty x^{t-1} e^{-x}\, dx, \qquad t > 0. \tag{3.2.36}$$

It can be shown that

$$\Gamma(n+1) = n!, \qquad n = 0, 1, 2, \ldots \tag{3.2.37}$$

and that

$$\Gamma(t+1) = t\Gamma(t) \qquad \text{for all} \quad t > 0. \tag{3.2.38}$$

For an excellent discussion of the gamma function see Parzen [7].

If X has a gamma distribution with parameters α and λ, then its moment generating function $\psi(\cdot)$ is given by

$$\psi(\theta) = E[e^{\theta X}] = \int_0^\infty \frac{e^{\theta x} \lambda^\alpha x^{\alpha-1} e^{-\lambda x}}{\Gamma(\alpha)}\, dx$$

$$= \frac{\lambda^\alpha}{\Gamma(\alpha)} \int_0^\infty x^{\alpha-1} e^{-(\lambda-\theta)x}\, dx. \tag{3.2.39}$$

This integral converges if $\theta < \lambda$. Making the substitution $y = (\lambda - \theta)x$ in (3.2.39), yields

$$\psi(\theta) = \frac{\lambda^\alpha}{\Gamma(\alpha)(\lambda - \theta)^\alpha} \int_0^\infty y^{\alpha - 1} e^{-y} \, dy$$

$$= \frac{\lambda^\alpha}{\Gamma(\alpha)(\lambda - \theta)^\alpha} \Gamma(\alpha) = \frac{\lambda^\alpha}{(\lambda - \theta)^\alpha}, \qquad \theta < \lambda. \qquad (3.2.40)$$

Similarly, the Laplace–Stieltjes transform, $X^*(\theta)$, is given by

$$X^*(\theta) = \left(\frac{\lambda}{\lambda + \theta}\right)^\alpha, \qquad \theta < \lambda. \qquad (3.2.41)$$

We summarize some of the important properties of gamma random variables in the next theorem.

THEOREM 3.2.4   (*Properties of Gamma Random Variables*)   Suppose X is a gamma random variable with parameters α and λ, that is, its density function f is given by

$$f(x) = \begin{cases} \dfrac{\lambda(\lambda x)^{\alpha - 1}}{\Gamma(\alpha)} e^{-\lambda x}, & x > 0 \\ 0, & x \le 0. \end{cases} \qquad (3.2.42)$$

Then the following are true.

(a)   The moment generating function $\psi(\cdot)$ is defined for all $\theta < \lambda$ by

$$\psi(\theta) = \frac{\lambda^\alpha}{(\lambda - \theta)^\alpha}; \qquad (3.2.43)$$

with Laplace–Stieltjes transform

$$X^*(\theta) = \left(\frac{\lambda}{\lambda + \theta}\right)^\alpha. \qquad (3.2.44)$$

(b) $E[X] = \alpha/\lambda$, $\text{Var}[X] = \alpha/\lambda^2$.

(c)   If Y is independent of X and has a gamma distribution with parameters β and λ then $Z = X + Y$ has a gamma distribution with parameters $\alpha + \beta$ and λ. (Gamma random variables are reproductive with respect to α.)

(d)   If $X_1, X_2, \ldots, X_n$ are mutually independent random variables, each with an exponential distribution with parameter λ, then their sum Y has a gamma distribution with parameters n and λ. Furthermore, the distribution function of Y is given by

$$F_Y(x) = G_n(x) = 1 - e^{-\lambda x}\left\{1 + \lambda x + \frac{(\lambda x)^2}{2!} + \cdots\right.$$

$$\left. + \frac{(\lambda x)^{n-1}}{(n-1)!}\right\}, \qquad x \ge 0. \qquad (3.2.45)$$

Proof (a) was proved by (3.2.40). The proofs of (b) and (c) are simple exercises in the use of Theorem 2.9.1 (see Exercise 15). For the proof of (d), see Feller [10, Sect. 1.3].

Example 3.2.4 Suppose the time, X, between inquiries in the Cutrate Construction on-line system has an exponential distribution with an average value of 10 seconds. Let t be an arbitrary point in time and T the elapsed time until the fifth inquiry arrives (after time t). Find the expected value and variance of T. What is the probability that T does not exceed 60 seconds? Exceeds 90 seconds?

Solution By Theorem 3.2.4d, since $T = X_1 + X_2 + X_3 + X_4 + X_5$ where X_1, X_2, X_3, X_4, X_5 are independent, identically distributed, exponential random variables, each with an average value of 10 seconds, T is a gamma random variable with parameters $\alpha = 5$ and $\lambda = 1/10$. Hence,

$$E[T] = \alpha/\lambda = 50 \quad \text{seconds} \quad \text{and} \quad \text{Var}[T] = \alpha/\lambda^2 = 500 \quad \text{seconds}^2.$$

By (3.2.45),

$$P[T \le 60] = G_5(60) = 1 - e^{-6}\left(1 + 6 + \frac{6^2}{2!} + \frac{6^3}{3!} + \frac{6^4}{4!}\right) = 0.7149.$$

By the same formula,

$$P[T > 90] = 1 - P[T \le 90] = e^{-9}\left(1 + 9 + \frac{9^2}{2!} + \frac{9^3}{3!} + \frac{9^4}{4!}\right) = 0.055.$$

If X is a gamma random variable with parameter $\alpha = n$, where n is a small positive integer, then the values of the distribution function of X, G_n, can be calculated fairly easily using (3.2.45). However, if the parameters α and λ are arbitrary positive numbers, probability calculations are more difficult. Sometimes, the formula $P[X \le t] = 1 - Q(2\lambda t \,|\, 2\alpha)$, can be used, where $Q(2\lambda t \,|\, 2\alpha)$ is obtained from Table 26.7 of Abramowitz and Stegun [8]. In other cases approximate formulas such as those found in Abramowitz and Stegun [8] must be used.

Example 3.2.5 The response time at the terminal of the Hopdup Autos on-line system (the time from the keying of the last character of the inquiry until the first character of the response is received) has a gamma distribution with an average value of 5 seconds and a variance of 10 seconds. What is the probability that the response time of a randomly selected inquiry will not exceed 7.2 seconds? 10 seconds?

Solution Let X be the response time. Since X has a gamma distribution, we must have $\alpha/\lambda = 5$ and $\alpha/\lambda^2 = 10$. Solving for α and λ yields $\alpha = 2.5$ and $\lambda = 0.5$. Hence, since $2\lambda t = t$,

$$P[X \le 7.2] = 1 - Q(7.2 \,|\, 5) = 0.79381,$$

and

$$P[X \le 10] = 1 - Q(10/5) = 0.92476,$$

by Table 26.7 of Abramowitz and Stegun [8].

3.2.5 Erlang-k Random Variables

The Danish mathematician A. K. Erlang used a special class of gamma random variables, now often called Erlang-k random variables, in his study of delays in telephone traffic. A random variable, T, is said to be an *Erlang-k random variable with parameter λ* or to have an *Erlang distribution with parameters k and λ* if T is a gamma random variable with the density function f given by

$$f(x) = \begin{cases} \dfrac{\lambda k (\lambda k x)^{k-1}}{(k-1)!} e^{-\lambda k x} & \text{for} \quad x > 0 \\ 0 & \text{for} \quad x \le 0. \end{cases} \qquad (3.2.46)$$

The physical model that Erlang had in mind was a service facility consisting of k identical independent stages, each with an exponential distribution of service time. He wanted this special facility to have the same average service time as a single service facility whose service time was exponential with parameter λ. Thus the service time, T, for the facility with k stages could be written as the sum of k exponential random variables, each with parameter λk. Hence, by Theorem 3.2.4d, T has a gamma distribution with parameters k and λk. Thus,

$$E[T] = 1/\lambda, \qquad \text{Var}[T] = 1/k\lambda^2 = E[T]^2/k,$$

and

$$F(t) = P[T \le t] = 1 - e^{-yt} \left[1 + \frac{yt}{1!} + \frac{(yt)^2}{2!} + \cdots + \frac{(yt)^{k-1}}{(k-1)!} \right],$$

where $y = \lambda k$.

It can also be shown (see Exercise 16) that

$$E[T^n] = \frac{k(k+1)\cdots(k+n-1)}{(k\lambda)^n},$$

and thus

$$E[T^2] = \frac{(k+1)(E[T])^2}{k} \qquad \text{and} \qquad E[T^3] = \frac{(k+1)(k+2)(E[T])^3}{k^2}.$$

$$(3.2.47)$$

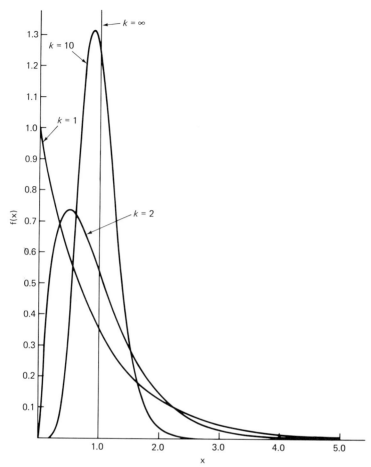

Fig. 3.2.5 Density functions of Erlang-k random variables for $k = 1, 2, 10,$ $\infty\,(\lambda = 1)$.

It should be noted that, for a fixed average value $E[T]$, the variance of T decreases as k increases and, in the limit, goes to 0. Thus an Erlang-k distribution can be used to approximate any nonnegative random variable whose variance does not exceed the square of its mean. Figure 3.2.5 shows the density functions of some selected Erlang-k random variables.

The random variable T in Example 3.2.4 has an Erlang-5 distribution with parameter $\lambda = 1/50 = 0.02$ (mean value 50).

Example 3.2.6   There are five independent stages in the repair of a certain piece of computer equipment. The repair time for each stage is exponentially distributed with an average value of 10 minutes. What is the

probability that a customer engineer can repair the equipment in an hour or
less? Not more than 90 minutes?

Solution   The repair time, T, has an Erlang-5 distribution with average
value 50 minutes (parameter $\lambda = 0.02$) and a variance of $E[T]^2/k = 2500/5 = 500$ minutes^2. (T has the same distribution as the random variable
T in Example 3.2.4, except that there the unit of time was seconds.) Thus the
distribution function F is given by

$$F(t) = 1 - e^{-t/10}\left[1 + \frac{t}{10} + \frac{1}{2}\left(\frac{t}{10}\right)^2 + \frac{1}{6}\left(\frac{t}{10}\right)^3 + \frac{1}{24}\left(\frac{t}{10}\right)^4\right],$$

$$t \text{ in minutes.} \quad (3.2.48)$$

Hence, the probability that the repair time will not exceed 60 minutes is
$F(60) = 0.7149$. (The details of this calculation are shown in Example 3.2.4.)
The probability that the repair time will not exceed 90 minutes is

$$F(90) = 1 - e^{-9}\left[1 + 9 + \frac{9^2}{2} + \frac{9^3}{6} + \frac{9^4}{24}\right] = 0.945.$$

Random variables which are useful in computer science can often be
approximated by Erlang-k random variables, thereby simplifying calculations. In addition, some useful mathematical models, particularly in
queueing theory, assume Erlang-k probability distributions. Thus, if an empirically determined random variable can be approximated by an Erlang-k
distributed random variable, well-known mathematical models can be
applied to make useful predictions.

The usual procedure for selecting an Erlang-k random variable Y to
approximate a given random variable X is as follows.

(1)   Let $\lambda = 1/E[X]$.
(2)   Let k be the largest integer less than or equal to $E[X]^2/\text{Var}[X]$ (the
"floor" of this quantity). Then Y is the Erlang-k random variable with
parameters k and λ.

This method is called the "method-of-moments" (see Section 7.1.1).

Example 3.2.7   Message lengths for the Euphoria State on-line system
have the distribution shown in Table 3.2.1. Approximate this message length
distribution by an Erlang-k random variable Y.

Solution   Let X be the message length. Then

$$E[X] = 25 \times 0.4 + 50 \times 0.3 + 70 \times 0.1 + 100 \times 0.15$$

$$+ 140 \times 0.05 = 54 \quad \text{characters}$$

TABLE 3.2.1

Message Length Distribution

Message length in characters:	25	50	70	100	140
Fraction with this length:	0.4	0.3	0.1	0.15	0.05

and

$$\text{Var}[X] = (25 - 54)^2 \times 0.4 + (50 - 54)^2 \times 0.3 + (70 - 54)^2 \times 0.1$$

$$+ (100 - 54)^2 \times 0.15 + (140 - 54)^2 \times 0.05 = 1054.$$

Since $E[X]^2/\text{Var}[X] = 2.77$, we let Y be the Erlang-2 random variable with average value $54(\lambda = 1/54)$. Thus Y has the same mean as X but its variance is slightly different—1458 rather than the correct 1054. A gamma distribution U with exactly the same mean and variance as X could be used to approximate the message length distribution but probability calculations with U would be much more complex than with Y. In addition the mathematical model under consideration may require an Erlang-k distributed random variable.

3.2.6 Chi-square Random Variables

A random variable Y is said to have a *chi-square distribution with n degrees of freedom* if it can be represented as

$$Y = X_1^2 + X_2^2 + \cdots + X_n^2, \tag{3.2.49}$$

where $X_1, X_2, \ldots, X_n$ are independent standard normal random variables. Thus it is evident that Y can assume only nonnegative values. To discuss the properties of a chi-square random variable we need the following theorem.

THEOREM 3.2.5   Let X be a continuous random variable with density function f and distribution function F. Then the density function g of the random variable $Y = X^2$ is defined by

$$g(y) = \begin{cases} \dfrac{1}{2\sqrt{y}}(f(\sqrt{y}) + f(-\sqrt{y}), & \text{for } y > 0 \\ 0 & \text{for } y \leq 0, \end{cases} \tag{3.2.50}$$

and the distribution function G is given by

$$G(y) = \begin{cases} F(\sqrt{y}) - F(-\sqrt{y}) & \text{for } y \geq 0 \\ 0 & \text{for } y < 0. \end{cases} \tag{3.2.51}$$

Proof Since Y cannot assume negative values, $G(y) = 0$ for $y \leq 0$. For $y > 0$, $Y = X^2 \leq y$ is equivalent to $-\sqrt{y} \leq X \leq \sqrt{y}$. Hence

$$G(y) = P[Y \leq y] = P[-\sqrt{y} \leq X \leq \sqrt{y}]$$
$$= F(\sqrt{y}) - F(-\sqrt{y}).$$

This proves (3.2.51). By differentiation we calculate

$$g(y) = G'(y) = \frac{1}{2\sqrt{y}}(F'(\sqrt{y}) + F'(-\sqrt{y})) = \frac{1}{2\sqrt{y}}(f(\sqrt{y}) + f(-\sqrt{y}))$$

which completes the proof.

Suppose now that Y has a chi-square distribution with one degree of freedom, that is, that $Y = X^2$ where X is a standard normal random variable. Then, by Theorem 3.2.5, the density function g for Y is given by

$$g(y) = \frac{1}{2\sqrt{y}}\left(\frac{e^{-y/2}}{\sqrt{2\pi}} + \frac{e^{-y/2}}{\sqrt{2\pi}}\right) = \frac{e^{-y/2}}{\sqrt{2}\sqrt{\pi}\sqrt{y}}$$

$$= \frac{(\frac{1}{2})(y/2)^{(1/2)-1}e^{-y/2}}{\Gamma(1/2)} \qquad \text{for} \quad y > 0 \qquad (3.2.52)$$

since $\Gamma(1/2) = \sqrt{\pi}$. (For a proof that $\Gamma(1/2) = \sqrt{\pi}$ see Cramér [11, p. 127].) A comparison of (3.2.52) with (3.2.35) shows that Y is a gamma random variable with parameters $\alpha = 1/2$ and $\lambda = 1/2$. Hence the moment generating function of Y is given by

$$\psi(\theta) = (1 - 2\theta)^{-1/2}. \qquad (3.2.53)$$

If now Y is a chi-square random variable with n degrees of freedom we can apply Theorem 2.91 to conclude that the moment generating function of Y is

$$\psi(\theta) = ((1 - 2\theta)^{-1/2})^n = (1 - 2\theta)^{-n/2}. \qquad (3.2.54)$$

The moment generating function given by (3.2.54) is that of a gamma random variable with parameters $\alpha = n/2$ and $\lambda = 1/2$. We summarize the properties of chi-square random variables in the following theorem.

THEOREM 3.2.6 (*Properties of a Chi-square Random Variable*) Let X be a chi-square random variable with n degrees of freedom (and thus a gamma random variable with parameters $\alpha = n/2$ and $\lambda = 1/2$). Then the following statements are true.

(a) $\psi(\theta) = (1 - 2\theta)^{-n/2}$, $E[X] = n$, $\text{Var}[X] = 2n$ and the density function of X is given by

$$f(x) = \begin{cases} \dfrac{x^{(n/2)-1}e^{-x/2}}{2^{n/2}\Gamma(n/2)} & \text{for} \quad x > 0 \\[2mm] 0 & \text{for} \quad x \leq 0. \end{cases} \qquad (3.2.55)$$

(b) If Y is an independent chi-square random variable with m degrees of freedom then $X + Y$ is a chi-square random variable with $n + m$ degrees of freedom (the chi-square distribution has the reproductive property).

(c) As n increases X approaches a normal distribution; that is, for large n, X is approximately $N(n, 2n)$.

Proof (a) follows immediately from the fact that X has a gamma distribution with parameters $\alpha = n/2$ and $\lambda = 1/2$, as we showed above. Now

$$\psi_{X+Y}(\theta) = (1 - 2\theta)^{n/2}(1 - 2\theta)^{m/2} = (1 - 2\theta)^{(n+m)/2}$$

and therefore (b) holds by the uniqueness of the moment generating function (Theorem 2.9.1).

(c) is a consequence of the central limit theorem, which is discussed in Section 3.3. This completes the proof.

The chi-square distribution is best known for its use in "chi-square tests," which are used to test various hypotheses about observed random variables. Some of these tests are discussed in Chapter 8.

Table 4 of Appendix A gives critical values, χ_α^2, of the chi-square distributed random variable χ^2, defined by $P[\chi^2 > \chi_\alpha^2] = \alpha$. For example, the table shows that, if X has a chi-square distribution with 25 degrees of freedom, then the probability that X assumes a value greater than 37.653 is 0.05.

3.2.7 Student-t Random Variables

A continuous random variable X is said to have a *Student-t distribution with n degrees of freedom* if its density function is given by

$$f_n(x) = \frac{1}{\sqrt{n\pi}} \frac{\Gamma((n+1)/2)}{\Gamma(n/2)} \left(1 + \frac{x^2}{n}\right)^{-(n+1)/2} \qquad \text{for all real} \quad x. \quad (3.2.56)$$

This distribution was discovered in 1908 by William S. Gosset who used the pen name "A. Student" [12]. (Gosset worked for the Guiness brewery in Dublin which, at that time, did not allow its employees to publish research papers.) It is evident from (3.2.56) that f_n is symmetric about $x = 0$, and it is easy to show that it assumes a maximum value there (see Exercise 23).

THEOREM 3.2.7 (*Properties of a Student-t Random Variable*) Let X be a Student-t random variable with n degrees of freedom as defined above. Then the following statements are true.

(a) For $n = 1$, X has no expected value; for $n > 1$, $E[X] = 0$.
(b) For $n = 1, 2$ the second moment does not exist; for $n > 2$, $\text{Var}[X] = n/(n - 2)$.
(c) For large values of n, X can be approximated by a standard normal random variable, that is, $\lim_{n \to \infty} X = Y$ where Y is $N(0, 1)$.

Proof   The proof will be omitted but can be found in Kendall and Stuart [13].

The Student-*t* distribution is used, primarily, in dealing with small samples from a normal population. This is discussed in Chapters 7 and 8.

Table 5 of Appendix A gives critical values, t_α, of a Student-*t* distributed random variable X, defined by $P[X > t_\alpha] = \alpha$.

3.2.8   *F*-Distributed Random Variables
(Sometimes called Snedcor-*F*)

A continuous random variable X is said to have an *F distribution with* (*n, m*) *degrees of freedom* if it has the density function f_{nm} given by

$$f_{nm}(x) = \begin{cases} \dfrac{(n/m)^{n/2}\Gamma((n+m)/2)x^{(n/2)-1}}{\Gamma(n/2)\Gamma(m/2)(1+(n/m)x)^{(n+m)/2}} & \text{for} \quad x > 0 \\ 0 & \text{for} \quad x \le 0. \end{cases} \qquad (3.2.57)$$

THEOREM 3.2.8   (*Properties of the F Distribution*)   Suppose U has a chi-square distribution with n degrees of freedom and V a chi-square distribution with m degrees of freedom, with U and V independent. Then $Y = (U/n)/(V/m)$ has an F distribution with (*n, m*) degrees of freedom and thus has the density (3.2.57). If we define $f_\alpha(n, m)$ to be the unique number such that $P[Y > f_\alpha(n, m)] = \alpha$, then

$$f_{1-\alpha}(n, m) = \frac{1}{f_\alpha(m, n)}.$$

Selected values of $f_\alpha(n, m)$ are given in Table 6 of Appendix A. If $m > 2$, then $E[Y] = m/(m-2)$. If $m > 4$, then

$$\text{Var}[Y] = \frac{m^2(2n + 2m - 4)}{n(m-2)^2(m-4)}.$$

The proof of this theorem can be found in Kendall and Stuart [13]. The *F* distribution is important for statistical inference; some examples of its use are given in Chapters 7 and 8.

3.2.9   Hyperexponential Random Variables

If the service time of a queueing system has a large standard deviation relative to the mean value, it can often be approximated by a hyperexponential distribution. Hyperexponential, in this case, means "super exponential." It would seem entirely proper to call a distribution for which the standard deviation is less than the mean "hypoexponential." Thus the constant distribution is the most hypoexponential of all! A hyperexponential random variable may, of course, represent many other interesting phenomena besides the service time of a queueing system; however, this provides an intuitively appealing way of describing a hyperexponential distribution, so we use it.

The model representing the simplest hyperexponential distribution is shown in Fig. 3.2.6. This model shows two parallel stages to the facility; the

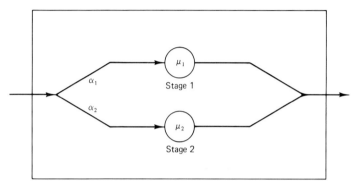

Fig. 3.2.6 Model to represent a two-stage hyperexponential service facility.

top one having exponential service with parameter μ_1, and the bottom stage having exponential service with parameter μ_2. A customer entering the service facility chooses the top stage with probability α_1 or the bottom stage with probability α_2 where $\alpha_1 + \alpha_2 = 1$. After receiving service at the chosen stage, the service time being exponentially distributed with average service rate μ_i (average service time $1/\mu_i$), the customer leaves the service facility. A new customer is not allowed to enter the facility until the original customer has completed service. Thus the density function for the service time is given by

$$f_s(t) = \alpha_1 \mu_1 e^{-\mu_1 t} + \alpha_2 \mu_2 e^{-\mu_2 t}, \qquad t \geq 0. \tag{3.2.58}$$

Therefore, by integration,

$$W_s = E[s] = (\alpha_1/\mu_1) + (\alpha_2/\mu_2), \tag{3.2.59}$$

and

$$E[s^2] = (2\alpha_1/\mu_1^2) + (2\alpha_2/\mu_2^2). \tag{3.2.60}$$

Hence, we calculate

$$\mathrm{Var}[s] = E[s^2] - E[s]^2 = \frac{2\alpha_1}{\mu_1^2} + \frac{2\alpha_2}{\mu_2^2} - \left(\frac{\alpha_1}{\mu_1} + \frac{\alpha_2}{\mu_2}\right)^2. \tag{3.2.61}$$

The distribution function of a two-stage hyperexponential service time can be calculated by integrating (3.2.58) yielding

$$W_s(t) = P[s \leq t] = 1 - \alpha_1 e^{-\mu_1 t} - \alpha_2 e^{-\mu_2 t}. \qquad (3.2.62)$$

Similarly, the moment generating function of the service time, $\psi_s(\cdot)$, is

$$\psi_s(\theta) = \frac{\alpha_1 \mu_1}{\mu_1 - \theta} + \frac{\alpha_2 \mu_2}{\mu_2 - \theta} \qquad \text{if} \quad \theta < \mu_1 \quad \text{and} \quad \theta < \mu_2. \qquad (3.2.63)$$

Likewise, the Laplace–Stieltjes transform, $W_s^*(\theta)$, is

$$W_s^*(\theta) = \frac{\alpha_1 \mu_1}{\mu_1 + \theta} + \frac{\alpha_2 \mu_2}{\mu_2 + \theta} \qquad \text{if} \quad \theta < \mu_1 \quad \text{and} \quad \theta < \mu_2, \qquad (3.2.64)$$

and the third moment is given by

$$E[s^3] = (6\alpha_1/\mu_1^3) + (6\alpha_2/\mu_2^3). \qquad (3.2.65)$$

Algorithm 6.2.1, of Chapter 6, shows how to construct a two-stage hyperexponential random variable with a given squared coefficient of variation $C^2 \geq 1$ and mean $1/\mu$.

The two-stage hyperexponential distribution can be generalized to k stages for any positive integer ≥ 2. The model to represent a k-stage hyperexponential service facility is shown in Fig. 3.2.7.

In this model the customer chooses stage i for $i = 1, 2, \ldots, k$ with probability α_i, where $\alpha_1 + \alpha_2 + \cdots + \alpha_k = 1$. As in the two-stage facility the customer receives exponentially distributed service with average service time of

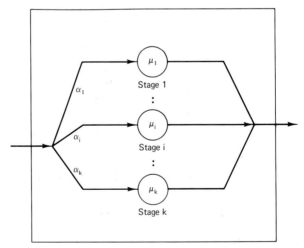

Fig. 3.2.7 Model to represent a k-stage hyperexponential service facility.

$1/\mu_i$ and a new customer cannot enter the facility until the previous customer completes service. Of course a server in a queueing system may provide k-stage hyperexponential service without any physical realization in terms of parallel stages, but the intuitive picture of k parallel exponential stages is helpful in visualizing the service time distribution. For the k-stage hyperexponential distribution we have

$$f_s(t) = \sum_{i=1}^{k} \alpha_i \mu_i e^{-\mu_i t}, \qquad t \geq 0, \tag{3.2.66}$$

$$W_s = E[s] = \sum_{i=1}^{k} \frac{\alpha_i}{\mu_i}, \tag{3.2.67}$$

$$E[s^2] = 2 \sum_{i=1}^{k} \frac{\alpha_i}{\mu_i^2}, \tag{3.2.68}$$

and

$$\text{Var}[s] = E[s^2] - E[s]^2, \tag{3.2.69}$$

where (3.2.68) and (3.2.67) are substituted into the right side of (3.2.69). The third moment, $E[s^3]$, is given by

$$E[s^3] = 6 \sum_{i=1}^{k} \frac{\alpha_i}{\mu_i^3}, \qquad \text{(see Exercise 22, Chapter 5)} \tag{3.2.70}$$

and the distribution function, $W_s(t)$, by

$$W_s(t) = P[s \leq t] = 1 - \sum_{i=1}^{k} \alpha_i e^{-\mu_i t}, \qquad t \geq 0. \tag{3.2.71}$$

Finally, the k-stage hyperexponential distribution has the moment generating function

$$\psi_s(\theta) = \sum_{i=1}^{k} \frac{\alpha_i \mu_i}{\mu_i - \theta}, \qquad \text{if} \quad \theta < \mu_i, \quad i = 1, 2, \ldots, k, \tag{3.2.72}$$

with the Laplace–Stieltjes transform, $W_s^*(\theta)$, given by

$$W_s^*(\theta) = \sum_{i=1}^{k} \frac{\alpha_i \mu_i}{\mu_i + \theta}. \tag{3.2.73}$$

3.3 CENTRAL LIMIT THEOREM

In Chapter 2 we discussed the weak law of large numbers which indicates, roughly speaking, that the probability $P[A]$ of an event A can be estimated by S_n/n where S_n is the number of times the event A occurs in n

independent trials of the basic experiment. Unfortunately, the law of large numbers does not provide a method for estimating how close we are to the true probability, although we saw in Section 2.10, that, by using Chebychev's inequality, we could make a crude estimate of how large n need be so that

$$P\left[\left|\frac{S_n}{n} - p\right| \geq \delta\right] \leq \varepsilon$$

for given positive δ and ε.

The central limit theorem allows us to improve this estimate. It also allows us to make probability judgments about other types of estimates. This theorem is one of the most important in applied probability theory.

THEOREM 3.3.1   (*Central Limit Theorem*)   Let X_1, $X_2, \ldots$ be independent, identically distributed random variables, each having mean μ and standard deviation $\sigma > 0$. Let $S_n = X_1 + \cdots + X_n$. Then for each $x < y$

$$\lim_{n \to \infty} P\left[x \leq \frac{S_n - n\mu}{\sigma\sqrt{n}} \leq y\right] = \Phi(y) - \Phi(x) \qquad (3.3.1)$$

where Φ is the standard normal distribution function.

The proof of this theorem may be found in Parzen [7]. This theorem is truly remarkable in that no special assumptions need be made about the character of X_1. It can be discrete, continuous, or of mixed type. No matter what the form of X_1, the sum S_n approaches a normal distribution with mean $n\mu$ and variance $n\sigma^2$ (S_n is approximately $N(n\mu, n\sigma^2)$). Of course the rate of convergence of S_n to a normal distribution depends upon X_1. For example, if X_1 is normally distributed, then, by Theorem 3.2.3, S_n *is* normally distributed for all n—no approximation is involved. However, if X_1 is a discrete uniform distribution, then n must be somewhat large before S_n can be reasonably approximated by a normal random variable. The result of the central limit theorem is true, under rather general conditions, even if each X_k has a different distribution with mean μ_k and standard deviation σ_k if $\sum_{k=1}^{n} \mu_k$ is substituted for $n\mu$ and $(\sum_{k=1}^{n} \sigma_k^2)^{1/2}$ is substituted for $\sigma\sqrt{n}$ in (3.3.1), that is, (3.3.1) becomes

$$\lim_{n \to \infty} P\left[x \leq \frac{S_n - E[S_n]}{(\text{Var}[S_n])^{1/2}} \leq y\right] = \Phi(y) - \Phi(x). \qquad (3.3.2)$$

This version of the central limit theorem is the basis for an explanation of the observed fact that many random variables such as the height and weight of humans, the yields of crops, the temperature at a certain geographical location for a given day of the year, etc., tend to be normally distributed. Each of these random variables can be represented as the sum of a large number of independent random variables.

There are many applications of the central limit theorem. The first application we discuss is the following approximation theorem.

THEOREM 3.3.2 (*De Moivre-Laplace Limit Theorem*) Let S_n be a binomial random variable with parameters n and p. Then for any nonnegative integers a and b it is approximately true that

$$\lim_{n \to \infty} P[a \leq S_n \leq b] = \Phi\left(\frac{b - np + \frac{1}{2}}{\sqrt{npq}}\right) - \Phi\left(\frac{a - np - \frac{1}{2}}{\sqrt{npq}}\right).$$

$$(3.3.3)$$

The proof is immediate from the central limit theorem, since $E[S_n] = np$ and $\mathrm{Var}[S_n] = npq$, except for the $\frac{1}{2}$ terms on the right side of (3.3.3), which are called the "continuity corrections." The reason for the continuity corrections is that, if we use the normal distribution to approximate the discrete binomial distribution, we are, in effect, fitting a continuous distribution to a discrete distribution as suggested by Fig. 3.3.1. In this figure the step function gives the probabilities of k successes in eight Bernoulli trials with $p = 0.25$. That is, for each k, the area under the binomial graph between $k - \frac{1}{2}$ and $k + \frac{1}{2}$ is the probability of k successes $k = 0, 1, \ldots, 8$. The density function for the approximating normal random variable has mean $np = 2$ and standard deviation $\sqrt{npq} = \sqrt{1.5} = 1.225$. The true probability that S_n

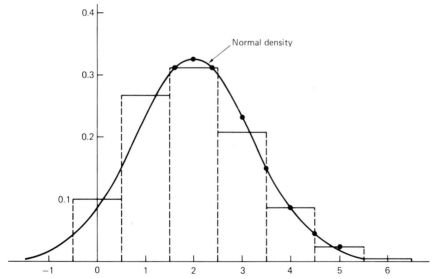

Fig. 3.3.1 Approximation of binomial random variable by normal random variable when $n = 8$, $p = 0.25$.

[1]The accuracy of this approximation improves with the size of n and in the limit goes to zero. It also improves with the degree of symmetry of S_n. One good rule of thumb is the requirement that $npq \geqslant 10$; another is that both $np > 5$ and $nq > 5$ hold. See Parzen [7, pp. 237–245] for a further discussion.

is between 1 and 3 inclusive is $\sum_{k=1}^{3} \binom{8}{k}(0.25)^k(0.75)^{8-k} = 0.7861$. If we use (3.3.3) with the continuity correction, we approximate this probability by the area under the normal density curve from 0.5 to 3.5, which is

$$\Phi\left(\frac{3.5 - 2}{1.225}\right) - \Phi\left(\frac{0.5 - 2}{1.225}\right) = \Phi(1.224) - \Phi(-1.224) = 2\Phi(1.224) - 1$$

$$= 2 \times 0.88952 - 1 = 0.77904.$$

However, if we ignore the continuity correction, we take, as our probability, the area under the normal density curve from 1 to 3, which gives the value

$$\Phi\left(\frac{3 - 2}{1.225}\right) - \Phi\left(\frac{1 - 2}{1.225}\right)$$

$$= 2\Phi(0.816) - 1 = 2 \times 0.79275 - 1 = 0.58549.$$

This is a rather poor approximation because we have neglected the area under the normal density curve from 0.5 to 1 and from 3.0 to 3.5.

Example 3.3.1　In Example 3.1.2 we considered a binomial random variable with parameters 20 and 0.6 which described the number of communication lines in use. The probability that 10 or more lines are in operation was found to be 0.872479. The normal approximation for this probability is

$$\Phi\left(\frac{20 - 12 + 0.5}{2.19}\right) - \Phi\left(\frac{10 - 12 - 0.5}{2.19}\right) = \Phi(3.881) + \Phi(1.142) - 1$$

$$= 0.99995 + 0.87327 - 1 = 0.87322,$$

a fairly good approximation.

The normal distribution can also be used to approximate a Poisson distribution. This follows from the fact that both the Poisson and normal distributions may be used to approximate the binomial distribution.

Example 3.3.2　In Example 3.1.5, X is a Poisson random variable with $\lambda = 10$. We calculated $P[X \leq 15]$ to be 0.95126 by using the APL function POISSONΔSUM. Approximate the answer using the normal distribution.

Solution　We use the normal distribution with $\mu = 10$ and $\sigma = \sqrt{10}$. Then

$$P[X \leq 15] \approx P\left[z \leq \frac{15.5 - 10}{\sqrt{10}}\right] = P[z \leq 1.739] = 0.95898.$$

Suppose we want to estimate $p = P[A]$ for some event A, where we know that $0 < P[A] < 1$. We can let X be the Bernoulli random variable which is 1 when the event A occurs on a particular trial of the experiment and 0 otherwise. Successive independent trials of the experiment yield the sequence of independent Bernoulli random variables X_1, X_2, X_3, ... If we let $S_n = X_1 + X_2 + \cdots + X_n$, then S_n counts the number of times that the event A occurred in n trials of the experiment. The weak law of large numbers, Theorem 2.10.4, indicated that the ratio S_n/n converges to $p = P[A]$. In Example 2.10.5 we saw that Chebychev's inequality enabled us to make a crude estimate of the required value of n such that

$$P\left[\left|\frac{S_n}{n} - p\right| \geq \delta\right] \leq \varepsilon \qquad (3.3.4)$$

where δ and ε are given positive numbers. We will now show how the central limit theorem allows us to improve that estimate. Let $p = P[A]$ and $q = 1 - p$. S_n is a binomial random variable with parameters n and p so $E[S_n] = np$ and $\text{Var}[S_n] = npq$. Suppose δ and ε are given positive numbers. Then

$$P\left[\left|\frac{S_n}{n} - p\right| \geq \delta\right] = P\left[\frac{S_n}{n} - p \leq -\delta\right] + P\left[\frac{S_n}{n} - p \geq \delta\right]. \qquad (3.3.5)$$

Some algebraic manipulation of (3.3.5) yields

$$P\left[\left|\frac{S_n}{n} - p\right| \geq \delta\right] = P\left[\frac{S_n - np}{\sqrt{npq}} \leq -\frac{\delta\sqrt{n}}{\sqrt{pq}}\right] + P\left[\frac{S_n - np}{\sqrt{npq}} \geq \frac{\delta\sqrt{n}}{\sqrt{pq}}\right]. \qquad (3.3.6)$$

The right side of (3.3.6) is now in the form for which we can apply the central limit theorem (the mean of X_1 is p and the standard deviation is $\sqrt{pq}$). Hence, we conclude that

$$P\left[\left|\frac{S_n}{n} - p\right| \geq \delta\right] \approx \Phi\left(\frac{-\delta\sqrt{n}}{\sqrt{pq}}\right) + 1 - \Phi\frac{\delta\sqrt{n}}{\sqrt{pq}} = 2\left(1 - \Phi\left(\frac{\delta\sqrt{n}}{\sqrt{pq}}\right)\right). \qquad (3.3.7)$$

To find n such that (3.3.4) is valid we set the right side of (3.3.7) to ε and $r = \delta\sqrt{n}/\sqrt{pq}$ to arrive at the equation

$$2(1 - \Phi(r)) = \varepsilon, \qquad (3.3.8)$$

or

$$\Phi(r) = (2 - \varepsilon)/2. \qquad (3.3.9)$$

The value of r which makes (3.3.9) true can be found from Table 3 of Appendix A. The definition of r then yields the following estimate for n:

$$n = r^2 pq/\delta^2 \leq r^2/4\delta^2, \qquad (3.3.10)$$

since $pq = p(1 - p)$ has a maximum value of $1/4$, achieved when $p = q = 1/2$ (see Exercise 2 of Chapter 2).

Example 3.3.3   In Example 2.10.5 we wanted to estimate the probability, p, that a randomly selected terminal chosen during the peak period was busy. The estimation method was to randomly choose a terminal n times during this period, count the number of times S_n that a selected terminal was in use, and use the ratio S_n/n as the estimate of p. Chebychev's inequality was used to find the smallest n such that (3.3.4) would be true with $\delta = 0.1$ and $\varepsilon = 0.05$. The value of n was 500, if no knowledge of p was assumed, and 320, if it were known that p was approximately 0.2. If we apply (3.3.9) and (3.3.10) for the first case we get $r = 1.96$ and $n = 96$. If we assumed that p was about 0.2 then (3.3.10) yields $n = 62$. Thus, for the given requirements, 100 samples should suffice. However, an error of 0.1 in a quantity with a magnitude of only 0.2 is a large relative error! Let us turn the question about and ask, "If we make 500 observations to estimate p and let $\varepsilon = 0.05$ in (3.3.4), what is the value of δ?" "That is, what is the maximum error in the estimate at the 5% level of uncertainty?" As before, (3.3.9) yields $r = 1.96$ and

$$\delta = \frac{r\sqrt{pq}}{\sqrt{n}} = \frac{1.96}{\sqrt{500}} \sqrt{pq}$$

which has the value 0.0438 or 0.0351, depending upon our assumption about the value of p. For 100 observations these values of δ are 0.098 and 0.0784, respectively.

3.4 SUMMARY

The name of this chapter is "Probability Distributions" which is meant to suggest that the most important property of a random variable X is how it "distributes probability." That is, how the probability associated with values of X is distributed over these values. For a discrete random variable the most convenient way to do this, usually, is an analytical formula for the probability mass function. That is, given a mass point x_n of X, the pmf $p(\cdot)$ describes how to calculate the associated probability. For example, if X has a Poisson distribution with parameter λ, and k is a nonnegative integer, then $p(k) = P[X = k] = e^{-\lambda}\lambda^k/k!$. For a continuous random variable the most convenient method of describing the probability distribution is by means of the distribution function F, defined for all real x by $F(x) = P[X \leq x]$. Thus, if X is an exponential random variable with parameter λ, then $F(x) = 1 - e^{-\lambda x} = 1 - \exp(-x/E[X])$ for $x \geq 0$.

In the first two sections of this chapter we considered some discrete and continuous random variables which have been found to be especially useful in applied probability theory, and of special importance to computer science applications. Each of these random variables is determined by either one or two parameters; that is, given the parameter or parameters, the entire probability distribution is known. This makes it relatively easy to fit one of these distributions to an empirical distribution. In Chapter 7 we discuss the problem of how to estimate the parameters necessary to fit a well-known distribution to an empirically derived one and, in Chapter 8, we address the problem of judging how good the fit is.

A summary of the properties of the random variables discussed in this chapter is given in Tables 1 and 2 of Appendix A. Examples are given in the text of the use of most of these random variables.

In the third section of this chapter we discussed the central limit theorem and some of its applications. The basic idea of the theorem is that the sum of independent random variables tends toward a normal random variable under very weak restrictions. This explains the special importance of the normal distribution. Several examples were given of the use of the central limit theorem.

Allen [14] presents some of the topics from this chapter in a more condensed form.

Student Sayings

Socrates took Poisson.

Monique is exponentially distributed!

No μs is good μs.

Keep your hyperexponential away from me!

Exercises

1. [15] One-fourth of the source programs submitted by Jumpin Jack compile successfully. What is the probability that exactly one of Jumpin's next five programs will compile? That three out of five will?

2. [20] Six programmers from Amalmogated Malmogates decide to toss coins on an "odd person out" basis to determine who will buy the coffee. Thus there will be a loser if exactly one of the coins falls heads or exactly one falls tails. What is the probability that the outcome will be decided on the first toss? What is the probability that exactly four trials will be required? Not more than four?

3. [HM22] Some authors modify our definition of a geometric random variable X so that it counts the number of trials *including* the trial at which the first success occurs. Thus X can assume the values 1, 2, 3, ... For this modified geometric random variable find the pmf $p(\cdot)$, the expected value, and the variance in terms of the probability of success on each trial, p, and of $q = 1 - p$.

4. [15] Jumpin Jill finds that, when she is developing a program module, syntax errors are discovered by the compiler on 60% of the runs she makes. Furthermore this percentage is independent of the number of runs made on the same module. (Is this reasonable?) How many runs does she need to make on one module, on the average, to get a run with no syntax errors? What is the probability that more than 4 runs will be required? [*Hint:* Use the result of Exercise 3.]

5. [15] Let X be a Poisson random variable with parameter λ. Prove that
(a) $P[X \leq \lambda/2] \leq 4/(\lambda + 4) < 4/\lambda$ and
(b) $P[X \geq 2\lambda] \leq 1/(1 + \lambda) < 1/\lambda$.
[*Hint:* Use one-sided inequality, Theorem 2.10.3.]

6. [15] About 1% of the population is left handed. What is the probability that at least four out of 200 people at Kysquare Testing will be left handed?

7. [12] There are 125 misprints in a 250 page user manual for the EZYASPI system. What is the probability that there are at least two misprints on a given page?

8. [C15] The number of telephone calls per minute arriving at the main switchboard of Binomial Distributors has a Poisson distribution with an average rate of 20 calls per minute. If the switchboard can handle only 30 calls per minute, what is the probability that in the next minute, starting with all lines clear, the switchboard will be overloaded? [*Hint:* The Poisson distribution can be approximated by the normal distribution if computational facilities for the Poisson calculations are not available.]

9. [10] Let X be a discrete uniform random variable assuming only the value c (X is a constant random variable). Show that for each positive integer n, $E[X^n] = E[X]^n$.

10. [15] Suppose a discrete uniform random variable, X, assumes only the values $C + L, C + 2L, \ldots, C + nL$, where C and L are constants. Show that

$$E[X] = C + \frac{(n + 1)}{2} L, \qquad E[X^2] = C^2 + (n + 1)LC + \frac{(n + 1)(2n + 1)}{6} L^2$$

and

$$\mathrm{Var}[X] = \frac{(n^2 - 1)}{12} L^2.$$

[*Hint:*

$$\sum_{i=1}^{n} i = \frac{n(n + 1)}{2} \qquad \text{and} \qquad \sum_{i=1}^{n} i^2 = \frac{n(n + 1)(2n + 1)}{6}.]$$

11. [12] The simulation model of proposed computer system for Students Gosset uses a discrete approximation to a continuous uniform distribution on the interval 10 to 30. Find the mean and variance of the continuous uniform distribution and compare these values to those for the discrete approximation if
(a) the eleven values 10, 12, 14, 16, ..., 20, ..., 30 are used for the discrete distribution,
(b) the 101 values 10, 10.2, ..., 20, 20.2, ..., 30 are used. [*Hint:* See Exercise 10.]

12. [15] Inquiries for the Poisson Cannery on-line system arrive in a Poisson pattern at an average rate of 12 per minute. What is the probability that the time interval between the next two inquiries will be less than 7.5 seconds? More than 10 seconds? What is the 90th percentile value for interarrival time?

13. [HM15] Suppose X is uniformly distributed on the interval a to b. Show that

$$E[X] = (a + b)/2 \quad \text{and} \quad \text{Var}[X] = (b - a)^2/12.$$

14. [20] Suppose entries to an order-entry system of the SHOOTEMUP Arms Company arrive at the central processor with a Poisson pattern at an average rate of 30 per minute. Given that an order entry transaction has just arrived, what is the average time until the fourth succeeding transaction arrives? What is the probability that it will take longer than 10 seconds for this entry to arrive? Less than 5 seconds? Will the answers to the above questions change if the point in time at which measurement begins is 1 second after a transaction has arrived?

15. [HM18] Prove Theorems 3.2.4b and 3.2.4c using Theorem 2.9.1.

16. [HM22] Prove that for an Erlang random variable X with parameters k and λ the moments are given by

$$E[X^n] = \frac{k(k + 1) \cdots (k + n - 1)}{(k\lambda)^n}, \quad n = 1, 2, 3, \ldots.$$

17. [15] The message length distribution for the incoming messages of an on-line system for the SOCKITUEM Finance Company has a mean of 90 characters and a variance of 1500. Fit an Erlang distribution to this message length distribution.

18. [HM20] Show that the density function for a chi-square random variable with n degrees of freedom has a unique maximum at $x = n$, if $n > 2$.

19. [18] A large-scale on-line system at FLYBYNIGHT Airlines has 200 terminals each provided with a dial-up capability to the main computer center. Each terminal k, independently, has probability 0.05 of being in use. What is the probability that 20 or more terminals are in use? Use the normal approximation and compare it to the estimate given by the one-sided inequality, Theorem 2.10.3.

20. [15] The arrival pattern of order messages of the on-line order entry system of the HUBAHUBA Record Company has a Poisson distribution with an average of 20 arrivals per minute during the peak period. What is the probability that more than 30 orders will arrive in one minute of the peak period? Use the normal approximation.

21. [18] Every fifth customer arriving at POURBOY FINANCE is given a prize. If customers arrive in a Poisson pattern with parameter λ, that is, if the number of customers who arrive in any one minute interval has a Poisson distribution with average value λ, describe the interarrival time distribution for the customers who receive gifts. If λ is 5 customers per minute what is the probability that the time between two successive winners exceeds 1 minute?

22. [15] A simulation model of a proposed new computer system for the HUNKYDORY Boat Company has been constructed. The model provides an estimate of the utilization, ρ, of the central processing unit (CPU) by testing every millisecond to determine whether or not it is busy and using the formula

$$\rho = P[\text{CPU is busy}] = S_n/n,$$

where n is the number of samples and S_n the number of times the CPU is busy. How many samples should be made if $\delta = 0.005$ and $\varepsilon = 0.001$ in the formula

$$P\left[\left|\frac{S_n}{n} - \rho\right| \geq \delta\right] \leq \varepsilon?$$

Assume ρ is near 0.5.

23. [HM18]   Show that the density function f_n given by (3.2.56) for a Student-t random variable with n degrees of freedom assumes a unique maximum at $x = 0$.

References

1. H. Bateman, On the probability distribution of α particles, *Philos. Mag.* Ser. 6, **20** (1910), 704–707.
2. E. Rutherford and H. Geiger, The probability variations in the distribution of α particles, *Philos. Mag.* Ser. 6, **20** (1910), 700–704.
3. S. A. Lippman, *Elements of Probability and Statistics.* Holt, New York, 1971.
4. R. D. Clarke, An application of the Poisson distribution, *J. Inst. Actuaries* **72** (1946), 48.
5. W. Feller, *An Introduction to Probability Theory and Its Applications*, Vol. 1, 3d ed. Wiley, New York, 1968.
6. W. A. Wallis, The Poisson distribution and the Supreme Court, *J. Amer. Statist. Assoc.* **31** (1936), 376–380.
7. E. Parzen, *Modern Probability Theory and Its Applications.* Wiley, New York, 1960.
8. M. Abramowitz and I. Stegun, *Handbook of Mathematical Functions.* National Bureau of Standards, Washington, D.C., 1964.
9. T. M. Apostol, *Calculus*, Vol. II, 2nd ed. Ginn (Blaisdell), Boston, 1969.
10. W. Feller, *An Introduction to Probability Theory and Its Applications*, Vol. 2, 2nd ed. Wiley, New York, 1971.
11. H. Cramér, *Mathematical Methods of Statistics.* Princeton University Press, Princeton, New Jersey, 1946.
12. "A. Student" (W. S. Gosset), The probable error of a mean, *Biometrika* **6** (1908).
13. M. G. Kendall and A. Stuart, *The Advanced Theory of Statistics*, Vol. I. Hafner, New York, 1969.
14. A. O. Allen, Elements of probability for system design, *IBM Syst. J.* **13**, 4 (1974) 325–348.

Chapter Four

STOCHASTIC PROCESSES

INTRODUCTION

In Chapter 3 we considered some common random variables which are useful in investigating probabilistic computer science phenomena such as the number of computer jobs waiting to be processed, the response time in an on-line inquiry system, the time between messages in an order entry system, etc. When we considered a random variable such as the number of jobs, N, waiting to be processed, we did not allow for the fact that the probability distribution of N changes with time. That is, if we let N_1 be the number of jobs in the job queue at 8 A.M. and N_2 the corresponding number at 11 A.M., then N_1 and N_2 probably have different probability distributions. (To investigate the nature of the distribution of N_1, we could note the number of jobs at 8 A.M. each day for a number of days; for N_2 we could do the same thing at 11 A.M.) Thus we really have a family of random variables $\{N_t, t \in T\}$, where T is the set of all times during the day that the computer center is in operation. Such a family of random variables is called a *stochastic process.*

4.1   STOCHASTIC PROCESS DEFINITIONS

A family of random variables $\{X(t), t \in T\}$ is called a *stochastic process*. Thus for each $t \in T$, where T is the *index set* of the process, $X(t)$ is a random variable. An element t of T is usually referred to as a time parameter and we will often refer to t as time. The *state space* of the process is the set of all possible values that the random variables $X(t)$ can assume. Each of these values is called a *state* of the process.

Stochastic processes are classified in a number of ways, such as by the index set and by the state space. If

$$T = \{0, 1, 2, 3, \ldots\} \qquad \text{or} \qquad T = \{0, \pm 1, \pm 2, \ldots\},$$

the stochastic process is said to be a *discrete parameter process* and we will usually indicate the process by $\{X_n\}$. If $T = \{t : -\infty < t < \infty\}$ or $T = \{t : t \geq 0\}$, the stochastic process is said to be a *continuous parameter process* and will be indicated by $\{X(t), -\infty < t < \infty\}$ or $\{X(t), t \geq 0\}$. The state space is classified as *discrete* if it is finite or countable; it is *continuous* if it consists of an interval (finite or infinite) of the real line.

Example 4.1.1   The waiting time of an arriving inquiry message until processing is begun, $\{W(t), t \geq 0\}$. The arrival time, t, of the message is the continuous parameter. The state space is also continuous.

Example 4.1.2   The number of messages which arrive in the time period from 0 to t, $\{N(t), t \geq 0\}$. This is a continuous parameter discrete state space process.

Example 4.1.3   Let $\{X_n, n = 1, 2, 3, 4, 5, 6, 7\}$ denote the average time to run a job at a computer center on the nth day of the week. Thus X_1 is the average job time on Sunday, X_2 is the average job time on Monday, etc. Then $\{X_n\}$ is a discrete parameter continuous state space process.

Example 4.1.4   Let $\{X_n, n = 1, \ldots, 365 \, (366)\}$ denote the number of jobs run at a computing center on the nth day of the year. This is a discrete parameter discrete state space process.

Consider random (unpredictable) events such as

(a)   the arrival of an inquiry at the central computer center of an on-line system,
(b)   a telephone call to an airline reservation center,
(c)   an end-of-file interrupt, or
(d)   the occurrence of a hardware or software failure in a computer system.

Such events can be described by a *counting function* $N(t)$, defined for all $t > 0$, as the number of events that have occurred after time 0 but not later

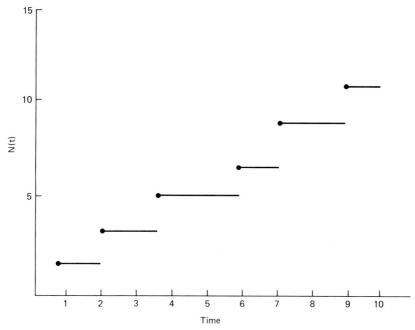

Fig. 4.1.1   Typical counting function $N(t)$.

than time t. (A typical counting function $N(\cdot)$ is shown in Fig. 4.1.1.) Here time 0 represents the time at which the counting is begun. For each t, the number $N(t)$ is an observed value of a random variable. Thus $\{N(t), t \geq 0\}$ is a stochastic process, called a *counting process*.

Definition 4.1.1 $\{N(t), t \geq 0\}$ constitutes a *counting process* provided

(1) $N(0) = 0$,
(2) $N(t)$ assumes only nonnegative integer values,
(3) $s < t$ implies that $N(s) \leq N(t)$, and
(4) $N(t) - N(s)$ is the number of events that have occurred after s but not later than t, that is, in the time interval $(s, t]$.

The next definition formalizes the idea that "one quantity is small relative to another quantity" and makes it possible to indicate this fact without specifying the exact relationship between the two quantities.

Definition 4.1.2   The function f is $o(h)$ (read "f is little-oh of h" and written "$f = o(h)$") if

$$\lim_{h \to 0} \frac{f(h)}{h} = 0,$$

that is, if given $\varepsilon > 0$, there exists $\delta > 0$ such that $0 < |h| < \delta$ implies that[1]

$$\left| \frac{f(h)}{h} \right| < \varepsilon.$$

Example 4.1.5   (a)   The function $f(x) = x$ is *not* $o(h)$ since

$$\lim_{h \to 0} \frac{f(h)}{h} = \lim_{h \to 0} \frac{h}{h} = 1 \neq 0.$$

(b)   The function $f(x) = x^2$ is $o(h)$ since

$$\lim_{h \to 0} \frac{f(h)}{h} = \lim_{h \to 0} \frac{h^2}{h} = \lim_{h \to 0} h = 0.$$

(c)   The function $f(x) = x^r$, where $r > 1$, is $o(h)$ since

$$\lim_{h \to 0} \frac{f(h)}{h} = \lim_{h \to 0} h^{r-1} = 0.$$

This generalizes (b).

(d)   If f is $o(h)$ and g is $o(h)$, then $f + g$ is $o(h)$ since

$$\lim_{h \to 0} \frac{f(h) + g(h)}{h} = \lim_{h \to 0} \frac{f(h)}{h} + \lim_{h \to 0} \frac{g(h)}{h} = 0 + 0 = 0.$$

(e)   If f is $o(h)$ and c is a constant, then cf is $o(h)$ since

$$\lim_{h \to 0} \frac{cf(h)}{h} = c \lim_{h \to 0} \frac{f(h)}{h} = c \times 0 = 0.$$

(f)   It follows from (d) and (e), by mathematical induction that any finite linear combination of functions, each of which is $o(h)$, is also $o(h)$. That is, if $c_1, c_2, \ldots, c_n$ are n constants and $f_1, f_2, \ldots, f_n$ are n functions, each of which is $o(h)$, then $\sum_{i=1}^{n} c_i f_i$ is $o(h)$.

(g)   Suppose X is an exponential random variable with parameter λ and that $h > 0$. Then

$$P[X \leq t + h \,|\, X > t] = P[X \leq h]$$

by the Markov property of the exponential distribution (Theorem 3.2.1d). But

$$P[X \leq h] = 1 - e^{-\lambda h} = 1 - \left[1 - \lambda h + \sum_{n=2}^{\infty} \frac{(-\lambda h)^n}{n!} \right]$$

$$= \lambda h - (\lambda h)^2 \sum_{n=2}^{\infty} \frac{(-\lambda h)^{n-2}}{n!} = \lambda h + o(h),$$

---

[1] If f is not defined for $h < 0$ then $f = o(h)$ means that for each $\varepsilon > 0$ there exists $\delta > 0$ such that $0 < h < \delta$ implies $|f(h)/h| < \varepsilon$.

so that

$$P[X \leq t + h \mid X > t] = \lambda h + o(h).$$

A continuous parameter stochastic process $\{X(t), t \geq 0\}$ has *independent increments* if events occurring in nonoverlapping time intervals are independent; that is, if $(a_1, b_1), \ldots, (a_n, b_n)$ are n nonoverlapping intervals, then thr n random variables

$$X1b_1) - X1a_1), X1b_2) - X1a_2), \ldots, X1b_n) - X1a_n)$$

are independent. The process has *stationary increments* if $X(t + h) - X(s + h)$ has the same distribution as $X(t) - X(s)$ for each choice of indices s and t (with $s < t$) and for every $h > 0$; that is, the distribution $X(t) - X(s)$ depends only upon the length of the interval from s to t and not on the particular value of s.

Example 4.1.6 Suppose $X_1, X_2, \ldots$ are independent identically distributed Bernouli random variables, each with probability p of success (that is, of assuming the value 1). Let $S_n = X_1 + \cdots + X_n$, that is, the number of successes in n Bernoulli trials. Then $\{S_n, n = 1, 2, 3, \ldots\}$ is called a *Bernoulli process*. It has the state space $\{0, 1, 2, 3, \ldots\}$ so it is a discrete parameter discrete state space process. For each n, S_n has a binomial distribution with pmf $p(\cdot)$ defined by $p(k) = P[S_n = k] = \binom{n}{k} p^k q^{n-k}$, $k = 0, 1, \ldots, n$, where $q = 1 - p$. As we saw in Section 3.1.3, starting at any particular Bernoulli trial, the number of succeeding trials, Y, before the next success has a geometric distribution; that is,

$$P[Y = k] = q^k p, \qquad k = 0, 1, 2, \ldots.$$

4.2 THE POISSON PROCESS

Definition 4.2.1 A counting process $\{N(t), t \geq 0\}$ (see Definition 4.1.1) is a *Poisson process with rate* $\lambda > 0$, if (1)–(4), below, are true.

(1) The process has independent increments. (Events occurring in nonoverlapping intervals of time are independent of each other.)

(2) The increments of the process are stationary. (The distribution of the number of events in any interval of time depends only upon the length of the interval and not upon when the interval begins.)

(3) The probability that exactly one event occurs in any time interval of length h is $\lambda h + o(h)$, that is,

$$P[N(h) = 1] = \lambda h + o(h).$$

(4) The probability that more than one event occurs in any time interval of length h is $o(h)$, that is,

$$P[N(h) \geq 2] = o(h).$$

Note that (3) and (4) together imply that

$$P[N(h) = 0] = 1 - \lambda h + o(h),$$

since

$$P[N(h) = 0] = 1 - P[N(h) = 1] - P[N(h) \geq 2]$$
$$= 1 - \lambda h - o(h) - o(h) = 1 - \lambda h + o(h). \qquad (4.2.1)$$

The last equality follows from Example 4.1.5d and e.

The following theorem shows that the Definition 4.2.1 of a Poisson process is descriptive.

THEOREM 4.2.1   Let $\{N(t), t \geq 0\}$ be a Poisson process with rate $\lambda > 0$. Then the random variable Y describing the number of events in any time interval of length $t > 0$ has a Poisson distribution with parameter λt. That is,

$$P[Y = k] = e^{-\lambda t} \frac{(\lambda t)^k}{k!}, \qquad k = 0, 1, 2, \ldots. \qquad (4.2.2)$$

Thus the average number of events occurring in any time interval of length t is λt.

Proof   Let $t > 0$. By the definition of a Poisson process the number of events occurring in any time interval is independent of those in any nonoverlapping interval and depends only upon the length of the given interval. Therefore, we can assume, without loss of generality, that the interval of interest extends from 0 to t. We define

$$P_n(t) = P[N(t) = n] \qquad \text{for each nonnegative integer} \quad n. \qquad (4.2.3)$$

It is true that no events occur by time $t + h$ only if no events occur by time t and no events occur in the interval from t to $t + h$. Hence,

$$P_0(t + h) = P_0(t)P[N(t + h) - N(t) = 0]$$
$$= P_0(t)P[N(h) = 0] = P_0(t)(1 - \lambda h + o(h)). \qquad (4.2.4)$$

The first equality in (4.2.4) follows from the fact that the process has independent increments; the second equality from the stationarity of the increments. The last equality follows from (4.2.1).

Equation (4.2.4) and Example 4.1.5e, together, yield

$$\frac{P_0(t + h) - P_0(t)}{h} = -\lambda P_0(t) + \frac{o(h)}{h}. \qquad (4.2.5)$$

Letting $h \to 0$ in (4.2.5), we arrive at the differential equation

$$\frac{dP_0(t)}{dt} = -\lambda P_0(t). \qquad (4.2.6)$$

The solution of (4.2.6) with the initial condition $P_0(0) = P[N(0) = 0] = 1$ is given by

$$P_0(t) = e^{-\lambda t},$$

as can be verified by direct substitution in (4.2.6).

Now suppose $n > 0$. Then n events can occur by time $t + h$ $(N(t + h) = n)$ in three mutually exclusive ways.

(1) n events occur by time t and no event occurs in the interval from t to $t + h$.

(2) $n - 1$ events occur by time t and exactly one event occurs in the interval between t and $t + h$.

(3) $n - k$ events occur by time t for some k from the set $\{2, 3, \ldots, n\}$ and exactly k events occur in the interval from t to $t + h$.

Hence, summing up the probabilities associated with (1), (2), and (3) yields

$$P_n(t + h) = P_n(t)(1 - \lambda h + o(h)) + \lambda h P_{n-1}(t) + o(h). \tag{4.2.7}$$

Therefore,

$$\frac{P_n(t + h) - P_n(t)}{h} = -\lambda P_n(t) + \lambda P_{n-1}(t) + \frac{o(h)}{h}, \tag{4.2.8}$$

and taking the limit as $h \to \infty$ gives

$$\frac{dP_n(t)}{dt} = -\lambda P_n(t) + \lambda P_{n-1}(t). \tag{4.2.9}$$

The solution to (4.2.9) subject to the initial condition $P_n(0) = 0$ is given by

$$P_n(t) = \frac{e^{-\lambda t}(\lambda t)^n}{n!}, \tag{4.2.10}$$

as can be checked by direct substitution.

Since $P[Y = k] = P_k(t) = e^{-\lambda t}(\lambda t)^k/k!$, $k = 0, 1, 2, \ldots$, this completes the proof.

It is important to note that, according to Theorem 4.2.1, the number of events occurring in *each* time interval of length t has a Poisson distribution with an average value of λt. Hence, the average number of events occurring per unit time is $\lambda t/t = \lambda$. The next theorem gives another important attribute of a Poisson process. It was previously stated as Theorem 3.2.1f.

THEOREM 4.2.2 Let $\{N(t), t \geq 0\}$ be a Poisson process with rate λ. Let $0 < t_1 < t_2 < t_3 < \cdots$ be the successive occurrence times of events, and let the interarrival (interoccurrence) times $\{\tau_n\}$ be defined by $\tau_1 = t_1$, $\tau_2 = t_2 - t_1, \ldots, \tau_k = t_k - t_{k-1}, \ldots$. Then the interarrival times $\{\tau_n\}$ are mutually independent identically distributed exponential random variables each with mean $1/\lambda$.

Proof Since a Poisson process has independent increments ((1) of Definition 4.2.1), events occurring after t_n are independent of those occurring before t_n, $n = 1, 2, \ldots$. This proves that $\tau_1, \tau_2, \ldots$ are independent random variables. For any $s \geq 0$ and any $n \geq 1$ the events $\{\tau_n > s\}$ and $\{N(t_{n-1} + s) - N(t_{n-1}) = 0\}$ are equivalent (we define t_0 to be zero so that t_{n-1} is defined when $n = 1$). The events in brackets are equivalent since $\{\tau_n > s\}$ is true, if and only if the nth event has not yet occurred s time units after the occurrence of the $(n - 1)$th event; but this is the same as the requirement that $\{N(t_{n-1} + s) - N(t_{n-1}) = 0\}$ is true. Thus it is true that

$$P[\tau_n > s] = P[N(t_{n-1} + s) - N(t_{n-1}) = 0] = P[N(s) = 0] = e^{-\lambda s},$$
(4.2.11)

by Theorem 4.2.1 and the fact that the process has stationary increments. Therefore,

$$P[\tau_n \leq s] = 1 - e^{-\lambda s}, \qquad s \geq 0.$$
(4.2.12)

This completes the proof.

The next theorem shows that the converse of Theorem 4.2.2 also is true.

THEOREM 4.2.3 Let $\{N(t), t \geq 0\}$ be a counting process such that the interarrival times of events, $\{\tau_n\}$, are independent identically distributed exponential random variables, each with the average value $1/\lambda$. Then $\{N(t), t \geq 0\}$ is a Poisson process with rate λ.

Proof We omit the proof of this theorem. The proof can be found in Chung [1, pp. 200–202].

When the occurrence of some event such as the arrival of an inquiry to an inquiry system, the arrival of customers at a bank, the arrival of messages to a message switching center, etc., is described by a Poisson process we often hear the events described as "random." From a technical standpoint we usually associate the word "random" with some sort of selection process in which either

(a)   each of a finite number of elements has the same probability of selection or

(b)   a time is chosen in some interval of time so that each subinterval of the same length has the same probability of containing the selected point.

The next theorem shows us that the word "random" *is* appropriate for describing a Poisson process.

THEOREM 4.2.4   Suppose $\{N(t), t \geq 0\}$ is a Poisson process and one event has taken place in the interval from 0 to t. Then Y, the random variable describing the time of occurrence of this Poisson event, has a continuous uniform distribution on the interval from 0 to t, that is, if $0 < \delta < t$, then any subinterval of $(0, t]$ of length δ has probability δ/t of containing the time of occurrence of the event.

Proof   Let $0 < x < t$. By the definition of Y

$$P[Y \leq x] = P[\tau_1 \leq x \mid N(t) = 1]. \tag{4.2.13}$$

But, by the definition of conditional probability

$$P[\tau_1 \leq x \mid N(t) = 1] = \frac{P[(N(x) = 1) \text{ and } (N(t) - N(x) = 0)]}{P[N(t) = 1]}$$

$$= \frac{P[N(x) = 1]P[N(t - x) = 0]}{P[N(t) = 1]}$$

$$= \frac{\lambda x e^{-\lambda x} e^{-\lambda(t - x)}}{\lambda t e^{-\lambda t}} = \frac{x}{t}, \tag{4.2.14}$$

where the next to last equality in (4.2.14) follows from Theorem 4.2.1. This completes the proof.

The Poisson process is a special case of a general type of stochastic process which is important to queueing theory. In the next section we study this type of process called a "birth-and-death process."

4.3   BIRTH-AND-DEATH PROCESS

In the last section we studied a Poisson process $\{N(t), t \geq 0\}$ which counted the number of occurrences of some type of event which could also be interpreted as an arrival of some entity at an average rate λ. We could think of such an arrival as a birth. For a Poisson process the probability of one birth in a short time interval h is $\lambda h e^{-\lambda h} = \lambda h + o(h)$, and this probability is independent of how many births have occurred. λ can be thought of as the birth rate. For some systems, such as a biological species or a queueing system, it might be reasonable to suppose that the birth rate depends upon the number of the population present, that is, that the probability of a birth in a short time interval h might be $\lambda_n h + o(h)$, where n is the size of the population, and the birth rate λ_n depends upon this number n. It is also reasonable to allow deaths or decreases in the population with the probability of a death in an interval of length h equal to $\mu_n h + o(h)$. Thus the intuitive idea behind a birth-and-death process is that of some type of a "population" which is simultaneously gaining new members through births and losing old members through deaths—such as the human population of the earth. The population we have in mind for most applications of birth-and-death processes to computer science is that of customers in a queueing system. Of course "customer" is a generic word, here, and could correspond to a computer job to be processed, an I/O request, a message arrival to a communication system, etc. Customer arrivals correspond to births, and customer departures (after receiving service) correspond to deaths.

Definition 4.3.1   Consider a continuous parameter stochastic process $\{X(t), t \geq 0\}$ with the discrete state space 0, 1, 2, ... Suppose this process describes a system which is in state E_n, $n = 0, 1, 2, \ldots$ at time t, if and only if, $X(t) = n$ (the system has a population of n elements or customers at time t). Then the system is said to be described by a *birth-and-death process* if there exist nonnegative birth rates $\{\lambda_n, n = 0, 1, 2, \ldots\}$ and nonnegative death rates $\{\mu_n, n = 1, 2, \ldots\}$ such that the following postulates (sometimes called the "nearest-neighbor" assumptions) are satisfied.

(1)   State changes are only allowed from state E_n to state E_{n+1} or from state E_n to E_{n-1} if $n \geq 1$, but from state E_0 to state E_1 only.

(2)   If at time t the system is in state E_n, the probability that between time t and time $t + h$ the transition from state E_n to state E_{n+1} occurs (indicated by $E_n \rightarrow E_{n+1}$) equals $\lambda_n h + o(h)$, and the probability that the transition $E_n \rightarrow E_{n-1}$ occurs (if $n \geq 1$) equals $\mu_n h + o(h)$.

(3)   The probability that, in the time interval from t to $t + h$, more than one transition occurs is $o(h)$.

Postulate (1) allows only one birth or death to occur at a time and states that no death can occur if the system is empty. Postulate (2) gives the transition probabilities, that is, the probability of a birth or of a death in a small time interval when the system population is n. The last postulate states that the probability of more than one birth or death in a short time interval is negligible.

When we describe a queueing system as a birth-and-death process we think of state E_n as corresponding to n customers in the system, either waiting for or receiving service.

We will now derive the differential-difference equations for $P_n(t) = P[X(t) = n]$, the probability that the system is in state E_n at time t. The procedure is very similar to the method we used in the proof of Theorem

4.2.1. In fact the differential-difference equations we derived there are a special case of the equations we derive here.

If $n \geq 1$ the probability $P_n(t + h)$ that at time $t + h$ the system will be in state E_n has four components.

(1) The probability that it was in state n at time t and no transitions occurred, either births or deaths. This probability is the product of (a) $P_n(t)$, (b) the probability that the transition $E_n \rightarrow E_{n+1}$ did *not* occur or $1 - \lambda_n h + o(h)$, and (c) the probability that the transition $E_n \rightarrow E_{n-1}$ did *not* occur or $1 - \mu_n h + o(h)$. Hence, the required probability is

$$P_n(t)(1 - \lambda_n h + o(h))(1 - \mu_n h + o(h))$$
$$= P_n(t)[1 - \mu_n h + o(h) - \lambda_n h + \lambda_n \mu_n h^2 - \lambda_n h o(h) + o(h)]$$
$$= P_n(t)[1 - \mu_n h - \lambda_n h + o(h)] = P_n(t)(1 - \lambda_n h - \mu_n h) + o(h), \qquad (4.3.1)$$

since, by Example 4.1.5,

$$o(h)(1 - \mu_n h + o(h)) = o(h),$$
$$\lambda_n \mu_n h^2 - \lambda_n h o(h) + o(h) = o(h), \qquad \text{and} \qquad P_n(t)o(h) = o(h).$$

(2) The probability $P_{n-1}(t)$ that the system was in state E_{n-1} at time t, times the probability that the transition $E_{n-1} \rightarrow E_n$ occurred in the interval from t to $t + h$. This latter probability equals $\lambda_{n-1} h + o(h)$ so the total contribution is

$$P_{n-1}(t)(\lambda_{n-1} h + o(h)) = P_{n-1}(t)\lambda_{n-1} h + o(h). \qquad (4.3.2)$$

(3) The probability $P_{n+1}(t)$ that the system was in state E_{n+1} at time t, multiplied by the probability that the transition $E_{n+1} \rightarrow E_n$ occurred during the interval from t to $t + h$. The contribution is thus

$$P_{n+1}(t)\mu_{n+1} h + o(h). \qquad (4.3.3)$$

(4) The probability that two or more transitions occur between times t and $t + h$ which leave the system in state E_n. (For example, the two transitions $E_{n+2} \rightarrow E_{n+1}$ and $E_{n+1} \rightarrow E_n$.) By hypothesis this probability is $o(h)$.

Since the events leading to the four components are mutually exclusive the result is that

$$P_n(t + h) = [1 - \lambda_n h - \mu_n h]P_n(t) + \lambda_{n-1} h P_{n-1}(t)$$
$$+ \mu_{n+1} h P_{n+1}(t) + o(h). \qquad (4.3.4)$$

Transposing the term $P_n(t)$ and dividing by h we get

$$\frac{P_n(t + h) - P_n(t)}{h} = -(\lambda_n + \mu_n)P_n(t)$$

$$+ \lambda_{n-1}P_{n-1}(t) + \mu_{n+1}P_{n+1}(t) + \frac{o(h)}{h}. \quad (4.3.5)$$

Taking the limit as $h \to 0$ gives us the equation

$$\frac{dP_n(t)}{dt} = -(\lambda_n + \mu_n)P_n(t) + \lambda_{n-1}P_{n-1}(t) + \mu_{n+1}P_{n+1}(t). \quad (4.3.6)$$

This equation is valid for $n \geq 1$. For $n = 0$, by similar reasoning, we get

$$\frac{dP_0(t)}{dt} = -\lambda_0 P_0(t) + \mu_1 P_1(t). \quad (4.3.7)$$

If the initial state is E_i then the initial conditions are given by

$$P_i(0) = 1 \quad \text{and} \quad P_j(0) = 0 \quad \text{for} \quad j \neq i. \quad (4.3.8)$$

The birth-and-death process depends upon the infinite set of differential-difference equations (4.3.6) and (4.3.7) with initial conditions (4.3.8). It can be shown that this set of equations has a solution $P_n(t)$ for all n and t under very general conditions. However, the solutions are very difficult to obtain, analytically, except for some very special cases.

One such special case is the pure-birth process with $\lambda_n = \lambda > 0$ for all n, $\mu_n = 0$ for all n, and the initial conditions $P_0(0) = 1$, $P_j(0) = 0$ for $j \neq 0$. (Any process for which all the μ_n are zero is called a *pure-birth process* while any for which all the λ_n are zero is called a *pure-death process*.) This leads to the set of equations

$$\frac{dP_n(t)}{dt} = -\lambda P_n(t) + \lambda P_{n-1}(t), \quad n \geq 1$$

$$\frac{dP_0(t)}{dt} = -\lambda P_0(t). \quad (4.3.9)$$

As we saw in the proof of Theorem 4.2.1, the solution of (4.3.9) satisfying the given initial conditions is given by

$$P_n(t) = \frac{e^{-\lambda t}(\lambda t)^n}{n!}, \quad n \geq 0, \quad t \geq 0. \quad (4.3.10)$$

Thus the process is a Poisson process and we have a new characterization of a Poisson process. It is a pure-birth process with a constant birth rate.

In general, finding the time-dependent solutions of a birth-and-death process is very difficult. However, if $P_n(t)$ approaches a constant value p_n as $t \to \infty$ for each n, then we say that the system is in *statistical equilibrium*. Under very general conditions these limits exist and are independent of the initial conditions. When a system is in statistical equilibrium we sometimes say the system is in the *steady state* or that the system is *stationary*, because the state of the system does not depend upon time. If we could obtain the time-dependent solutions $\{P_n(t)\}$ we could solve for the steady state solutions $\{p_n\}$ by the equations $\lim_{t \to \infty} P_n(t) = p_n$, $n = 0, 1, 2, \dots$. Since we cannot, in general, find the time-dependent or transient solutions to the birth-and-death differential-difference equations (4.3.6) and (4.3.7), analytically, we will take limits as $t \to \infty$ on both sides of these equations and, using the fact that $\lim_{t \to \infty} dP_n(t)/dt = 0$ for all n and $\lim_{t \to \infty} P_n(t) = p_n$ (we assume that the steady state solutions do exist), we obtain the set of difference equations

$$0 = \lambda_{n-1} p_{n-1} + \mu_{n+1} p_{n+1} - (\lambda_n + \mu_n) p_n, \qquad n \geq 1 \qquad (4.3.11)$$

$$0 = \mu_1 p_1 - \lambda_0 p_0, \qquad n = 0. \qquad (4.3.12)$$

The last equation yields

$$p_1 = (\lambda_0/\mu_1) p_0. \qquad (4.3.13)$$

Equation (4.3.11) can be written as

$$\mu_{n+1} p_{n+1} - \lambda_n p_n = \mu_n p_n - \lambda_{n-1} p_{n-1}, \qquad n \geq 1. \qquad (4.3.14)$$

If we define $g_n = \mu_n p_n - \lambda_{n-1} p_{n-1}$ for $n = 1, 2, 3, \dots$ we see that (4.3.14) can be written as

$$g_{n+1} = g_n, \qquad n \geq 1. \qquad (4.3.15)$$

Hence $g_n = $ constant and, by (4.3.12), $g_1 = $ constant $= 0$. Hence $g_n = 0$ for all n or (assuming $\mu_n > 0$ for all n)

$$p_{n+1} = \frac{\lambda_n}{\mu_{n+1}} p_n, \qquad n \geq 0. \qquad (4.3.16)$$

Thus we compute successively

$$p_1 = \frac{\lambda_0}{\mu_1} p_0, \qquad p_2 = \frac{\lambda_1}{\mu_2} p_1 = \frac{\lambda_0 \lambda_1}{\mu_1 \mu_2} p_0, \qquad p_3 = \frac{\lambda_2}{\mu_3} p_2 = \frac{\lambda_0 \lambda_1 \lambda_2}{\mu_1 \mu_2 \mu_3} p_0.$$

Continuing, for $n = 4, 5, 6, \dots$, we see, by induction, that

$$p_n = \frac{\lambda_0 \lambda_1 \cdots \lambda_{n-1}}{\mu_1 \mu_2 \cdots \mu_n} p_0, \qquad n \geq 1. \qquad (4.3.17)$$

This gives the solutions in terms of p_0, the probability that the system is in state E_0 (the system is empty). p_0 is determined by the condition

$$\sum_{n=0}^{\infty} p_n = p_0 + p_1 + p_2 + \cdots = 1. \tag{4.3.18}$$

If we substitute (4.3.17) into (4.3.18) we obtain

$$p_0 \left(1 + \frac{\lambda_0}{\mu_1} + \frac{\lambda_0 \lambda_1}{\mu_1 \mu_2} + \cdots + \frac{\lambda_0 \lambda_1 \cdots \lambda_{n-1}}{\mu_1 \mu_2 \cdots \mu_n} + \cdots \right) = 1. \tag{4.3.19}$$

Hence, the steady state probabilities (4.3.17) exist if the series

$$S = 1 + \frac{\lambda_0}{\mu_1} + \frac{\lambda_0 \lambda_1}{\mu_1 \mu_2} + \cdots + \frac{\lambda_0 \lambda_1 \cdots \lambda_{n-1}}{\mu_1 \mu_2 \cdots \mu_n} + \cdots < \infty. \tag{4.3.20}$$

(We assume the λ_n and μ_n are nonnegative.) When this is true, $p_0 = 1/S > 0$, or the probability that the system is empty is positive. In the case of a queueing system this means that the service facility sometimes "catches up" or gets all the customers processed. On the other hand, if the series for S diverges, this is an indication that the queueing system is unstable because arrivals are occurring faster, on the average, than departures. For actual real-life queueing systems described by birth-and-death processes we may safely assume that the steady state probabilities $\{p_n\}$ exist, if and only if the series for S converges and then they are given by (4.3.17) with $p_0 = 1/S$.

Before we study the steady state solutions to some important birth-and-death processes, we consider a simple queueing system for which we *can* calculate the time dependent functions $P_n(t)$ for all n.

Example 4.3.1 Consider a queueing system with one server and no waiting line. We assume a Poisson arrival process with parameter λ and an exponential service time distribution with parameter μ. The former means, by definition, that the probability of an arrival in the interval $(0, h]$ is $\lambda h + o(h)$. If the server is busy at time t, then the probability that service for the customer will be completed by time $t + h$ is $1 - e^{-\mu h} = \mu h + o(h)$. (Here we have used the "lack of memory" or "Markov property" of the exponential distribution.) Thus our birth-and-death model has only states E_0 and E_1, with E_0 corresponding to the server being idle and E_1 to the server being busy. An arrival that occurs when the server is busy is turned away and thus has no effect on the system. Therefore, an arrival will cause a state transition (from E_0 to E_1) only if the arrival occurs when the server is idle. Hence, $\lambda_0 = \lambda$, $\lambda_n = 0$ for $n \neq 0$, $\mu_1 = \mu$ and $\mu_n = 0$ for $n \neq 1$. The birth-and-death differential-difference equations (4.3.6) and (4.3.7) become

$$\frac{dP_1(t)}{dt} = \lambda P_0(t) - \mu P_1(t), \qquad \frac{dP_0(t)}{dt} = -\lambda P_0(t) + \mu P_1(t). \tag{4.3.21}$$

We can set the initial conditions to be $P_0(0) + P_1(0) = 1$. Since

$$\frac{d(P_0(t) + P_1(t))}{dt} = 0$$

by (4.3.21), we have

$$P_0(t) + P_1(t) = 1, \qquad t \geq 0. \tag{4.3.22}$$

If we substitute (4.3.22) into the second equation of (4.3.21), we obtain

$$\frac{dP_0(t)}{dt} + (\lambda + \mu)P_0(t) = \mu. \tag{4.3.23}$$

By elementary differential equation theory (see, for example, Coddington [2, p. 41]) we have

$$P_0(t) = \frac{\mu}{\lambda + \mu} + \left(P_0(0) - \frac{\mu}{\lambda + \mu}\right)e^{-(\lambda + \mu)t}. \tag{4.3.24}$$

By symmetry we also have

$$P_1(t) = \frac{\lambda}{\lambda + \mu} + \left(P_1(0) - \frac{\lambda}{\lambda + \mu}\right)e^{-(\lambda + \mu)t}. \tag{4.3.25}$$

Now, if we take the limit as $t \to \infty$ in (4.3.24) and (4.3.25), we obtain the steady state probabilities p_0 and p_1 as

$$p_0 = \lim_{t \to \infty} P_0(t) = \frac{\mu}{\lambda + \mu}, \tag{4.3.26}$$

and

$$p_1 = \lim_{t \to \infty} P_1(t) = \frac{\lambda}{\lambda + \mu}. \tag{4.3.27}$$

On the other hand, if we set the time derivatives to 0 in (4.3.21) and replace $P_0(t)$ by p_0, $P_1(t)$ by p_1, we obtain

$$0 = \lambda p_0 - \mu p_1, \qquad 0 = -\lambda p_0 + \mu p_1. \tag{4.3.28}$$

Either equation of (4.3.28) gives

$$p_1 = (\lambda/\mu)p_0 \tag{4.3.29}$$

(the equations of (4.3.28) are equivalent). Substituting (4.3.29) into the equation $p_0 + p_1 = 1$ gives $p_0(1 + \lambda/\mu) = 1$ or

$$p_0 = \mu/(\lambda + \mu). \tag{4.3.30}$$

Substituting (4.3.30) into (4.3.29) gives

$$p_1 = \lambda/(\lambda + \mu). \tag{4.3.31}$$

The latter method of obtaining the solution was obviously more straight-forward than the former.

A useful, intuitively appealing technique has been devised to derive the steady state difference equations (4.3.11) and (4.3.12). It involves the use of a state-transition rate diagram which graphically illustrates the postulates for a birth-and-death system. Figure 4.3.1 is a general *state-transition rate dia-*

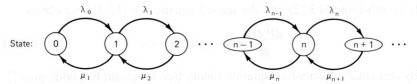

Fig. 4.3.1 State-transition rate diagram for a general birth-and-death process.

gram for a birth-and-death system. In this diagram a state E_n is represented by a circle or oval labeled with the number n. The arrows in this diagram show the only state transitions allowed and are labeled with the mean transition rates (either birth or death). Since we assume the system is in the steady state we assume the following principle: *FLOW RATE IN = FLOW RATE OUT PRINCIPLE.* If a birth-and-death system has reached the steady state (equilibrium) condition, then for every state of the system n $(n = 0, 1, 2, \ldots)$ the mean flow rate of the population into the state must equal the mean flow rate out. The equations expressing this condition are called the *balance equations.*

Consider first a state E_n, with $n \geq 1$. Then, by Fig. 4.3.1, we see that the mean flow rate of population into the state is $\lambda_{n-1} p_{n-1} + \mu_{n+1} p_{n+1}$ while the mean flow rate of population out of state n is $\mu_n p_n + \lambda_n p_n = (\mu_n + \lambda_n) p_n$. Therefore, by the FLOW RATE IN = FLOW RATE OUT PRINCIPLE we have the balance equation

$$\lambda_{n-1} p_{n-1} + \mu_{n+1} p_{n+1} = (\lambda_n + \mu_n) p_n, \qquad n \geq 1. \qquad (4.3.32)$$

Equation (4.3.32) is equivalent to (4.3.11). For state E_0 the above PRINCIPLE immediately gives

$$\mu_1 p_1 = \lambda_0 p_0, \qquad (4.3.33)$$

which is the same as (4.3.12).

Example 4.3.2   The queueing system of Example 4.3.1 has the state-transition rate diagram of Fig. 4.3.2. We can write the following balance equation by inspection:

$$\mu p_1 = \lambda p_0. \qquad (4.3.34)$$

Hence,

$$p_1 = (\lambda/\mu) p_0. \qquad (4.3.35)$$

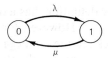

Fig. 4.3.2 State-transition rate diagram for birth-and-death system of Example 4.3.1.

But,

$$1 = p_0 + p_1 = p_0 \left(1 + \frac{\lambda}{\mu}\right), \qquad (4.3.36)$$

so

$$p_0 = \mu/(\lambda + \mu), \qquad (4.3.37)$$

and

$$p_1 = (\lambda/\mu)p_0 = \lambda/(\lambda + \mu), \qquad (4.3.38)$$

as we got in Example 4.3.1.

4.4 MARKOV CHAINS

A stochastic process $\{X(t), t \in T\}$ is a *Markov process* if for any set of $n + 1$ values $t_1 < t_2 < \cdots < t_n < t_{n+1}$ in the index set and any set $\{x_1, x_2, \ldots, x_{n+1}\}$ of $n + 1$ states, we have

$$P[X(t_{n+1}) = x_{n+1} \,|\, X(t_1) = x_1, X(t_2) = x_2, \ldots, X(t_n) = x_n]$$

$$= P[X(t_{n+1}) = x_{n+1} \,|\, X(t_n) = x_n]. \qquad (4.4.1)$$

Intuitively, (4.4.1) indicates that the future of the process depends only upon the present state and not upon the history of the process. That is, the entire history of the process is summarized in the present state.

A Markov process is called a *Markov chain* if its state space is discrete. Markov processes can thus be classified as in Table 4.4.1, where we assume the index set is time.

TABLE 4.4.1
Classification of Markov Processes

Type of parameter	State space	
	Discrete	Continuous
Discrete time	Discrete time Markov chain	Discrete time Markov process
Continuous time	Continuous time Markov chain	Continuous time Markov process

Since a Markov chain, by definition, has a discrete state space, we could label the states $\{E_0, E_1, E_2, \ldots\}$. However, for notational convenience, we will usually assume the states are the nonnegative integers $\{0, 1, 2, \ldots\}$. For the general case it is easy to associate each integer i with the corresponding state E_i.

For a discrete time Markov chain it is fruitful to think of the process as making *state transitions* at times t_n, $n = 1, 2, 3, \ldots$ (possibly into the same state). Thus the discrete time Markov chain $\{X_n\}$ (we write X_n for $X(t_n)$) starts in an initial state, say i, when $t = t_0$ ($X_0 = i$), and makes a state transition at the next step (time in the sequence); that is, when $t = t_1$, so that $X_1 = j$, etc. The one-step transition probabilities defined by

$$P[X_{n+1} = j \,|\, X_n = i], \qquad n, i, j = 0, 1, 2, \ldots$$

generally depend upon the index n. However, we are primarily interested in Markov chains for which the one-step transition probabilities are independent of n and thus can be denoted by P_{ij}. Such a Markov chain is said to have *stationary transition probabilities* or to be *homogeneous in time*. All discrete Markov chains in the sequel are assumed to be time homogeneous, unless the contrary is stated. The transition probabilities P_{ij} can be exhibited as a square matrix

$$P = \begin{bmatrix} P_{00} & P_{01} & P_{02} & P_{03} & \cdots \\ P_{10} & P_{11} & P_{12} & P_{13} & \cdots \\ P_{20} & P_{21} & P_{22} & P_{23} & \cdots \\ \vdots & \vdots & \vdots & \vdots & \\ P_{i0} & P_{i1} & P_{i2} & P_{i3} & \cdots \\ \vdots & \vdots & \vdots & \vdots & \end{bmatrix}$$

called the *transition probability matrix* of the chain. If the number of states is finite, say n, then there will be n rows and n columns in the matrix P; otherwise the matrix will be infinite. We must have

$$P_{ij} \geq 0, \qquad i, j = 0, 1, 2, \ldots, \tag{4.4.2}$$

and

$$\sum_{j=0}^{\infty} P_{ij} = 1, \qquad i = 0, 1, 2, \ldots. \tag{4.4.3}$$

Equation (4.4.2) is true because each P_{ij} is a transition probability. The $(i + 1)$st row of P represents the probabilities that the process will make the transition from state i to state j, $j = 0, 1, 2, \ldots$, at the next step. That is, this row gives the probability distribution of X_{n+1} given that $X_n = i$. Thus (4.4.3) must hold since some transition occurs (possibly back to state i).

Example 4.4.1 Consider a sequence of Bernoulli trials in which the probability of success on each trial is p and of failure is q, where $p + q = 1$ and $0 < p < 1$. Let the state of the process at trial n be the number of uninterrupted successes that have been completed at this point (a sequence of such successes is called a success run). Thus, if the first 5 outcomes were SFSSF (a success followed by a failure, two successes, and another failure), we would have $X_0 = 1$, $X_1 = 0$, $X_2 = 1$, $X_3 = 2$, and $X_4 = 0$. The state transition matrix is given by

$$P = [P_{ij}] = \begin{bmatrix} q & p & 0 & 0 & 0 & \cdots \\ q & 0 & p & 0 & 0 & \cdots \\ q & 0 & 0 & p & 0 & \cdots \\ q & 0 & 0 & 0 & p & \cdots \\ \vdots & & & & & \end{bmatrix}.$$

The state 0 can be reached in one transition from any state while the state $i + 1$ can only be reached (in one transition) from state i, $i = 0, 1, 2, \ldots$. This Markov chain is clearly homogeneous in time.

We now define the n-step transition probabilities P_{ij}^n by

$$P_{ij}^n = P[X_n = j \mid X_0 = i]. \tag{4.4.4}$$

Since the Markov chain $\{X_n\}$ has stationary transition probabilities we have

$$P_{ij}^n = P[X_{m+n} = j \mid X_m = i] \qquad \text{for all} \quad m \geq 0 \quad \text{and} \quad n > 0. \tag{4.4.5}$$

It is true, of course, that $P_{ij} = P_{ij}^1 = P_{ij}$.

For convenience, we define

$$P_{ij}^0 = \delta_{ij} = \begin{cases} 1 & \text{if} \quad i = j \\ 0 & \text{if} \quad i \neq j. \end{cases}$$

(δ_{ij} is called the Kronecker delta function.) The n-step transition probabilities can be computed using the *Chapman–Kolmogorv equations*, which are

$$P_{ij}^{n+m} = \sum_{k=0}^{\infty} P_{ik}^n P_{kj}^m \qquad \text{for all} \quad n, m, i, j \geq 0. \tag{4.4.6}$$

In particular,

$$P_{ij}^n = \sum_{k=0}^{\infty} P_{ik}^{(n-1)} P_{kj}, \qquad n = 2, 3, \ldots. \tag{4.4.7}$$

If we denote the matrix of n-step transition probabilities, P_{ij}^n, by $P^{(n)}$, then (4.4.7) shows that $P^{(n)} = P^{(n-1)}P = P^n$, that is, the matrix product of $P^{(n-1)}$ by P. Hence $P^{(n)}$ can be calculated as the nth power of the matrix P.

Example 4.4.2   Consider a communication system which transmits the digits 0 and 1 through several stages. At each stage the probability that the same digit will be received by the next stage, as transmitted, is 0.75. What is the probability that a 0 that is entered at the first stage is received as a 0 by the 5th stage?

Solution   We want to find P_{00}^4. The state transition matrix P is given by

$$P = \begin{bmatrix} 0.75 & 0.25 \\ 0.25 & 0.75 \end{bmatrix}.$$

Hence

$$P^2 = \begin{bmatrix} 0.625 & 0.375 \\ 0.375 & 0.625 \end{bmatrix}, \quad \text{and} \quad P^4 = P^2 P^2 = \begin{bmatrix} 0.53125 & 0.46875 \\ 0.46875 & 0.53125 \end{bmatrix}.$$

Therefore, the probability that a zero will be transmitted through four stages as a zero is

$$P_{00}^4 = 0.53125.$$

State j of a Markov chain $\{X_n\}$ is said to be *reachable* from state i if it is possible for the chain to proceed from state i to state j in a finite number of transitions; that is, $P_{ij}^n > 0$ for some $n \geq 0$. If every state is reachable from every other state, the chain is said to be *irreducible*. We shall be concerned only with irreducible Markov chains.

Example 4.4.3   The Markov chain of Example 4.4.1 is irreducible because there is a one-step transition from any state i to state 0, while if $j > 0$ there is clearly a j-step transition from state 0 to state j.

The state i is said to be *periodic with period d* if $P_{ii}^n > 0$ only for $n = d, 2d, 3d, \ldots$, where d is the largest such integer and $d > 1$; if $d = 1$ then i is an *aperiodic state*. That is, i is an aperiodic state if $P_{ii}^n > 0$ for $n = 1, 2, 3, \ldots$. For each state i we define $f_i^{(n)}$ to be the probability that the first return to state i occurs n steps or transitions after leaving i. That is, $f_i^{(n)} = P[X_n = i, X_k \neq i$ for $k = 1, 2, \ldots, n - 1 \mid X_0 = i]$. (We define $f_i^{(0)} = 1$ for all i.) The probability of ever returning to state i is given by $f_i = \sum_{n=1}^{\infty} f_i^{(n)}$. If $f_i < 1$, then i is said to be a *transient state*. If $f_i = 1$, then state i is said to be *recurrent*. If state i is recurrent, we define the *mean recurrence time of i* as m_i given by

$$m_i = \sum_{n=1}^{\infty} n f_i^{(n)}. \tag{4.4.8}$$

Thus m_i is the average time to return to state i. If $m_i = \infty$, then state i is said to be *recurrent null*, while, if $m_i < \infty$, then state i is said to be *positive recurrent* or *recurrent nonnull*. Thus each recurrent state i is either positive recurrent or recurrent null.

Two states i and j are said to *communicate* if i is reachable from j and j is reachable from i, that is, there exist integers m and n such that $P_{ji}^m > 0$ and $P_{ij}^n > 0$. We indicate that states i and j communicate by writing $i \leftrightarrow j$. We state without proof, some important theorems about the states of a Markov chain. The proofs can be found in [3] and [4].

THEOREM 4.4.1 The relation $i \leftrightarrow j$ is an equivalence relation. That is, (a) $i \leftrightarrow i$ for each state i, (b) if $i \leftrightarrow j$, then $j \leftrightarrow i$, and (c) if $i \leftrightarrow j$ and $j \leftrightarrow k$, then $i \leftrightarrow k$. Thus the states of Markov state can be partitioned into equivalence classes of states such that two states i and j are in the same class, if and only if $i \leftrightarrow j$. A Markov chain is irreducible, if and only if there is exactly one equivalence class. Furthermore, if $i \leftrightarrow j$, then i and j have the same period. Finally, if $i \leftrightarrow j$, and i is recurrent, then so is j.

THEOREM 4.4.2 Let $\{X_n\}$ be an irreducible Markov chain. Then exactly one of the following holds:

(1) all states are positive recurrent,
(2) all states are recurrent null,
(3) all states are transient.

Let us define the probability that the discrete Markov chain $\{X_n\}$ is in the state j at the nth step by $\pi_j^{(n)}$. That is,

$$\pi_j^{(n)} = P[X_n = j].$$

Thus our *initial distribution* of the states $0, 1, 2, \ldots$ is given by

$$\pi_j^{(0)} = P[X_0 = j], \qquad j = 0, 1, \ldots.$$

A discrete Markov chain is said to have a *stationary probability distribution* $\pi = (\pi_1, \pi_2, \ldots)$ if the matrix equation $\pi = \pi P$ is satisfied, where each $\pi_i \geq 0$ and $\sum_i \pi_i = 1$. The matrix equation $\pi = \pi P$ can be written as the set of equations

$$\pi_j = \sum_i \pi_i P_{ij}, \qquad j = 0, 1, 2, \ldots. \tag{4.4.9}$$

The reason such a probability distribution π is called a stationary distribution is that, if $\pi_j^{(0)} = \pi_j, j = 0, 1, \ldots$, for such a distribution, then $\pi_j^{(n)} = \pi_j$ for all n and j; the probabilities $\pi_j^{(n)}$ do not change with time, but are stationary. A Markov chain is said to have a *long-run* or *limiting* probability distribution $\pi = (\pi_1, \pi_2, \ldots)$ if

$$\lim_{n \to \infty} \pi_j^{(n)} = \lim_{n \to \infty} P[X_n = j] = \pi_j, \qquad j = 0, 1, \ldots.$$

A discrete Markov chain which is irreducible, aperiodic, and for which all states are positive recurrent is said to be *ergodic*. The next theorem is important for the application of Markov chains to queueing theory and is stated without proof. (The proof can be found in Feller [8].)

THEOREM 4.4.3   If $\{X_n\}$ is an irreducible, aperiodic, time homogeneous Markov chain then the limiting probabilities

$$\pi_j = \lim_{n \to \infty} \pi_j^{(n)}, \qquad j = 0, 1, \ldots$$

always exist and are independent of the initial state probability distribution $\pi^{(0)} = (\pi_1^{(0)}, \pi_2^{(0)}, \ldots)$. If all the states are *not* positive recurrent (and thus either all states are recurrent null or all are transient), then $\pi_j = 0$ for all j and no stationary probability distribution exists. However, if all the states of $\{X_n\}$ are positive recurrent so the chain is ergodic, then $\pi_j > 0$ for all j and $\pi = (\pi_1, \pi_2, \ldots)$ forms a stationary probability distribution where

$$\pi_j = 1/m_j, \qquad j = 0, 1, \ldots. \tag{4.4.10}$$

In this latter case the limiting distribution is the unique solution of the set of equations

$$\sum_i \pi_i = 1 \tag{4.4.11}$$

$$\pi_j = \sum_i \pi_i P_{ij}, \qquad j = 0, 1, 2, \ldots. \tag{4.4.12}$$

Note that part of the conclusion is that for an ergodic Markov chain the stationary probability distribution and the long-run (limiting) probability distribution are the same. Such probability distributions are called *equilibrium* or *steady state* distributions. In Markov chain applications to queueing theory these are the distributions of most interest. Thus it is important to know under what conditions a Markov chain is ergodic and thus has a steady state distribution. The next two theorems provide some answers.

THEOREM 4.4.4   The irreducible, aperiodic Markov chain $\{X_n\}$ is positive recurrent (and thus ergodic) if there exists a nonnegative solution of the system

$$\sum_j P_{ij} x_j \le x_i - 1 \qquad (i \ne 0) \tag{4.4.13}$$

such that

$$\sum_j P_{0j} x_j < \infty. \tag{4.4.14}$$

THEOREM 4.4.5   The irreducible, aperiodic Markov chain $\{X_n\}$ is positive recurrent (and thus ergodic), if and only if there exists a nonnull solution of the equations

$$\sum_j x_j P_{ji} = x_i, \qquad i = 0, 1, 2, \ldots \tag{4.4.15}$$

such that

$$\sum_j |x_j| < \infty. \tag{4.4.16}$$

The proofs of the last two theorems are given in Foster [5].

THEOREM 4.4.6 A finite-state Markov chain $\{x_n\}$ which is irreducible and aperiodic is ergodic.

Proof The theorem follows immediately from Theorem 4 of Feller [8, p. 392].

Example 4.4.4 Consider again the Markov chain of Example 4.4.2 with state-transition matrix

$$P = \begin{bmatrix} 0.75 & 0.25 \\ 0.25 & 0.75 \end{bmatrix}.$$

It is clear that this Markov chain is irreducible and aperiodic. Hence, by Theorem 4.4.6 it is also ergodic and thus has a limiting probability distribution which is also a stationary distribution. We can apply Theorem 4.4.3 to calculate the equilibrium probability distribution $\pi = (\pi_0, \pi_1)$.
  We have the equations

$$\pi_0 + \pi_1 = 1, \qquad \pi_0 = 0.75\pi_0 + 0.25\pi_1, \qquad \pi_1 = 0.25\pi_0 + 0.75\pi_1.$$

The unique solution of these equations is $\pi_0 = 0.5$, $\pi_1 = 0.5$. This means that if data are passed through a large number of stages, the output is independent of the original input and each digit received is equally likely to be a 0 or a 1. This also means that

$$\lim_{n \to 0} P^n = \begin{bmatrix} 0.5 & 0.5 \\ 0.5 & 0.5 \end{bmatrix}.$$

(We saw, in Example 4.4.2, that

$$P^4 = \begin{bmatrix} 0.53125 & 0.46875 \\ 0.46875 & 0.53125 \end{bmatrix}$$

so that

$$P^8 = \begin{bmatrix} 0.501953125 & 0.498046875 \\ 0.498046875 & 0.501953125 \end{bmatrix}$$

and the convergence is rapid.)
  Note also that

$$\pi P = (0.5, 0.5) = \pi,$$

so π *is* a stationary distribution, as advertised.

For completeness we list another theorem on recurrence, although it is not as useful as the preceding three theorems.

THEOREM 4.4.7   The irreducible Markov chain $\{X_n\}$ is recurrent if there exists a sequence $\{y_i\}$ such that

$$\sum_j P_{ij} y_j \leq y_i \qquad \text{for} \quad i \neq 0 \qquad\qquad (4.4.17)$$

with $\lim_{i \to \infty} y_i = \infty$.

The proof of this theorem is given in Karlin and Taylor [4] and Foster [5]. Note that the recurrent chain could be recurrent null.

Sometimes we are interested in showing that a particular Markov chain is *not* recurrent but rather transient. The next theorem gives a necessary and sufficient condition for this.

THEOREM 4.4.8   An irreducible Markov chain is transient, if and only if there exists a bounded nonconstant solution of the equations

$$\sum_j P_{ij} y_j = y_i, \qquad i \neq 0. \qquad\qquad (4.4.18)$$

For a proof see Foster [5].

Example 4.4.5   In Fig. 4.4.1 we represent a computer system consisting of N independent processors (CPUs) and M independent memory modules. It is assumed that each processor can access each memory module. In such a system independent programs may simultaneously request access to the same memory module and interference will occur. The model we consider here was proposed by Baskett and Smith [6]. Mills [7] shows how the model can be used to estimate the performance of some computer systems and to decide what additions of hardware (CPUs or memory modules) can be cost justified. Certain simplifying assumptions are made to make the analysis tractable but the model can be used to at least approximate the performance of some computer systems. Thus it is assumed that the N processors and M

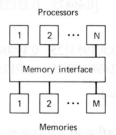

Processors

Fig. 4.4.1   A computer system of N processors (CPUs) and M memory modules.

memory modules are synchronized. Each processor always has a memory request ready for a memory module to accept as soon as it is able. At the beginning of each memory cycle each of the processors whose request from the previous cycle was satisfied makes a new request. This request is made to a memory module chosen at random; that is, each memory module is equally likely to be chosen. Each processor whose memory request from the last cycle was not honored must wait at least one more cycle before it is allowed to make another memory request; such processors are the victims of "memory interference." On each memory cycle several requests may be made to the same memory module. Each module will service one request per cycle, if any are pending; those remaining are queued for future memory cycles. All N processors simultaneously make their memory requests and all those which were successful on the last cycle receive their data at the same time.

Using the notation of Baskett and Smith [6], the state of the system shown in Fig. 4.4.1 can be represented as an M-tuple $K = (k_1, k_2, \ldots, k_M)$ where k_i is the number of access requests queued for memory module i. Since each processor always has exactly one request pending, $\sum_{i=1}^{M} k_i = N$. The number of states is thus equivalent to the number of ways N indistinguishable requests can be distributed to M memory modules. As Feller [8, p. 38] shows, this number is

$$\binom{M + N - 1}{M - 1} = \binom{M + N - 1}{N}.$$

At the end of a memory cycle (thus just before new memory requests are made) the state of the system is represented by $H = (h_1, h_2, \ldots, h_M)$, where $h_i = k_i - 1$ if $k_i > 0$, and $h_i = 0$ otherwise. (Thus $\sum_{i=1}^{M} h_i = N - h$, where h is the number of memory requests serviced during the memory cycle and $N - h$ is the number of requests *not* serviced.) Baskett and Smith [6] show that the following are true.

(a) The state $G = (g_1, g_2, \ldots, g_M)$ is reachable in one step from $K = (k_1, k_2, \ldots, k_M)$, if and only if $g_i \geq h_i$ for all i.

(b) Suppose G is reachable from K in one step, that $d_i = g_i - h_i$, and that $x = \sum_{i=1}^{M} d_i$. Then $P(K, G)$, the probability of transition from state K to G (in one step), is

$$P(K, G) = ((x!/(d_1! \, d_2! \cdots d_M!))(1/M)^x. \qquad (4.4.19)$$

Some other facts follow.

(c) The system is described by a Markov chain since the next state is completely determined by the current state.

(d) The system is aperiodic since a one-step transition from any state to itself is possible.

(e)   The system is irreducible since it can reach any state from any other state in a finite number of steps.

We can now apply Theorem 4.4.6 to conclude that the system is ergodic and thus, by Theorem 4.4.3, has an equilibrium or steady state probability distribution $\pi = (\pi_1, \pi_2, \ldots, \pi_J)$, where

$$J = \binom{M + N - 1}{N}$$

is the number of states. Thus π satisfies the matrix equation $\pi = \pi P$ subject to the conditions

$$\pi_i \geq 0, \quad i = 1, 2, \ldots, J, \quad \text{and} \quad \sum_{i=1}^{J} \pi_i = 1.$$

Here P is the state-transition matrix $[P_{ij}]$. The interpretation of π is that

$\pi_i = P[\text{the system is in state } i], \quad i = 1, 2, \ldots, J$, or

π_i is the fraction of the time that the system is in state i.

We can use the steady state distribution π to calculate what Mills [7] calls the "effective processor power, EP(N, M)." EP(N, M) compares the processing power of a computer system with N processors and M memory modules, as described above, to the processing power of a single processor with one memory module. Thus the formula for EP(N, M) must be

$$EP(N, M) = \sum_{i=1}^{J} PROC(i)\pi_i, \tag{4.4.20}$$

where PROC(i) is the number of processors that are in operation just after a memory cycle in which the system was in state i. Thus PROC(i) is the number of memory requests serviced during a memory cycle in which the system is in state i. We illustrate by calculating EP$(2, 2)$.

For $N = 2$, $M = 2$ there are

$$\binom{M + N - 1}{N} = \binom{3}{2} = 3 \quad \text{states.}$$

It is easy to see that the three states are

state 1: $(2, 0)$;   state 2: $(1, 1)$;   state 3: $(0, 2)$.

(The states could be numbered in other ways.) We claim the state transition matrix is given by

$$P = \begin{bmatrix} \frac{1}{2} & \frac{1}{2} & 0 \\ \frac{1}{4} & \frac{1}{2} & \frac{1}{4} \\ 0 & \frac{1}{2} & \frac{1}{2} \end{bmatrix}.$$

We illustrate the calculation of the first row and leave the calculation of the second and third rows to the reader. For row one, $i = 1$ and just after a memory cycle $(2, 0)$ is transformed to $(1, 0)$. Hence, by (a), above, only the states 1 and 2 can be reached from state 1 in one step. To calculate P_{11} we see that $d_1 = 1$, $d_2 = 0$, $x = d_1 + d_2 = 1$, so that

$$P_{11} = \frac{x!}{d_1! \, d_2!} \left(\frac{1}{M}\right)^x = \frac{1!}{1! \, 0!} \left(\frac{1}{2}\right)^1 = \frac{1}{2}.$$

For P_{12} we see that $d_1 = 0$, $d_2 = 1$, $x = d_1 + d_2 = 1$ so

$$P_{12} = \frac{1!}{0! \, 1!} \left(\frac{1}{2}\right)^1 = \frac{1}{2}.$$

$P_{13} = 0$ since no one-step transition is possible from state 1 to state 3 by (a), above.

The matrix equation $\pi = \pi P$ yields the equations

$$\pi_1 = \frac{\pi_1}{2} + \frac{\pi_2}{4}, \tag{4.2.21}$$

$$\pi_2 = \frac{\pi_1}{2} + \frac{\pi_2}{2} + \frac{\pi_3}{2}, \tag{4.2.22}$$

$$\pi_3 = \frac{\pi_2}{4} + \frac{\pi_3}{2}. \tag{4.2.23}$$

The first of these equations yields $\pi_2 = 2\pi_1$. When this value of π_2 is substituted into (4.2.22) we get $\pi_3 = \pi_1$. The equation

$$1 = \pi_1 + \pi_2 + \pi_3 = 4\pi_1$$

shows that $\pi_1 = 0.25$, $\pi_2 = 0.50$, and $\pi_3 = 0.25$. Clearly, PROC(1) = 1, PROC(2) = 2, and PROC(3) = 1. Therefore, by (4.2.20),

$$EP(2, 2) = \sum_{i=1}^{3} PROC(i)\pi_i = 0.25 + 2 \times 0.5 + 0.25 = 1.50. \tag{4.2.24}$$

The values of $EP(N, M)$ calculated by Mills [7] for various values of N and M are shown in Table 4.4.2. The reader is invited to calculate $EP(2, 3)$ in Exercise 2.

In reviewing Table 4.4.2 for relevance to a particular multiprocessor system the assumptions made in the model leading to the table should be kept in mind. For example, the model does not take into account the memory interference caused by I/O operations. The model also assumes that the processors and memory modules are synchronized; as Basket and Smith [6] point out, this is not true for some multiprocessor systems. In addition it

TABLE 4.4.2

Effective Processor Power by the Model
of Example 4.4.5[a]

M	N Processors			
Memory modules	2	3	4	5
2	1.500			
3	1.667	2.048		
4	1.750	2.269	2.621	
5	1.800	2.409	2.863	3.199
6	1.833	2.505	3.036	3.453
7	1.857	2.575	3.166	3.648
8	1.875	2.627	3.265	3.801
9	1.889	2.668	3.344	3.925
10	1.900	2.701	3.407	4.025

[a] Adapted from Figure 2 of Mills [7] by permission of the author and the publisher, ACM SIGMETRICS.

is assumed that each processor makes a memory request on each cycle. In some systems memory accesses may be as low as 75% because some instructions require more than one cycle for execution. In spite of these deviations from the model, Baskett and Smith [6] found by trace driven simulations that "...our analytic results are quite accurate and relatively insensitive to departures from our model." We may have an example here of the observation of E. P. Wigner [9], "...the enormous usefulness of mathematics in the natural sciences is something bordering on the mysterious and there is no rational explanation for it." (I have seen acquaintances grossly misapply queueing theory models they got from me with spectacularly good results.) The reason for the good results here is probably that different inadequacies of the model tend to offset each other.

Example 4.4.6 Suppose a multiprocessor system such as described in Example 4.4.5 is to process a certain job stream. The job stream requires 250 hours of CPU processing time on a single processor plus 100 hours of I/O time. It is assumed that the degree of multiprogramming is four (four jobs are in the system at all times) and there are five memory modules. The problem is to estimate the mean running time for one, two, or three processors. Assume CPU time and I/O time are exponentially distributed.

Solution To formulate the model following Mills [7], we represent the state of the system i as the number of jobs queueing for or receiving CPU service. This means there are $4 - i$ jobs receiving I/O service or queueing for it. Thus the system can be viewed graphically as in Fig. 4.4.2, and thus is both a Markov process and a birth-and-death process. Thus, by (4.3.17), we

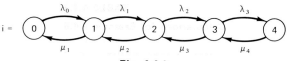

Fig. 4.4.2

see that

$$\frac{p_n}{p_0} = \frac{\lambda_0 \lambda_1 \cdots \lambda_{n-1}}{\mu_1 \mu_2 \cdots \mu_n}, \qquad n = 1, 2, 3, 4. \qquad (4.4.25)$$

Then p_0 can be calculated by the formula

$$p_0 = \left(1 + \sum_{n=1}^{4} \frac{p_n}{p_0}\right)^{-1}. \qquad (4.4.26)$$

In this formulation each λ_i refers to the mean I/O rate and each μ_i to the mean processing rate for the state i. To calculate (4.4.25) we must estimate the λ's and μ's. If I is the total number of I/O requests, then $\lambda_i = I/100$ for all i. I is also the total number of CPU requests because a CPU service continues until the job under service requires I/O. Thus $\mu_1 = I/250$, since only one processor is in use in state 1. When more than one processor is in use we estimate the CPU service rate by

$$\frac{I \times EP(N, 5)}{250}$$

where N is the number of processors in use and $EP(N, 5)$ is taken from Table 4.4.2. We make the calculations for $N = 1$ first. Thus, we calculate

$$\frac{p_1}{p_0} = \left(\frac{I/100}{I/250}\right) = 2.5, \qquad \frac{p_2}{p_0} = \left(\frac{I/100}{I/250}\right)^2 = 2.5^2 = 6.25,$$

$$\frac{p_3}{p_0} = 2.5^3 = 15.625, \qquad \text{and} \qquad \frac{p_4}{p_0} = 2.5^4 = 39.0625.$$

This yields $p_0 = (1 + \sum_{n=1}^{4} 2.5^n)^{-1} = 0.015518914$. The p_n are then as shown in the third column of Table 4.4.3.

For $N = 2$ processors the results are a little more complex. We have

$$\frac{p_1}{p_0} = \frac{I/100}{I/250} = 2.5,$$

but

$$\frac{p_2}{p_0} = \left(\frac{I/100}{I \times EP(2, 5)/250}\right) 2.5 = \frac{2.5^2}{1.8} = 3.47222.$$

TABLE 4.4.3

Calculations for Example 4.4.6 with $N = 1$

n	p_n/p_0	p_n
0	1	0.015518914
1	2.5	0.038797284
2	6.25	0.096993211
3	15.625	0.242483026
4	39.0625	0.606207566
Sum	64.4375	1.0

Similarly,

$$\frac{p_3}{p_0} = \left(\frac{1/100}{I \times EP(2,5)/250}\right) \times 3.47222 = 3.47222 \times \frac{2.5}{1.8} = 4.82253,$$

and

$$\frac{p_4}{p_0} = 4.82253 \times \frac{2.5}{1.8} = 6.69796.$$

Hence,

$$p_0 = \left(1 + \sum_{n=1}^{4} \frac{p_n}{p_0}\right)^{-1} = 18.49271^{-1} = 0.0.054075.$$

The results for $N = 2$ are shown in Table 4.4.4 and for $N = 3$ in Table 4.4.5.

We now have all the information we need to calculate the expected running time $E[t]$, assuming an expected processing time $E[t_1] = 250$ hours for one processor. For the single processor case we use the formula

$$E[t] = \frac{E[t_1]}{\rho} = \frac{E[t_1]}{1 - p_0}. \tag{4.4.27}$$

TABLE 4.4.4

Calculations for Example 4.4.6 with $N = 2$

n	p_n/p_0	p_n
0	1	0.054075
1	2.5	0.135188
2	3.47222	0.187762
3	4.82253	0.260780
4	6.69796	0.362195
Sum	18.49271	1.0

TABLE 4.4.5

**Calculations for Example 4.4.6
with** $N = 3$

n	p_n/p_0	p_n
0	1	0.069856
1	2.5	0.174641
2	3.47222	0.242556
3	3.60339	0.251719
4	3.73950	0.261228
Sum	14.31511	1.0

The reasoning behind (4.4.27) is that $E[t_1]$ time units of CPU processing time is required, and the processor is actually processing with probability ρ. Thus

$$E[t_1] = \rho E[t] \quad \text{or} \quad E[t] = \frac{E[t_1]}{\rho} = \frac{E[t_1]}{1 - p_0}.$$

The reason the central processor is not busy all the time is that, occasionally, all four of the jobs in the system are receiving I/O or queueing for it. Thus, for the problem of this example, with one processor, we have

$$E[t] = \frac{250}{1 - 0.015518914} = 253.94 \quad \text{hours.}$$

When more than one processor is available (that is, $N > 1$), the formula for expected run time is given by

$$E[t] = \frac{E[t_1]}{\text{WP}}, \tag{4.4.28}$$

where WP, the weighted processing power, is given by

$$\text{WP} = \sum_{i=1}^{N} \text{EP}(i, 5)p_i + \sum_{i=N+1}^{4} \text{EP}(N, 5)p_i. \tag{4.4.29}$$

Since $\text{EP}(1, 5) = 1$ so that, for the single-processor case we have

$$\text{WP} = \sum_{i=1}^{4} \text{EP}(1, 5)p_i = 1 - p_0,$$

we see that (4.4.28) is valid for the single-processor case, also. The final expected run times for various number of processors are shown in Table 4.4.6.

TABLE 4.4.6

Final Results for Example 4.4.6[a]

	Number of processors, N		
	1	2	3
WP	0.98448	1.5945	1.8469
$E[t]$	253.94	156.79	135.36

[a] All times in hours.

The calculation for the three-processor case, $N = 3$, is as follows:

$$WP = \sum_{i=1}^{3} EP(i, 5)p_i + EP(3, 5)p_4$$

$$= p_1 + 1.8p_2 + 2.409p_3 + 2.409p_4 = 1.8469,$$

so that

$$E[t] = \frac{250}{1.8469} = 135.36 \quad \text{hours.}$$

The same technique used in this example can be used to estimate the running time using processors of different processing powers. Thus, if all the processors are R times as powerful as the ones used in the example, we would replace the $E[t_1]$ of 250 hours by $250/R$ in all our calculations.

4.5 SUMMARY

This chapter has been an introduction to stochastic processes with emphasis on those that have immediate application to computer science problems. Thus we considered the Poisson process, which is fundamental to computer science as well as to all areas of applied probability. The birth-and-death process is a key element in queueing theory. Finally, Markov chains are used in queueing theory and in many other computer science applications. We have included a number of examples to illustrate the use of the concepts discussed in this chapter.

Student Sayings

This birth-and-death process is suffering from labor pains; it will be the death of me yet.

Exercises

1. [00] Is a constant function c $o(h)$?

2. [T22] Calculate EP(2, 3) in the notation of Example 4.4.5.

3. [T25] Consider Example 4.4.6. Suppose a job stream is given such that on a single processor, such as that considered in Example 4.4.6, 240 hours of processing are required plus 100 hours of I/O. Suppose the degree of multiprogramming is four, and there are five memory modules. If processors of twice the power are considered, estimate the mean running time for one, two, or three processors. Assume the I/O requires 100 hours as before.

4. [10] Recall that for a transition probability matrix P, by (4.4.3), $\sum_j P_{ij} = 1$ for each i; the rows add to one. Such a matrix is said to be doubly stochastic if the columns also add to one, that is, if $\sum_i P_{ij} = 1$ for each j. Show that, if P is doubly stochastic and the Markov process it represents is irreducible, aperiodic, and consists of the $M + 1$ states 0, 1, ..., M (and thus is ergodic by Theorem 4.4.6), then the limiting probabilities are given by

$$\pi_i = \frac{1}{M + 1}, \qquad i = 0, 1, \ldots, M.$$

[*Hint:* Try it (the above solution); maybe you'll like it.]

5. [HM15] Consider a birth-and-death process in which

$$\lambda_n = \alpha/(n + 1), \qquad n = 0, 1, 2, \ldots,$$

for some $\alpha > 0$, and

$$\mu_n = \mu > 0, \qquad n = 1, 2, \ldots.$$

(Such a system is called a "discouraged arrival" system because the arrival rate shrinks as the population grows.) Show that

$$p_n = e^{-\alpha/\mu}[(\alpha/\mu)^n/n!], \qquad n = 0, 1, 2, \ldots.$$

References

1. K. L. Chung, *Elementary Probability Theory with Stochastic Processes*. Springer-Verlag, New York, 1974.
2. E. A. Coddington, *An Introduction to Ordinary Differential Equations*. Prentice-Hall, Englewood Cliffs, New Jersey, 1961.
3. E. Parzen, *Stochastic Processes*. Holden-Day, San Francisco, 1962.
4. S. Karlin and H. M. Taylor, *A First Course in Stochastic Processes*, 2nd ed. Academic Press, New York, 1975.
5. F. G. Foster, On stochastic matrices associated with certain queueing processes, *Ann. Math. Statist.* **24** (1953), 355–360.
6. F. Baskett and A. J. Smith, Interference in multiprocessor computer systems with interleaved memory, *Commun. ACM* **19** (6), (June 1976), 327–334.
7. P. M. Mills, A simple model for cost considerations in a batch multiprocessor environment, *Perform. Evaluation Rev.* **5** (3), (Summer 1976), 19–27.

8. W. Feller, *An Introduction to Probability Theory and Its Applications*, Vol. I, 3rd ed. Wiley, New York, 1968.
9. E. Wigner, The unreasonable effectiveness of mathematics in the natural sciences, *Commun. Pure Appl. Math.* **13** (1), (February 1960). Also reprinted in *Symetries and Reflections Scientific Essays of Eugene P. Wigner.* The MIT Press, Cambridge, Massachusetts, 1970, and in *The Spirit and Uses of the Mathematical Sciences* (T. L. Saaty and F. Joachim Weyl, eds.). McGraw-Hill, New York, 1969.

They also serve who only stand and wait.
Milton

PART TWO

QUEUEING THEORY

Chapter Five

QUEUEING THEORY

INTRODUCTION

One of the most fruitful areas of applied probability theory for computer science applications is that of queueing theory or the study of waiting line phenomena (a queue is a waiting line). We are all familiar with queues in our daily lives. We must join a queue when we want to cash a check, buy stamps, pay for our groceries, purchase a movie ticket, obtain a table in a crowded restaurant, etc. Queues are also common in computer systems. Thus there are queues of people waiting to use a computer terminal, queues of inquiry messages waiting for processing by the central computer system, queues of channel requests, queues of I/O (input/output) requests, etc.

Figure 5.1.1 represents the elements of a queueing system, pictorially. (The reader may note that queueing is spelled "queuing" in some publications but I prefer "queueing" because (1) that is the way most queueing theory authorities spell it, and (2) it is a delightful and rare word having five consecutive vowels.) *Customers* from a *population* or *source* enter a queueing system to receive some type of service. "Customer" is used in the generic

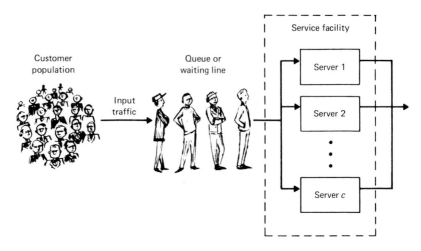

Fig. 5.1.1   Elements of a queueing system.

sense and thus may be an inquiry message requiring transmission and processing, a program requiring I/O service, a program in a multi-programming computer system requiring CPU service, etc. The *service facility* of the queueing system has one or more *servers* (sometimes called *channels*). A server is an entity capable of performing the required service for a customer. If all servers are busy when a customer enters the queueing system, the customer must join the queue until a server is free.

In Table 5.1.1 we list some typical computer science queueing systems. For some computer systems, such as an on-line inquiry system, different parts of the system may be viewed as the essential queueing system, depending upon where bottlenecks are felt to occur. Thus, if input and output message transmission times are negligible compared to the processing time at the central computer system, then the model of an on-line computer system may consist of the central computer system as the single server with the incoming inquiries playing the role of customers.

In any type of system which can be modeled as a queueing system there are trade-offs to be considered. If the service facility of the system has such a large capacity that queues rarely form, then the service facility is likely to be idle a large fraction of the time so that unused capacity exists. Contrariwise, if almost all customers must join a queue and the servers are rarely idle, there may be customer dissatisfaction and possibly lost customers. Queueing theory, in many cases, enables a designer to ensure that the proper level of service is provided in terms of response time requirements (response time is the sum of customer waiting time and service time) while avoiding excessive cost. The designer can do this by considering several alternative systems and

TABLE 5.1.1

Typical Computer Science Queueing Systems

Queueing system	Customer	Server(s)
Airline reservation system	Traveler wanting information	Agent plus terminal to computer reservation system
On-line inquiry system	Inquiry from terminal	Communication line plus central computer
On-line order entry system	Order	Communication line plus central computer
On-line data entry system	Data record	Communication line plus central computer
DASD (direct access storage device) queueing system	Request for record(s) from DASD	Channel plus DASD unit
Message buffering system	Incoming or outgoing message	Message buffer(s) (all of them, together, form the service facility)

evaluating them by analytic queueing theory models. The future performance of an existing system can be predicted so that upgrading of the system
can be done on a timely basis. For example, an analytical model of an
on-line system may indicate that the expected load two years into the future
will swamp the present system; the model may make it possible to evaluate
different alternatives for increased capacity, such as adding more main
memory, getting a faster CPU, providing more auxiliary storage, replacing
some disk drives by drums, etc. We shall give a number of practical examples
of how queueing theory can help one explore the alternatives available in an
informed way, rather than, as is often the case, making "seat-of-the-pants"
judgments.

In this chapter we discuss the elements of queueing theory and study
some basic queueing models which are of great utility in the study of computer systems. In the next chapter (Chapter 6) we show how some of these
basic queueing models can be combined to study more complex systems in
which the output of one queue may be the input to another (queues in
tandem or networks of queues).

5.1   DESCRIBING A QUEUEING SYSTEM

Figure 5.1.2 illustrates the primary random variables in a queueing
system. The basic queueing theory definitions and notation are listed in

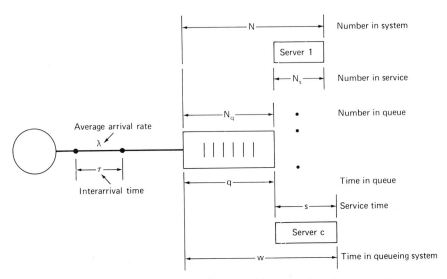

Fig. 5.1.2   Some random variables used in queueing theory models.

Table 5.1.2. A more complete set of definitions and notation is given in Table 1 of Appendix C.

There are some obvious relationships between some of the random variables shown in Fig. 5.1.2. With respect to the number of customers in various parts of the queueing system, we must have

$$N(t) = N_q(t) + N_s(t), \qquad (5.1.1)$$

and

$$N = N_q + N_s. \qquad (5.1.2)$$

In (5.1.2) we assume that the queueing system has reached the *steady state* condition. When a queueing system is first put into operation and for some time afterwards, the number in the queue and in service depends strongly upon both the initial conditions (such as the number of customers queued up waiting for the system to go into operation) and upon how long the system has been in operation (the time parameter t). The system is said to be in the *transient* state. However, after the system has been in operation for a long time, the influences of the initial conditions and of the time since start-up have "damped out" and the number of customers in the system and in the queue is independent of time—the system is in the "steady state." However, N, N_q, and N_s *are* random variables, that is, they are not constant but have probability distributions.

TABLE 5.1.2

Basic Queueing Theory Notation and Definitions

c	Number of identical servers.
L	Expected steady state number of customers in the queueing system, $E[N]$.
L_q	Expected steady state number of customers in the queue, not including those in service, $E[N_q]$.
λ	Average arrival rate of customers to the system.
μ	Average service rate per server, that is, the average rate of service completions while the server is busy.
$N(t)$	Random variable describing the number of customers in the system at time t.
N	Random variable describing the steady state number of customers in the system.
$N_q(t)$	Random variable describing the number of customers in queue at time t.
N_q	Random variable describing the steady state number in the queue (waiting line).
$N_s(t)$	Random variable describing the number of customers receiving service at time t.
N_s	Random variable describing the steady state number of customers in the service facility.
$P_n(t)$	Probability that there are n customers in the queueing system at time t, assuming some initial number at time 0.
p_n	Steady state probability that there are n customers in the system.
q	Random variable describing the time a customer spends in the queue (waiting line) before receiving service.
ρ	Server utilization $= \lambda/c\mu$.
s	Random variable describing the service time.
τ	Random variable describing interarrival time.
w	Random variable describing the total time a customer spends in the queueing system, including both time waiting in the queue for service and service time, $w = q + s$.
W	Expected steady state time in the system, $E[w] = E[q] + E[s]$.
W_q	Expected steady state time in the queue, excluding service time, $E[q] = W - E[s]$.
W_s	Expected service time, $E[s]$.

Equation (5.1.2) of course implies

$$E[N] = E[N_q] + E[N_s],\qquad(5.1.3)$$

which is often written as

$$L = L_q + L_s.\qquad(5.1.4)$$

There are some obvious relationships between the random variables describing time, also; clearly the total time in the queueing system for each customer is the sum of waiting time (time in the queue) and service time, that is,

$$w = q + s,\qquad(5.1.5)$$

and

$$E[w] = E[q] + E[s].$$ (5.1.6)

Equation (5.1.6) is often written as

$$W = W_q + W_s.$$ (5.1.7)

Equations (5.1.5)–(5.1.7) all refer to the steady state values.

Some common English words have a special meaning in queueing theory. A customer who refuses to enter a queueing system because the queue is too long is said to be *balking* while one who leaves the queue without receiving service because of excessive waiting time is said to have *reneged*. Customers may *jockey* from one queueing system to another with a shorter waiting line.

In order to describe a queueing system, analytically, a number of elements of the system must be known. We consider the most important of these.

Population or Source   The source or population of potential customers can be either finite or infinite. An infinite source system is often easier to describe mathematically than a finite source system. The reason for this is that, in a finite source system, the number of customers in the system affects the arrival rate; indeed, if every potential customer is either waiting for or receiving service, the arrival rate must be zero. If the customer population is finite but large we sometimes assume an infinite source to simplify the mathematics of the model.

Arrival Pattern   The ability of a queueing system to provide service for an arriving stream of customers depends not only upon the average rate, λ, at which the customers arrive but also upon the "pattern" in which they arrive. Thus, if customer arrivals are evenly spaced in time, say every h time units, the service facility can provide better service than if the customers arrive in clusters. We assume customer arrivals at times

$$0 \le t_0 < t_1 < t_2 < \cdots < t_n < \cdots$$

(we always assume that observation of a queueing system begins at time $t = 0$). The random variables $\tau_k = t_k - t_{k-1}$, $k = 1, 2, 3, \ldots$, are called *interarrival times*. We usually assume that $\tau_1, \tau_2, \tau_3, \ldots$ is a sequence of independent identically distributed random variables and use the symbol τ for an arbitrary interarrival time. The usual method of specifying the arrival pattern is to give the distribution function, $A(\cdot)$, of the interarrival time, that is, $A(t) = P[\tau \leq t]$. The arrival pattern most commonly assumed for queueing theory models (because of its pleasant mathematical properties) is the exponential pattern; $A(t) = 1 - e^{-\lambda t}$, where λ is the average arrival rate. The reader should review the properties of the exponential distribution by reading Section 3.2.2, especially Theorem 3.2.1.

Because of the properties of the exponential distribution, summarized in Theorem 3.2.1, if the interarrival time of customers to a queueing system has an exponential distribution, the arrival pattern is called a *Poisson arrival pattern (process)* or a *random arrival pattern* (or sometimes just said to be *random*). Other commonly assumed arrival patterns are constant, Erlang-k, and hyperexponential.

The symbol λ is reserved (except for finite queue systems) for the average arrival rate to the system; therefore, the average interarrival time, $E[\tau]$, equals $1/\lambda$. (For finite queue systems λ_a is used for average arrival rate into the queueing system.)

Service Time Distribution   The exponential distribution is often used to describe the service time of a server because of the Markov or "memoryless" property of this distribution (Theorem 3.2.1d). Thus, if the service time is exponential, the expected time remaining to complete a customer service is independent of the service already provided. Suppose now that the queueing system has several identical servers, each with an exponential service time with parameter μ, and that n of the servers are *now* busy. Let T_i be the remaining service time for server i $(i = 1, 2, \ldots, n)$. By the Markov property, each T_i has an exponential distribution with parameter μ. T, the time until the next service completion, is the minimum of $\{T_1, T_2, \ldots, T_n\}$. Hence, by Theorem 3.2.1h, T has an exponential distribution with parameter $n\mu$; the queueing system is presently performing like a single exponential server system with average service rate $n\mu$. In queueing theory, exponentially distributed service time is called *random service*, and the distribution function, $W_s(t)$, is given by

$$W_s(t) = P[s \leq t] = 1 - e^{-\mu t}.$$

Here μ is called the *average service rate*. The average service rate $\mu = 1/E[s]$ can be defined as the average rate at which a server processes customers when the server is busy. This definition is valid for all service time distributions. Other common service time distributions are Erlang-k, constant, and hyperexponential. The hyperexponential distribution is useful to describe a service time distribution with a large variance relative to the mean (see Section 3.2.9).

The *squared coefficient of variation*, C_X^2, defined for a random variable X by

$$C_X^2 = \frac{\text{Var}[X]}{E[X]^2}, \qquad (5.1.8)$$

is a useful parameter to measure the character of probability distributions used to represent service time or interarrival time. If X is a constant random variable, $C_X^2 = 0$; if X has an exponential distribution, then $C_X^2 = 1$; if X has an Erlang-k distribution, then $C_X^2 = 1/k$; and (see Exercise 18), if X has a k-stage hyperexponential distribution, $C_X^2 \geq 1$. (For example, if X has a two-stage hyperexponential distribution with $\alpha_1 = 0.4$, $\alpha_2 = 0.6$, $\mu_1 = 0.5$, and $\mu_2 = 0.01$; then $E[X] = 60.8$, $E[X^2] = 12{,}003.2$, $\text{Var}[X] = 8306.56$, and $C_X^2 = 2.25$.) We conclude that, for C_s^2 close to zero, the service time is almost constant; if C_s^2 is close to one, the service time is approximately random (exponential); if C_s^2 is close to $1/k$ for some positive integer k, then s can be approximated by an Erlang-k distribution; and, finally, if $C_s^2 > 1$ then s can probably be approximated by a k-stage hyperexponential distribution and has a great deal of variability. Similarly, if C_τ^2 is close to zero, the arrival process has a regular pattern; if C_τ^2 is close to one, the arrival pattern is nearly random; while, if C_τ^2 is greater than one, the arrivals tend to cluster.

Maximum Queueing System Capacity   In some systems the queue capacity is assumed to be infinite; that is, every arriving customer is allowed to wait until service can be provided. Other queueing systems, called "loss systems," have zero queue capacity; thus, if a customer arrives when the service facility is fully utilized (all the servers are busy), the customer is turned away. For example, some dial-up telephone systems are loss systems. Still other queueing systems, such as a message buffering system, have a positive but not infinite capacity. We use K to represent the maximum number of customers allowed in such a system.

Number of Servers   The simplest queueing system, in this sense, is the *single-server system* which can serve only one customer at a time. A *multi-server system* has c identical servers and can provide service to as many as c customers, simultaneously. In an *infinite-server system* each arriving customer is immediately provided with a server.

Queue Discipline   (*Service Discipline*)   This is the rule for selecting the next customer to receive service. The most common queue discipline is "first-come, first-served," abbreviated as FCFS (or more commonly termed "first-in, first-out," and abbreviated FIFO). Other queue disciplines include "last-come, first-served," LCFS, (or "last-in, first-out," LIFO); "random-selection-for-service," RSS (or "service-in-random-order," SIRO), which means each customer in the queue has the same probability of being selected for service; or "priority service," PRI. Priority service means that some customers get preferential treatment, just as in George Orwell's "Animal Farm" some animals were "more equal" than others. In a priority queueing system, customers are divided into priority classes with preferential treatment afforded by class. We study priority queueing systems in Section 5.4.

A shorthand notation, called the Kendall notation, after David Kendall, has been developed to describe queueing systems and has the form $A/B/c/K/m/Z$. Here A describes the interarrival time distribution, B the service time distribution, c the number of servers, K the system capacity (maximum number of customers allowed in the system), m the number in the source, and Z the queue discipline. Usually the shorter notation $A/B/c$ is used and it is assumed that there is no limit to the queue size, the customer source is infinite, and the queue discipline is FCFS. The symbols traditionally used for A and B are

GI   general independent interarrival time;
G   general service time distribution;
H_k k-stage hyperexponential interarrival or service time distribution;
E_k Erlang-k interarrival or service time distribution;
M   exponential interarrival or service time distribution;
D   deterministic (constant) interarrival or service time distribution.

When we say a queueing model, such as M/G/1, has a "general service time distribution," we mean the equations of the model are valid for very general service time distributions (make few assumptions about the service time distribution) and thus, in particular, the equations are valid for the M/M/1 system. However, equations developed specifically to describe an M/M/1 queueing system would give more information than the general equations developed for the M/G/1 model and applied, as a special case, to M/M/1. Similar remarks apply to the phrase "general independent interarrival time distribution."

As an example of the full Kendall notation, an $M/E_4/3/20/\infty/SIRO$ queueing system has exponential interarrival time, three servers with identical Erlang-4 service time distributions, a system capacity of 20 (3 in service and 17 in the queue), an infinite customer source, and the "service in random order" queue discipline (each waiting customer is equally likely to receive service next).

As the Kendall notation suggests, certain properties of a queueing system are assumed known; it is desired to calculate measures of performance of the queueing system from these known parameters. It is usually assumed that the average arrival rate λ and the average service rate per server μ are known. It is also assumed that something is known about the arrival and service time distributions—at least the averages and standard deviations are assumed known. One fundamental measure of queueing system performance is the *traffic intensity* $u = E[s]/E[\tau]$. It should be noted that $E[s]$ is the average service time per server, while $E[\tau]$ is the average interarrival time for all customers entering the queueing system and not for the customers who use a particular server (unless, of course, there is only one server). Since $\lambda = 1/E[\tau]$ and $\mu = 1/E[s]$, the traffic intensity can also be written as $\lambda E[s]$ or as λ/μ. $\rho = u/c = \lambda/c\mu$ is called the *server utilization* because it represents the average fraction of time that each server is busy (assuming the traffic is evenly distributed to the servers), that is, the probability that a given server is busy.

Consider a queueing system with a constant interarrival time of 20 seconds and a constant service time of 10 seconds. Then the server is busy half the time and $\rho = 10/20 = 0.5$. If the server is replaced by one which requires exactly 15 seconds to service a customer then $\rho = 15/20 = 0.75$ and this server is busy three-fourths of the time. Replacing this server with one requiring exactly 30 seconds to service a customer would yield a traffic intensity $u = 30/20 = 1.5$. One of these servers would be required to provide 30 seconds of service every 20 seconds! This is impossible; two servers must be provided! Thus the traffic intensity u is a measure of the minimum number of servers required to service the stream of arriving customers and $\rho = u/c$ is a measure of congestion. The following intuitive argument shows that u should give a measure of the minimum number of servers required without the assumption that both interarrival time and service time are constant. Let T be the length of a long time interval. During this interval of time about λT customers will arrive. Each customer will require approximately $E[s]$ time units of service so that about $(\lambda T)E[s]$ time units of service time will be required for all the customers. Hence, dividing by T, we see that $\lambda E[s]$ time units of service time are required per unit of time. If $u = \lambda E[s]$ is greater than one, more than one server will be required; the number of servers required is the smallest positive integer c such that $u/c < 1$. For example, if $\lambda = 7$ and $E[s] = 0.6$ then $u = 4.2$ so that 5 servers are required, each with an average service time of 0.6 time units. Although server utilization is one important measure of congestion, there are some other useful measures of queueing system performance including the following steady state values:

W the average waiting time in the system (including the service time);
W_q the average waiting time in the queue (excluding the service time).
$\pi_w(90)$ the 90th percentile value of waiting time in the system (90 percent of all customers spend less than this time in the system);

$\pi_q(90)$ the 90th percentile value of waiting time in the queue;
L the average number of customers in the queueing system (those in the queue and those in service);
L_q the average number of customers waiting in the queue;
p_n the probability that there are n customers in the queueing system.

J. D. C. Little [1] has shown that for a steady state queueing system, under very general conditions, the following formulas hold:

$$L = \lambda W, \tag{5.1.9}$$

and

$$L_q = \lambda W_q. \tag{5.1.10}$$

These formulas are called "Little's formulas" or "Little's result." These formulas together with (5.1.4) and (5.1.7) enable us to calculate the four performance measures W, W_q, L, and L_q if any one of them is known (assuming that we know λ and $W_s = E[s]$). For example, if W is known, then

$$L = \lambda W, \qquad W_q = W - W_s, \qquad \text{and} \qquad L_q = \lambda W_q.$$

Example 5.1.1 (*A Queueing Theory Paradox*) Taxis pass a certain corner with an average time between them of 20 seconds. What is the average time that one would expect to wait for a taxi?

Solution Intuitively it would seem that a taxi is just as likely to arrive at one point in time between arrivals as any other; that is, by symmetry, the distribution of the arrival time should be uniform on the interval from 0 to 20 seconds. Thus the average waiting time would be 10 seconds. This is true, however, only when the taxicabs arrive exactly 20 seconds apart. In fact, as is shown by Takács [2, p. 10], if W_t is the time until the next arrival, measured from time t, then

$$E[W_t] = \frac{1}{2}\left\{E[\tau] + \frac{\text{Var}[\tau]}{E[\tau]}\right\}. \tag{5.1.11}$$

Thus, with a Poisson arrival pattern we would expect to wait 20 seconds, on the average. If the arrival pattern is hyperexponential we would expect our average wait to exceed 20 seconds. This well-known queueing theory conundrum called "the inspection paradox" is discussed by Feller [3, pp. 11–14, 23, 187]. He shows that, if τ has an exponential distribution, then W_t has the same distribution as τ, that is,

$$P[W_t \leq x] = P[\tau \leq x] = 1 - e^{-\lambda x}. \tag{5.1.12}$$

This example highlights the fact that, in queueing theory, intuition is often misleading. One *can* get an appropriate intuitive picture of what is happening in this example by thinking of the taxicab arrivals as being appropriately scattered along the time axis and realizing that a randomly chosen point on this axis is more likely to fall in a long interval between two arrivals than in a short one.

5.2 BIRTH-AND-DEATH PROCESS QUEUEING MODELS

A number of important queueing theory models fit the birth-and-death process description of Section 4.3. A queueing system based on the birth-and-death process is in state E_n at time t if the number of customers in the system then is n, that is, $N(t) = n$. A birth is a customer arrival, and a death occurs when a customer leaves the system after completing service. We consider only steady state solutions to the queueing model. Thus, given the birth rates $\{\lambda_n\}$ and death rates $\{\mu_n\}$, and assuming that

$$S = 1 + C_1 + C_2 + C_3 + \cdots < \infty \qquad (5.2.1)$$

where

$$C_n = \frac{\lambda_0 \lambda_1 \cdots \lambda_{n-1}}{\mu_1 \mu_2 \cdots \mu_n}, \qquad n = 1, 2, 3, \ldots, \qquad (5.2.2)$$

we calculate

$$p_0 = 1/S, \qquad (5.2.3)$$

and

$$p_n = P[N = n] = C_n p_0, \qquad n = 1, 2, 3, \ldots. \qquad (5.2.4)$$

From the probabilities calculated by (5.2.4) we can generate measures of queueing system performance.

5.2.1 The M/M/1 Queueing System

This model assumes a random (Poisson) arrival pattern and a random (exponential) service time distribution. The arrival rate does not depend upon the number of customers in the system and, by Theorem 4.2.1, the probability of an arrival in a time interval of length $h > 0$ is given by

$$e^{-\lambda h}(\lambda h) = \lambda h \left(1 - \lambda h + \frac{(\lambda h)^2}{2!} - \cdots \right)$$

$$= \lambda h - (\lambda h)^2 + \frac{(\lambda h)^3}{2!} - \cdots + (-1)^{n+1} \frac{(\lambda h)^n}{(n-1)!} + \cdots$$

$$= \lambda h + o(h). \qquad (5.2.5)$$

Thus, we have

$$\lambda_n = \lambda, \qquad n = 0, 1, 2, \ldots. \tag{5.2.6}$$

By hypothesis, the service time distribution is given by

$$W_s(t) = P[s \leq t] = 1 - e^{-\mu t}, \qquad t \geq 0. \tag{5.2.7}$$

Hence, if a customer is receiving service, the probability of a service completion (death) in a short time interval, h, is given by

$$1 - e^{-\mu h} = 1 - \left(1 - \mu h + \frac{(\mu h)^2}{2!} - \cdots \right) = \mu h + o(h). \tag{5.2.8}$$

(Here, we have used the memoryless property of the exponential distribution in neglecting the service already completed.)

Thus,

$$\mu_n = \mu, \qquad n = 1, 2, 3, \ldots. \tag{5.2.9}$$

Thus the state-transition diagram for the M/M/1 queueing system is given by Fig. 5.2.1 and therefore, by (5.2.1), since $\lambda/\mu = \rho$, and each C_n is equal to ρ^n,

$$S = 1 + \rho + \rho^2 + \cdots + \rho^n + \cdots = 1/(1 - \rho). \tag{5.2.10}$$

(We have assumed that $\rho < 1$, for otherwise a steady state solution does not exist because customers arrive faster than the server can provide service for

Fig. 5.2.1 State-transition diagram of the M/M/1 queueing system.

them. To sum the series for S we have used the well-known summation formula for the geometric series, namely,

$$\sum_{n=0}^{\infty} x^n = 1 + x + x^2 + \cdots = 1/(1 - x) \qquad \text{if} \quad |x| < 1.)$$

Hence,

$$p_n = P[N = n] = (1 - \rho)\rho^n, \qquad n = 0, 1, 2, \ldots. \tag{5.2.11}$$

But (5.2.11) is the pmf for a geometric random variable, that is, N has a geometric distribution with $p = 1 - \rho$ and $q = \rho$. Hence, by Table 1 of Appendix A,

$$L = E[N] = q/p = \rho/(1 - \rho), \tag{5.2.12}$$

and

$$\sigma_N{}^2 = \rho/(1 - \rho)^2. \qquad (5.2.13)$$

By Little's formula,

$$W = E[w] = L/\lambda = E[s]/(1 - \rho), \qquad (5.2.14)$$

since $\rho = \lambda E[s]$.

Now,

$$W_q = E[q] = W - E[s] = \rho E[s]/(1 - \rho). \qquad (5.2.15)$$

Applying Little's formula, again, gives,

$$L_q = E[N_q] = \lambda W_q = \rho^2/(1 - \rho). \qquad (5.2.16)$$

By (5.2.11) we calculate

$$P[\text{server busy}] = 1 - P[N = 0] = 1 - (1 - \rho) = \rho.$$

By the law of large numbers this probability can be interpreted as the fraction of time that the server is busy; it *is* appropriate to call ρ the "server utilization."

We now have the four parameters most commonly used to measure the performance of a queueing system, W, W_q, L, and L_q, as well as the pmf, p_n, of the number in the system. For the M/M/1 system we can also derive the exact distribution of w and q.

If an arriving customer finds no customers in the system upon arrival ($N = 0$), then no waiting in the queue occurs so $W_q(0) = P[q = 0] = P[N = 0] = 1 - \rho$. Thus $W_q(\cdot)$ has a probability concentration or mass at $t = 0$. However, if a customer arrives when n customers are already in the system, then this arrival will have to wait through n exponential service times, that is, the conditional waiting time in the queue, given that n customers are in the system, is given by

$$q = s_1 + s_2 + \cdots + s_n, \qquad (5.2.17)$$

where $s_1, s_2, \ldots, s_n$ are n independent identically distributed exponential random variables, each with expected value $1/\mu$. (By the memoryless property of the exponential distribution there is no need to account for the service time already expended on the customer receiving service.) By Theorem 3.2.4d, q has a gamma distribution with parameters n and μ, which means that the conditional density function of q is given by

$$f_q(t) = \mu e^{-\mu t}\frac{(\mu t)^{n-1}}{(n - 1)!}, \qquad t \geq 0, \quad n \geq 1. \qquad (5.2.18)$$

Thus, if $n > 0$,

$$P[q \leq t \mid N = n] = \int_0^t \frac{\mu^n x^{n-1} e^{-\mu x}}{(n-1)!} \, dx. \tag{5.2.19}$$

Therefore, by the law of total probability, Theorem 2.4.2,

$$P[q \leq t \mid q > 0] = \sum_{n=1}^{\infty} P[q \leq t \mid N = n] P[N = n]. \tag{5.2.20}$$

Substituting (5.2.19) into (5.2.20) and using the fact that

$$p_n = P[N = n] = (1 - \rho)\rho^n = \left(1 - \frac{\lambda}{\mu}\right)\left(\frac{\lambda}{\mu}\right)^n, \tag{5.2.21}$$

yields

$$
\begin{aligned}
P[q \leq t \mid q > 0] &= \sum_{n=1}^{\infty} \int_0^t \frac{\mu^n x^{n-1} e^{-\mu x}}{(n-1)!} \left(\frac{\lambda}{\mu}\right)^n \left(1 - \frac{\lambda}{\mu}\right) dx \\
&= \int_0^t \lambda e^{-\mu x} \left(1 - \frac{\lambda}{\mu}\right) \sum_{n=1}^{\infty} \frac{(\lambda x)^{n-1}}{(n-1)!} \, dx \\
&= \int_0^t \lambda e^{-\mu x} \left(1 - \frac{\lambda}{\mu}\right) e^{\lambda x} \, dx \\
&= \int_0^t \lambda \left(1 - \frac{\lambda}{\mu}\right) e^{-x(\mu - \lambda)} \, dx = \frac{\lambda}{\mu} \int_0^t (\mu - \lambda) e^{-x(\mu - \lambda)} \, dx \\
&= \rho[1 - e^{-\mu(1 - \rho)t}] = \rho[1 - e^{-t/W}]. \tag{5.2.22}
\end{aligned}
$$

Hence, we have

$$
\begin{aligned}
W_q(t) &= P[q \leq t] = P[q = 0] + P[0 < q \leq t] \\
&= 1 - \rho + \rho[1 - e^{-\mu(1-\rho)t}] = 1 - \rho e^{-\mu(1-\rho)t} = 1 - \rho e^{-t/W}. \tag{5.2.23}
\end{aligned}
$$

Note that, even though q is discrete at the origin and continuous for $q > 0$, (5.2.23) is valid for all values of t.

Another quantity of interest is the average waiting time in queue for those who must wait. This quantity is of interest because a queueing system could have an acceptable average waiting time because many customers have no wait, while those who must wait have a very long wait.

By Theorem 2.8.1,

$$
\begin{aligned}
W_q &= E[q] = P[q = 0]E[q \mid q = 0] + P[q > 0]E[q \mid q > 0] \\
&= (1 - \rho) \times 0 + \rho E[q \mid q > 0] = \rho E[q \mid q > 0]. \tag{5.2.24}
\end{aligned}
$$

Hence,

$$E[q\,|\,q > 0] = E[q]/\rho = E[s]/(1 - \rho) = W. \qquad (5.2.25)$$

Since $W = E[q] + E[s]$, this means that, on the average, customers who must wait, wait one average service time longer than the average customer waits.

The derivation of the distribution function $W(\,\cdot\,)$ of total time in the system is similar to that for $W_q(\,\cdot\,)$. If a customer arrives when there are already n customers in the system, then the total time this arrival spends in the system is the sum of $n + 1$ independent exponential random variables, each with mean $1/\mu$. Hence, the density function, $f_w(\,\cdot\,)$, is the gamma density

$$f_w(t) = (\mu^{n+1} t^n e^{-\mu t})/n!, \qquad t \geq 0. \qquad (5.2.26)$$

By the law of total probability, we have

$$W(t) = P[w \leq t] = \sum_{n=0}^{\infty} P[w \leq t\,|\,N = n]P[N = n]$$

$$= \sum_{n=0}^{\infty} \int_0^t \frac{\mu^{n+1} x^n e^{-\mu x}}{n!} \left(\frac{\lambda}{\mu}\right)^n \left(1 - \frac{\lambda}{\mu}\right) dx$$

$$= \int_0^t \mu e^{-\mu x} \left(1 - \frac{\lambda}{\mu}\right) \sum_{n=0}^{\infty} \frac{x^n \lambda^n}{n!} dx$$

$$= \int_0^t \mu e^{-\mu x} \left(1 - \frac{\lambda}{\mu}\right) e^{\lambda x} dx = \int_0^t \mu \left(1 - \frac{\lambda}{\mu}\right) e^{-\mu(1 - \lambda/\mu)x} dx$$

$$= 1 - e^{-\mu(1 - \rho)t} = 1 - e^{-t/W}. \qquad (5.2.27)$$

This shows that w has an exponential distribution and that

$$W = E[w] = 1/\mu(1 - \rho) = E[s]/(1 - \rho), \qquad (5.2.28)$$

as we calculated before.

Because w has an exponential distribution we know that

$$\mathrm{Var}[w] = W^2 = (E[s]/(1 - \rho))^2. \qquad (5.2.29)$$

We also know, by Theorem 3.2.1, that the rth percentile value of w, $\pi_w(r)$, is given by

$$\pi_w(r) = E[w] \ln \left(\frac{100}{100 - r}\right), \qquad (5.2.30)$$

so that, in particular,

$$\pi_w(90) \approx 2.3W, \qquad (5.2.31)$$

and

$$\pi_w(95) \approx 3W. \qquad (5.2.32)$$

Similarly, using the distribution function of q, we calculate

$$\pi_q(r) = \frac{E[s]}{1 - \rho} \ln\left(\frac{100\rho}{100 - r}\right) = \frac{E[q]}{\rho} \ln\left(\frac{100\rho}{100 - r}\right). \qquad (5.2.33)$$

A number of formulas for an M/M/1 queueing system are shown in Table 3 of Appendix C and can be evaluated by the APL function MΔMΔ1 of Appendix B.

Example 5.2.1 For a small batch computing system the processing time per job is exponentially distributed with an average time of 3 minutes. Jobs arrive randomly at an average rate of one job every 4 minutes and are processed on a first-come-first-served basis. The manager of the installation has the following concerns.

(a) What is the probability that an arriving job will require more than 20 minutes to be processed (the job turn-around time exceeds 20 minutes)?

(b) A queue of jobs waiting to be processed will form, occasionally. What is the average number of jobs waiting in this queue?

(c) It is decided that, when the work load increases to the level such that the average time in the system reaches 30 minutes, the computer system capacity will be increased. What is the average arrival rate of jobs per hour at which this will occur? What is the percentage increase over the present job load? What is the average number of jobs in the system at this time?

(d) Suppose the criterion for upgrading the computer capacity is that not more than 10% of all jobs have a time in the system (turn-around time) exceeding 40 minutes. At the arrival rate at which this criterion is reached, what is the average number of jobs waiting to be processed?

Solution (a) $E[\tau] = 4$ minutes, so

$$\lambda = 1/E[\tau] = 0.25 \quad \text{jobs/minute,}$$

and

$$\rho = \lambda E[s] = 0.25 \times 3 = 0.75.$$

The average time in the system, $W = E[s]/(1 - \rho) = 12$ minutes, so, by (5.2.27),

$$W(t) = P[w \le t] = 1 - e^{-t/12} \qquad \text{or} \qquad P[w > t] = e^{-t/12}.$$

Therefore, the probability that w exceeds 20 minutes is $e^{-20/12} = e^{-5/3} = 0.1889$.

(b) If we assume a job queue has not formed unless there is a job in it, we use the formula

$$E[N_q | N_q > 0] = 1/(1 - \rho) = 4 \quad \text{jobs}$$

(see Exercise 3).

If the question is interpreted to mean the average job queue length, including queues of length zero, then we calculate

$$L_q = E[N_q] = \rho^2/(1 - \rho) = (0.75)^2/0.25 = 2.25 \quad \text{jobs.}$$

The most reasonable answer to the question, as stated, is 4 jobs.

(c)   When $W = 30$ minutes the system is to be upgraded, assuming the current $E[s]$ is 3 minutes. We solve the equation

$$30 = W = \frac{E[s]}{1 - \lambda E[s]} = \frac{3}{1 - 3\lambda}$$

or

$$\lambda = 27/90 = 3/10 \quad \text{jobs/minute} = 18 \quad \text{jobs/hour.}$$

The percentage increase is

$$100 \times (18 - 15)/15 = 100/5 = 20\%.$$

When $\lambda = 18$ jobs/hour $= 3/10$ jobs/minute, the average number of jobs in the system

$$L = \rho/(1 - \rho) = 0.9/(1 - 0.9) = 9 \quad \text{jobs.}$$

(d)   The criterion is that $\pi_w(90)$ reaches 40 minutes. We solve the equation

$$40 = \pi_w(90) = 2.3W = \frac{2.3 \times E[s]}{1 - \lambda E[s]} = \frac{2.3 \times 3}{1 - 3\lambda},$$

to obtain

$$\lambda = \frac{33.1}{120} \quad \text{jobs/minute} = 60 \times \frac{33.1}{120} = 16.55 \quad \text{jobs/hour.}$$

That is only a $[(16.55 - 15)/15] \times 100 = 10.3\%$ increase over the present arrival rate. At this arrival rate $\rho = \lambda E[s] = 0.8275$ and the average number of jobs in the queue is

$$L_q = E[N_q] = \rho^2/(1 - \rho) = 3.97.$$

This is an increase over the current value of 2.25 jobs. The average time in the system at this increased arrival rate is 17.39 minutes; it is only 12 minutes at the current arrival rate.

In part (c) of the above example we see that increasing the arrival rate by 20% increased the average time a job would spend in the system from 12 minutes to 30 minutes—a 150% increase! The reason for this phenomenon is shown graphically in Fig. 5.2.2. The curve of $E[w]/E[s]$ rises sharply as ρ approaches the value 1.

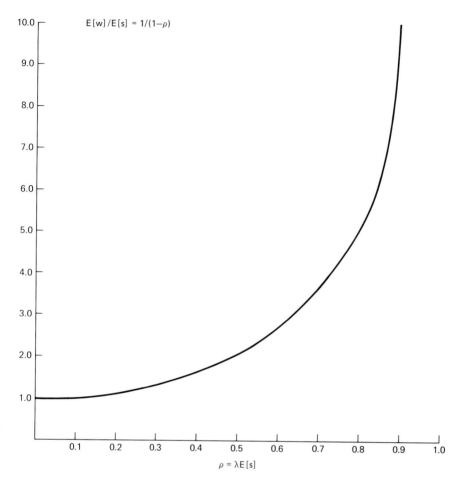

Fig. 5.2.2 Normalized average time in the system, $E[w]/E[s]$, for M/M/1 queueing system.

That is, the slope of the curve increases rapidly as ρ grows beyond about 0.8. Since

$$dW/d\rho = E[s](1 - \lambda E[s])^{-2},$$

a small change in ρ (due to a small change in λ, assuming $E[s]$ is fixed) causes a change in W given approximately by

$$(dW/d\rho)\Delta\rho = (dW/d\rho)E[s]\Delta\lambda = E[s]^2(1 - \lambda E[s])^{-2}\Delta\lambda.$$

Thus, if $\rho = 0.5$, a change $\Delta\lambda$ in λ will cause a change in W of about $4E[s]^2\Delta\lambda$, while, if $\rho = 0.9$, the change in W will be about $100E[s]^2\Delta\lambda$, or 25 times the size of the change that occurred for $\rho = 0.5$!

That is, when the system is operating at 90% server utilization, a small change in the system load (arrival rate) will cause 25 times as great an increase in the average system time as the same increase in load would cause if the system were operating at 50% utilization! This illustrates the danger of designing a system to operate at a high utilization level—a small increase in the load can have disastrous effects on the system performance.

Example 5.2.2 A computing facility has a large computer dedicated to a certain type of on-line application for users who are scattered about the country. The arrival pattern of requests to the central machine is random (Poisson), and the service time provided is random (exponential) also, so the system is an M/M/1 queueing system. A proposal is made that the workload be divided equally among n smaller machines—each with $1/n$ times the processing power of the original machine. It is claimed that the response time (time a request is in the system) will not change but the users will have a local computer. Are these claims justified?

Solution Let λ, μ be the average arrival and service rates, respectively, of the current system so that $\rho = \lambda/\mu$ is the computer utilization. For each of the proposed new systems the average arrival rate is λ/n and the average service rate is μ/n, so the server utilization is $(\lambda/n)/(\mu/n) = \lambda/\mu = \rho$, the same value as the present system. If we assume the small computers also provide random service, then

$$\frac{W_{\text{proposed}}}{W_{\text{current}}} = \left(\frac{n/\mu}{(1-\rho)}\right)\Big/\left(\frac{1/\mu}{1-\rho}\right) = n,$$

and

$$\frac{W_{q\text{proposed}}}{W_{q\text{current}}} = \left(\frac{\rho n/\mu}{1-\rho}\right)\Big/\left(\frac{\rho/\mu}{1-\rho}\right) = n.$$

Thus, the average time in the system and the average time in the queue would *increase n-fold* rather than remain the same! Of course the n new computer systems, together, process the same number of requests per hour as before, but each individual request requires n times as long to be processed, on the average, as in the present system. Thus, if the present system has an average service time of 2 seconds with a utilization of 0.7, then it has an average response time of 6.67 seconds; a proposed system of 10 computers, each providing 20-second service time, would yield a response time of 66.7 seconds! The effect discussed in this example is called the "scaling effect" and is discussed more fully by Streeter [4]. The result can be used to show that centralizing a computing facility can improve the response time while providing more computing capability for less money (economy of scale).

Example 5.2.3 A branch office of a large engineering firm has one on-line terminal connected to a central computer system for 16 hours each day. Engineers, who work throughout the city, drive to the branch office to use the terminal for making routine calculations. The arrival pattern of engineers is random (Poisson) with an average of 20 persons per day using the terminal. The distribution of time spent by an engineer at the terminal is exponential with an average time of 30 minutes. Thus the terminal is 5/8 utilized ($20 \times 1/2 = 10$ hours out of 16 hours available). The branch manager receives complaints from the staff about the length of time many of them have to wait to use the terminal. It does not seem reasonable to the manager to procure another terminal when the present one is only used five-eighths of the time, on the average. How can queueing theory help this manager?

Solution The M/M/1 queueing system is a reasonable model with $\rho = 5/8$, as we computed above. The M/M/1 formulas give the following.

$W = E[w] = E[s]/(1 - \rho) = 80$ minutes.   Average time an engineer spends at the branch office.

$L_q = \rho^2/(1 - \rho) = 1.0417.$ Average number of engineers waiting in the queue.

$E[N_q | N_q > 0] = 1/(1 - \rho) = 8/3.$ Average number of engineers in nonempty queues.

$W_q = E[q] = \rho E[s]/(1 - \rho) = 50$ minutes.   Average waiting time in queue.

$E[q | q > 0] = E[w] = 80$ minutes.   Average waiting time of those who must wait.

$\pi_q(90) = W \ln(10\rho) = 146.61$ minutes.   90th percentile of time in the queue.

$\pi_w(90) \approx 2.3W = 184$ minutes.   90th percentile time in the branch office.

Since $\rho = 5/8$, only three-eighths of the engineers who use the terminal need not wait. For those who must wait, the average wait for the terminal is 80 minutes—quite a long wait, by most standards! Ten percent of the engineers spend over 3 hours (actually 184 minutes) in the office to do an average of 30 minutes of computing. The probability of waiting more than an hour to use the terminal is

$$P[q > 60] = \tfrac{5}{8}e^{-60/80} = 0.295229,$$

or almost 30%.

These results may seem a little startling to those not acquainted with queueing theory. It might seem, intuitively, that adding another terminal would cut the average waiting time in half—from 50 minutes to 25 minutes (to 40 minutes for those who must wait). We shall see, in Example 5.2.6, that the improvement is much more dramatic than this. The queueing theory we have presented so far should suffice to convince the manager that an improvement is needed.

Example 5.2.4   Traffic to a message switching center for one of the outgoing communication lines arrives in a random pattern at an average rate of 240 messages per minute. The line has a transmission rate of 800 characters per second. The message length distribution (including control characters) is approximately exponential with an average length of 176 characters. Calculate the principal statistical measures of system performance assuming that a very large number of message buffers are provided. What is the probability that 10 or more messages are waiting to be transmitted?

Solution   The average service time is the average time to transmit a message or

$$E[s]$$

$$= \frac{\text{average message length}}{\text{line speed}} = \frac{176 \quad \text{characters}}{800 \quad \text{characters/second}} = 0.22 \quad \text{seconds}.$$

Hence, since the average arrival rate

$$\lambda = 240 \quad \text{messages/minute} = 4 \quad \text{messages/second},$$

the server utilization

$$\rho = \lambda E[s] = 4 \times 0.22 = 0.88,$$

that is, the communication line is transmitting outgoing messages 88% of the time. Using the M/M/1 formulas of Table 3, Appendix C we calculate the following.

$L = E[N] = \rho/(1 - \rho) = 7.33 \quad \text{messages.}$ Average number of messages in the system.

$L_q = E[N_q] = \rho^2/(1 - \rho) = 6.45 \quad \text{messages.}$ Average number of messages in the queue waiting to be transmitted.

$W = E[w] = E[s]/(1 - \rho) = 1.83 \quad \text{seconds.}$ Average time a message spends in the system.

$W_q = E[q] = \rho E[s]/(1 - \rho) = 1.61 \quad \text{seconds.}$ Average time a message waits for transmission.

$\pi_w(90) = 2.3W = 4.209 \quad \text{seconds.}$ 90th percentile time in the system.

$\pi_q(90) = W \ln(10\rho) = 3.98$ seconds.   90th percentile waiting
                                                     time in queue (90% of the
                                                     messages wait no longer
                                                     than 3.98 seconds.)

Since 10 or more messages are waiting if and only if 11 or more messages are in the system, the required probability is

$$P[11 \quad \text{or more messages in the system}] = \rho^{11} = 0.245.$$

Our discussion of the M/M/1 model has been more complete than it will be for many queueing models because it is an important but simple model. It is also a pleasant model to study because the probability distributions of the random variables w, q, N, and N_q can be calculated; for some queueing models only the averages W, W_q, L, and L_q can be computed, and these only with difficulty. A number of systems can be modeled, at least in a limiting sense, as an M/M/1 queueing system.

5.2.2 The M/M/1/K Queueing System

Example 5.2.4 was somewhat unrealistic in the sense that no message switching system can have an unlimited number of buffers. The M/M/1/K system is a more accurate model of this type of system in which a limit of K customers is allowed in the system. When the system contains K customers, arriving customers are turned away. Figure 5.2.3 is the state-transition diagram for this model. Thus, as a birth-and-death process, the coefficients are

$$\lambda_n = \begin{cases} \lambda & \text{for} \quad n = 0, 1, 2, \ldots, K-1 \\ 0 & \text{for} \quad n \geq K, \end{cases} \tag{5.2.34}$$

and

$$\mu_n = \begin{cases} \mu & \text{for} \quad n = 1, 2, \ldots, K \\ 0 & \text{for} \quad n > K. \end{cases} \tag{5.2.35}$$

This gives the steady state probabilities

$$p_n = \left(\frac{\lambda}{\mu}\right)^n p_0 = u^n p_0 \qquad \text{for} \quad n = 0, 1, 2, \ldots, K, \tag{5.2.36}$$

where

$$u = \lambda E[s] = \lambda/\mu.$$

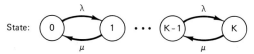

State: 0 ⟷ 1 $\cdots$ $K-1$ ⟷ K (λ above, μ below)

Fig. 5.2.3 State-transition diagram for the M/M/1/K queueing system.

Since

$$1 = p_0 + p_1 + \cdots + p_K = p_0 \sum_{n=0}^{K} u^n = \left(\frac{1 - u^{K+1}}{1 - u}\right) p_0,$$

if $\lambda \neq \mu$, we have,

$$p_0 = (1 - u)/(1 - u^{K+1}). \qquad (5.2.37)$$

Since there are never more than K customers in the system, the system reaches a steady state for all values of λ and μ. That is, we need not assume that $\lambda < \mu$ for the system to achieve a steady state. If $\lambda = \mu$ then $u = 1$ and

$$p_0 = 1/(K + 1) = p_n \qquad \text{for} \quad n = 1, 2, \ldots, K.$$

Thus the steady state probabilities are

$$p_n = \begin{cases} \dfrac{(1 - u)u^n}{1 - u^{K+1}} & \text{for} \quad \lambda \neq \mu, \quad n = 0, 1, \ldots, K \\[2mm] \dfrac{1}{K + 1} & \text{for} \quad \lambda = \mu, \quad n = 0, 1, \ldots, K. \end{cases} \qquad (5.2.38)$$

It should be noted that, if $\lambda < \mu$, as $K \to \infty$, each p_n in (5.2.38) approaches the value in (5.2.11), as it should.

If $\lambda \neq \mu$ then

$$
\begin{aligned}
L = E[N] &= \sum_{n=0}^{K} n p_n = \sum_{n=1}^{K} \left(\frac{1 - u}{1 - u^{K+1}}\right)(nu^n) = \left(\frac{1 - u}{1 - u^{K+1}}\right) u \sum_{n=1}^{K} n u^{n-1} \\
&= \left(\frac{1 - u}{1 - u^{K+1}}\right) u \sum_{n=1}^{K} \frac{du^n}{du} = \left(\frac{1 - u}{1 - u^{K+1}}\right) u \frac{d}{du} \sum_{n=0}^{K} u^n \\
&= \left(\frac{1 - u}{1 - u^{K+1}}\right) u \frac{d}{du}\left(\frac{1 - u^{K+1}}{1 - u}\right) \\
&= \left(\frac{1 - u}{1 - u^{K+1}}\right) u \left(\frac{-(K + 1)u^K}{1 - u} + (1 - u^{K+1})(1 - u)^{-2}\right) \\
&= \frac{u}{1 - u} - \frac{(K + 1)u^{K+1}}{1 - u^{K+1}}.
\end{aligned}
\qquad (5.2.39)
$$

Thus if $\lambda < \mu$, the expected number in the system, L, is always less than for the unlimited queue length case (where L is $u/(1 - u)$).

If $\lambda = \mu$ then $u = 1$ and

$$L = \sum_{n=0}^{K} n p_n = \frac{1}{K + 1}(1 + 2 + \cdots + K) = \frac{K(K + 1)}{2(K + 1)} = \frac{K}{2}. \qquad (5.2.40)$$

Thus (5.2.39) and (5.2.40) can be summarized by

$$L = \begin{cases} \dfrac{u}{1-u} - \dfrac{(K+1)u^{K+1}}{1-u^{K+1}} & \text{if } \lambda \neq \mu \\[2ex] \dfrac{K}{2} & \text{if } \lambda = \mu. \end{cases} \tag{5.2.41}$$

In either case,

$$L_q = L - (1 - p_0) \tag{5.2.42}$$

because

$$E[N_s] = P[N = 0]E[N_s \mid N = 0] + P[N > 0]E[N_s \mid N > 0]$$
$$= p_0 \times 0 + (1 - p_0) \times 1 = 1 - p_0.$$

All the traffic reaching the system does not enter the system because customers are not allowed admission when there are K customers in the system, that is, with probability p_K. Thus, if λ_a is the average rate of customers *into* the system,

$$\lambda_a = \lambda(1 - p_K). \tag{5.2.43}$$

We can then apply Little's formula to obtain

$$W = E[w] = L/\lambda_a, \tag{5.2.44}$$

and

$$W_q = E[q] = L_q/\lambda_a. \tag{5.2.45}$$

The true server utilization, ρ, which is the probability that the server is busy, is given by

$$\rho = \lambda_a E[s] = \lambda(1 - p_K)E[s] = (1 - p_K)u. \tag{5.2.46}$$

The derivation of the distribution functions of q and w is more complex for the M/M/1/K model than it was for the M/M/1 model. For $n = 0, 1, 2, \ldots,$ $K - 1$ let q_n be the probability that there are n customers in the system just before an arrival enters the system. Then $q_n \neq p_n$ because (1) no arrival enters the system when K customers are there so q_K is zero and (2) $\sum_{n=0}^{K-1} p_n \neq 1$. If it is assumed that $q_n = kp_n$, $n = 0, 1, \ldots, K - 1$, then it is easy to show (Exercise 8) that

$$q_n = p_n/(1 - p_K), \qquad n = 0, 1, \ldots, K - 1. \tag{5.2.47}$$

Proceeding as we did in deriving (5.2.27) for the $M/M/1$ model, we calculate (see Exercise 9)

$$W(t) = P[w \le t]$$

$$= \sum_{n=0}^{K-1} P[w \le t \mid N_a = n] P[N_a = n]$$

$$= \sum_{n=0}^{K-1} \left[\int_0^t \frac{\mu(\mu x)^n}{n!} e^{-\mu x} \, dx \right] q_n$$

$$= 1 - \sum_{n=0}^{K-1} q_n P[\mu t, n], \qquad (5.2.48)$$

where

$$P[\mu t, n] = \sum_{k=0}^n e^{-\mu t} \frac{(\mu t)^k}{k!}$$

is the Poisson distribution function and N_a is the random variable which counts the number of customers in an $M/M/1/K$ queueing system just before a customer arrives to enter the system. (Thus N_a assumes the values 0, 1, 2, ..., $K - 1$ and $P[N_a = n] = q_n$.) $W(t)$ can be calculated with the aid of tables of the Poisson distribution function or by using an APL function such as POISSONΔDIST. The APL function DMΔMΔ1ΔK computes the values $W(t)$ and $W_q(t)$ for the $M/M/1/K$ model. The same reasoning that led to (5.2.48) shows that $W_q(t)$ is given by

$$W_q(t) = P[q \le t] = W_q(0) + \sum_{n=1}^{K-1} P[q \le t \mid N_q = n] q_n$$

$$= q_0 + \sum_{n=1}^{K-1} q_n \int_0^t \frac{\mu(\mu x)^{n-1}}{(n-1)!} e^{-\mu x} \, dx = 1 - \sum_{n=0}^{K-2} q_{n+1} P[\mu t, n], \quad (5.2.49)$$

where

$$P[\mu t, n] = \sum_{k=0}^n \frac{e^{-\mu t}(\mu t)^k}{k!}.$$

Example 5.2.5   Consider Example 5.2.4. Suppose we have the same arrival pattern, message length distribution, and line speed as described in the example. Suppose, however, that it is desired to provide only enough message buffers so that the probability is less than one-half of a percent (probability < 0.005) that all the buffers are filled at any particular time. How many buffers should be provided? For the required number of buffers calculate L, L_q, W, and W_q. What is the probability that the time an arriving message spends in the system does not exceed 2.5 seconds? What is the probability that the queueing time of a message before transmission is begun does not exceed 2.5 seconds?

Solution The $M/M/1/K$ model fits this system with $u = \lambda E[s] = 0.88$ erlangs. The probability that all the buffers are filled, given $K - 1$ buffers are provided, is

$$p_K = \frac{(1 - u)u^K}{1 - u^{K+1}} \qquad \text{where} \quad u = 0.88. \qquad (5.2.50)$$

APL can be used to ease the burden of calculating the values of p_K. The number of buffers required is expected to be large. Let us try $K = 20$ (19 buffers). Using (5.2.50), we obtain

$$p_{20} = \frac{(1 - 0.88)(0.88)^{20}}{1 - (0.88)^{21}} = 0.00998936.$$

This is almost 1% so we try $K = 25$, which yields $p_{25} = 0.005095$; still not enough buffers.

Since $p_{26} = 0.004464 < 0.005$, we need 25 buffers, which allows 26 messages in the system (counting the one being transmitted).

Using the APL function $M\Delta M\Delta 1\Delta K$, which makes the calculations for the $M/M/1/K$ model (shown in Table 4 of Appendix C), we obtain the following.

$L = E[N] = 6.449$ messages. — Average number of messages in the system.

$\sigma_N = 6.0387$ messages. — Standard deviation of the number of messages in the system.

$L_q = E[N_q] = 5.573$ messages. — Average number of messages queued for the line.

$\sigma_{N_q} = 5.914$ messages. — Standard deviation of number of messages queued for the line.

$W = E[w] = 1.62$ seconds. — Average time a message spends in the system (queueing for the line and being transmitted).

$W_q = E[q] = 1.40$ seconds. — Average time a message queues for the line.

$E[q \mid q > 0] = 1.60$ seconds. — Average time in the queue for those messages delayed.

All of these numbers are smaller than the numbers for the $M/M/1$ model with the same λ and $E[s]$.

Using the APL function DMΔMΔ1ΔK we calculate the probability an arriving message is in the system for not more than 2.5 seconds is

$$W(2.5) = 0.77208, \quad \text{while} \quad W_q(2.5) = P[q \le 2.5] = 0.8039.$$

We also calculate the probability the system is empty, p_0, is 0.123928, while the server utilization, ρ, is 0.87607.

For the M/M/1 system of Example 5.2.4 with unlimited queue length, the 90th percentile of waiting time in the system, $\pi_w(90)$ is 4.216667 seconds, but, for the corresponding M/M/1/26 system, $P[w \le 4.216667] = 0.9328$; it appears that all the performance statistics for the M/M/1/26 system are superior to those for the system with unlimited queue length. The penalty for this improved performance, however, is that $100 \times p_K = 0.4464\%$ of the messages are refused and must be resent at a later time.

5.2.3 The M/M/c Queueing System

For this model we assume random (exponential) interarrival and service times with c identical servers. This system can be modeled as a birth-and-death process with the coefficients

$$\lambda_n = \lambda, \quad n = 0, 1, 2, \ldots, \tag{5.2.51}$$

and

$$\mu_n = \begin{cases} n\mu, & n = 1, 2, \ldots, c \\ c\mu, & n \ge c. \end{cases} \tag{5.2.52}$$

The state-transition diagram is shown in Fig. 5.2.4.

Thus, by (5.2.2), with $u = \lambda/\mu$ and $\rho = u/c$,

$$C_n = \begin{cases} \dfrac{u^n}{n!}, & n = 1, 2, 3, \ldots, c, \\ \dfrac{u^c}{c!}\left(\dfrac{u}{c}\right)^{n-c}, & n = c, c+1, \ldots. \end{cases} \tag{5.2.53}$$

Hence, if $\rho < 1$ so that the steady state exists, then

$$S = \frac{1}{p_0} = 1 + u + \frac{u^2}{2!} + \cdots + \frac{u^{c-1}}{(c-1)!} + \frac{u^c}{c!}\left(1 + \frac{u}{c} + \left(\frac{u}{c}\right)^2 + \cdots\right)$$

$$= \sum_{n=0}^{c-1} \frac{u^n}{n!} + \frac{u^c}{c!} \sum_{n=0}^{\infty} \rho^n = \sum_{n=0}^{c-1} \frac{u^n}{n!} + \frac{u^c}{c!\,(1-\rho)}. \tag{5.2.54}$$

Hence

$$p_0 = \left[\sum_{n=0}^{c-1} \frac{u^n}{n!} + \frac{u^c}{c!\,(1-\rho)}\right]^{-1} \tag{5.2.55}$$

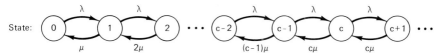

State:

Fig. 5.2.4 State-transition diagram for M/M/c queueing system.

and

$$
p_n = \begin{cases} \dfrac{u^n}{n!}\, p_0 & \text{if } n = 0, 1, \ldots, c, \\[2ex] \dfrac{u^n}{c!\, c^{n-c}}\, p_0 & \text{if } n \geq c. \end{cases}
\tag{5.2.56}
$$

We will now derive the primary measures of system performance, L_q, W_q, W, and L.

By definition,

$$
L_q = E[N_q] = \sum_{n=c}^{\infty} (n - c)p_n = \sum_{k=0}^{\infty} k p_{c+k} = \sum_{k=0}^{\infty} k \frac{u^c}{c!} \rho^k p_0 = p_0 \frac{u^c}{c!} \sum_{k=0}^{\infty} k \rho^k
$$

$$
= p_0 \frac{u^c}{c!} \{0 + 1\rho + 2\rho^2 + 3\rho^3 + \cdots\} = p_0 \frac{u^c}{c!} \rho \frac{d}{d\rho} \{1 + \rho + \rho^2 + \cdots\}
$$

$$
= p_0 \frac{u^c}{c!} \rho \frac{d}{d\rho}\left(\frac{1}{1 - \rho}\right) = \frac{p_0 u^c \rho}{c!\,(1 - \rho)^2}.
\tag{5.2.57}
$$

Having computed L_q by the formula (5.2.57), we can calculate

$$
W_q = L_q/\lambda,
\tag{5.2.58}
$$

$$
W = W_q + E[s] = W_q + (1/\mu),
\tag{5.2.59}
$$

and

$$
L = \lambda W.
\tag{5.2.60}
$$

While we have the pmf of the number in the system, N, and the expected values of the primary random variables, it is useful to have the distribution functions of w and q. We will derive $W_q(\cdot)$ and state the formula for $W(\cdot)$.

First, we note that

$$
W_q(0) = P[q = 0] = P[N \leq c - 1] = \sum_{n=0}^{c-1} p_n = p_0 \sum_{n=0}^{c-1} \frac{u^n}{n!}.
\tag{5.2.61}
$$

But, by (5.2.55),

$$
p_0 \left(\sum_{n=0}^{c-1} \frac{u^n}{n!}\right) + \frac{p_0 u^c}{c!\,(1 - \rho)} = 1,
$$

or

$$p_0 \left(\sum_{n=0}^{c-1} \frac{u^n}{n!} \right) = 1 - \frac{p_0 u^c}{c! \, (1 - \rho)}.$$

Therefore, we have

$$W_q(0) = 1 - \frac{p_0 u^c}{c! \, (1 - \rho)} = 1 - \frac{p_c}{1 - \rho}. \qquad (5.2.62)$$

Now suppose $N = n \geq c$ when a customer arrives. All c servers are busy so, as we explained earlier (in the discussion of "Service Time Distribution" in Section 5.1), the time between service completions has an exponential distribution with average value $1/c\mu$. There are c customers receiving service and $n - c$ customers waiting in the queue. Therefore the new arrival must wait for $n - c + 1$ service completions before receiving service. (If $n = c$, so no customer is waiting, the new arrival must wait for one service completion. If $n = c + 1$, two service completions are required, etc.) Hence, the waiting time in queue is the sum of $n - c + 1$ independent exponential random variables each with mean $1/c\mu$; that is, it is gamma with parameters $n - c + 1$ and $c\mu$. Hence, if $t > 0$, we can write, by (3.2.35), since $\Gamma(n - c + 1) = (n - c)!$, that

$$W_q(t) = W_q(0) + \sum_{n=c}^{\infty} p_n P[q \leq t \,|\, N = n]$$

$$= W_q(0) + \sum_{n=c}^{\infty} p_0 \frac{u^n}{c! \, c^{n-c}} \int_0^t \frac{c\mu(c\mu x)^{n-c}}{(n-c)!} e^{-c\mu x} \, dx$$

$$= W_q(0) + p_0 \frac{u^c}{(c-1)!} \int_0^t \mu e^{-c\mu x} \left(\sum_{n=c}^{\infty} \frac{(\mu u x)^{n-c}}{(n-c)!} \right) dx$$

$$= W_q(0) + \frac{p_0 u^c}{(c-1)!} \int_0^t \mu e^{-c\mu x} e^{\mu u x} \, dx$$

$$= W_q(0) + \frac{p_0 u^c}{(c-1)!} \int_0^t \mu e^{-\mu x(c-u)} \, dx$$

$$= W_q(0) + \frac{p_0 u^c}{(c-u)(c-1)!} (1 - e^{-\mu t(c-u)})$$

$$= 1 - \frac{p_0 u^c}{(c-u)(c-1)!} + \frac{p_0 u^c}{(c-u)(c-1)!} (1 - e^{-\mu t(c-u)})$$

$$= 1 - \frac{p_0 u^c}{(1-\rho)c!} e^{-\mu t(c-u)}$$

$$= 1 - \frac{p_c}{(1-\rho)} e^{-\mu t(c-u)} = 1 - \frac{p_c}{1-\rho} e^{-c\mu t(1-\rho)}. \qquad (5.2.63)$$

We note that (5.2.63) agrees with (5.2.23) when $c = 1$; that is, for the M/M/1 queueing system, as it should. The quantity $p_c/(1 - \rho)$ is an interesting quantity—in fact it is the probability that an arriving customer must wait; it is known as *Erlang's C formula* or Erlang's delay formula and written

$$C(c, u) = \frac{u^c}{c! (1 - \rho)} p_0 = \frac{u^c/c!}{(1 - \rho)\left[\left(\sum_{n=0}^{c-1} \frac{u^n}{n!}\right) + \frac{u^c}{c! (1 - \rho)}\right]}, \quad (5.2.64)$$

where, of course, $u = \lambda/\mu$ and $\rho = u/c$.

To see that Erlang's C formula, (5.2.64), does give the probability that an arriving customer must wait we note this probability is

$$\sum_{n=c}^{\infty} p_n = 1 - \sum_{n=0}^{c-1} p_n = 1 - W_q(0) = \frac{p_c}{1 - \rho},$$

by (5.2.62). Hence (5.2.63) can be rewritten as

$$W_q(t) = 1 - P[\text{arriving customer must queue}]e^{-c\mu t(1 - \rho)}$$

$$= 1 - C(c, u)e^{-c\mu t(1 - \rho)}, \quad t \geq 0. \quad (5.2.65)$$

Although q has a probability mass at the origin and is continuous for $t > 0$, formula (5.2.65) is valid for all values of t. Using (5.2.64) we can write (5.2.58) in the more pleasing form

$$W_q = E[q] = \frac{C(c, u)E[s]}{c(1 - \rho)} \quad (5.2.58')$$

which is reminiscent of (5.2.15), the formula for $E[q]$ when $c = 1$.

The distribution function $W(\cdot)$ for the waiting time in system is given by

$$W(t) = 1 + \frac{(u - c + W_q(0))}{c - 1 - u} e^{-\mu t} + \frac{C(c, u)}{c - 1 - u} e^{-c\mu t(1 - \rho)}, \quad (5.2.66)$$

if $u \neq c - 1$ and by

$$W(t) = 1 - [1 + C(c, u)\mu t]e^{-\mu t}, \quad (5.2.67)$$

if $u = c - 1$.

Formula (5.2.65) can be used to calculate the rth percentile value of q, $\pi_q(r)$, and yields

$$\pi_q(r) = \frac{E[s]}{c(1 - \rho)} \ln\left(\frac{100C(c, u)}{100 - r}\right). \quad (5.2.68)$$

In particular, this gives

$$\pi_q(90) = \frac{E[s]}{c(1 - \rho)} \ln(10C(c, u)). \quad (5.2.69)$$

The formulas for the M/M/c system are given in Table 5 of Appendix C. Table 6 gives the formulas for the special case that $c = 2$. Figure 1 of Appendix C is a graph of $C(c, u)$ as a function of traffic intensity u.

Example 5.2.6 In Example 5.2.3 the branch manager was mystified by the complaints of the staff concerning the availability of the computer terminal; it was idle 3/8 of the time yet the engineers were complaining about excessive waiting times. One of the engineers with an understanding of queueing theory explained to the manager that the theory showed that nearly 30% of the personnel using the terminal would have to wait more than an hour to gain access to it; that the average waiting time for the 5/8 of the users who must wait was 80 minutes. A committee of senior engineers met with the branch manager and jointly decided that the situation was intolerable. Moreover, it could not be solved by scheduling terminal usage. It was decided that enough terminals should be provided to ensure that the average queueing time should not exceed 10 minutes, that 90% of all engineers should queue less than 15 minutes, and that not more than 5% of all terminal users should have to queue more than 1 hour. After the specifications were agreed upon the manager had second thoughts. If the average value of queueing time is 50 minutes with one terminal, it seemed that it would require 5 terminals to reduce this value to 10 minutes; worse yet, to drop the 90th percentile value of waiting time in queue from 184 minutes to 15 minutes would require 13 terminals! How many terminals are actually needed?

Solution If more terminals are added at the branch office the M/M/c model applies. (We assume that adding a few terminals to the corporate on-line system will not affect the terminal response time of each terminal.) Let us first try two terminals and use the formulas of Table 6, Appendix C (we could also use the APL function MΔMΔC of Appendix B). With two terminals ($c = 2$) the server utilization ρ is 5/16. The average waiting time in the queue, W_q, is $\rho^2 E[s]/(1 - \rho^2) = 3.247$ minutes. The 90th percentile waiting time in the queue is

$$\pi_q(90) = \frac{E[s]}{2(1 - \rho)} \ln\left(\frac{20\rho^2}{1 + \rho}\right) = 8.673 \quad \text{minutes.}$$

The probability that the waiting time in queue exceeds 1 hour is

$$P[q > 60] = \frac{2\rho^2}{1 + \rho} e^{-2 \times 60(1 - \rho)/30} = 0.00951$$

or 0.951%. Thus all the specifications are satisfied with one additional terminal in the branch office and the branch manager relaxes. However, it would save commuting time and gasoline if the second terminal were placed in a convenient location across town so that half the engineering force could use it, thus having two M/M/1 systems, each with $u = 5/16$ erlang. Using the formulas for M/M/1 from Table 3, Appendix C, as implemented by the APL function MΔMΔ1 of Appendix B, we calculate the numbers in the third row of Table 5.2.1 and see that not one of the criteria is met. The fourth row shows that even with 4 terminals, at 4 different locations, the specification on 90th percentile queueing time fails. Thus the branch manager is right; it requires 5 terminals, if they are placed in 5 separate locations, to meet all the criteria. (It is not nice to fool your branch manager!)

TABLE 5.2.1

Summary of Calculations for Example 5.2.6[a]

System	ρ	W_q	$E[q \mid q > 0]$	$\pi_q(90)$	$P[q > 60]$
1 M/M/1	5/8	50	80	146.61	0.29523
1 M/M/2	5/16	3.25	21.82	8.67	0.00951
2 M/M/1's	5/16	13.64	43.64	49.72	0.07902
4 M/M/1's	5/32	5.56	35.56	15.87	0.02891
5 M/M/1's	1/8	4.29	34.29	7.65	0.02172

[a] All times in minutes.

Providing a second terminal does not cut the waiting time for service in half, as intuition might suggest, but rather to one-fourth when the new terminal is installed at a different location, and to one-sixteenth when the new terminal is placed in close proximity to the previous one. A cynic may note that the average queueing time for those who must queue is only cut in half (from 80 minutes to 44 minutes) if the second terminal is remotely located and to one-fourth if it is in proximity to the original terminal. However, it is difficult to improve this queueing time because of the relatively long service time (time using the terminal) which is exponentially distributed. Thus an engineer arriving at a one-terminal facility which is in use has to queue an average of at least one 30-minute service time, even if no one else is waiting, because of the memoryless property of the exponential distribution.

The manager, armed with the information in Table 5.2.1, is in a position to make an informed decision as to how many terminals to provide and where to put them. For example, suppose the average driving time to the branch office is now 30 minutes, but each terminal could be reached in 20 minutes if a new terminal is located across town. Then the average round trip to do some computing when both terminals are in the branch office (1 M/M/2 system) is $30 + 3.25 + 30 + 30 = 93.25$ minutes. With two separately located terminals (2 M/M/1 systems) it is $20 + 13.64 + 30 + 20 = 83.64$ minutes. However, round trips for engineers delayed, average about 112 minutes for the first system and 114 minutes for the second system; the latter system has a larger 90th percentile queueing time, and a larger (25.28% versus 9.2%) probability that an engineer must spend more than 1 hour in the office. On balance, the single M/M/2 system seems better.

Example 5.2.7 KAMAKAZY Airlines is planning a telephone reservations office. Each agent will have a reservations terminal and can service a typical caller, on the average, in 5 minutes, the time being exponentially distributed. Calls arrive randomly and the system will hold calls that arrive when no agent is free. Thirty-six calls per hour are expected during the peak period of the day, on the average. The three design criteria for the new office follow.

(1) The probability a caller will find all agents busy should not exceed 0.1 (10%).

(2) The average waiting time for those who must wait should be no greater than one minute.

(3) Less than 5% of all callers should have to wait more than one minute for an agent.

How many agents (and terminals) should be provided? How will this system perform if the number of callers per hour is 10% higher than anticipated?

Solution   The expected peak period average arrival rate, λ, is 36 calls per hour or $36/60 = 0.6$ calls per minute. Hence the traffic intensity is $\lambda E[s] = 0.6 \times 5 = 3$ erlangs. Thus a minimum of 4 agents (servers) are required to keep up with the inquiries. We seek the minimum c such that $C(c, 3) \leq 0.1$. Using Fig. 1 of Appendix C (the reader is warned that the use of such graphs induces vertigo in some people, including the author), we see that $C(6, 3)$ is very close to 0.1. Direct calculation using the formula for $C(c, u)$ in Table 5 of Appendix C or the APL function ERLANGAC of Appendix B, shows that $C(6, 3) = 0.0991$ while $C(5, 3) = 0.236$; six agents are required to satisfy the first design criterion. The formulas of Table 5 as implemented by the APL function MΔMΔC show that, for six agents, the average queueing time for callers delayed is 1.67 minutes. Thus six agents are not enough. Actually eight agents are required, since for seven servers, $E[q \mid q > 0] = 1.25$ minutes; for eight agents it is exactly 1 minute. With eight servers

$$P[q > 1] = C(8, 3)e^{-8(1 - \rho)/5} = 0.00476,$$

so the final design criterion is also satisfied. If the peak traffic is 10% higher than anticipated, the probability that all eight agents are busy is 0.022, the average queueing time for callers delayed is 1.06 minutes, and 97.8% of them will not have to wait at all. Thus the proposed system looks good, even if the traffic is slightly higher than estimated. Of course each agent is busy only three-eighths of the time during the peak hour—such is the price of good service. As shown by the figures in Table 5.2.2, with six agents only one of the design criteria is met and with 4 agents the performance is deplorable. Eight agents looks like a good choice. (We have shown that eight agents should be on duty during the peak period. More than this number may be needed to provide for coffee breaks, I/O breaks, etc., so that eight agents are available for duty.)

TABLE 5.2.2

Summary of Calculations for Example 5.2.7[a]

c	ρ	$C(c, 3)$	$E[q]$	$E[q \mid q > 0]$	$P[q > 1]$
8	0.3750	0.0129	0.013	1.00	0.00476
7	0.4286	0.0376	0.047	1.25	0.01692
6	0.5000	0.0991	0.165	1.67	0.05441
5	0.6000	0.2362	0.5904	2.50	0.15830
4	0.7500	0.5094	2.5472	5.00	0.41709

[a] All times in minutes.

Parzen [5] has suggested that an appropriate measure of effectiveness of a queueing system is the *customer loss ratio*, R, defined by

$$R = \frac{\text{average time spent by a customer waiting for service}}{\text{average time spent by a customer being served}}$$

$$= \frac{E[q]}{E[s]} = \frac{W_q}{W_s}. \tag{5.2.70}$$

For an $M/M/c$ system

$$R = \frac{C(c, u)}{c(1 - \rho)}. \tag{5.2.71}$$

Thus, for Example 5.2.7, if eight agents are provided, $R = 0.0026$; with six agents R increases to 0.0330, and for four agents R reaches the value of 0.509! The customer loss ratio for the original system of Example 5.2.6 (that is the system of Example 5.2.3) was 167% while the suggested $M/M/2$ system of Example 5.2.6 has an R value of 10.8%.

5.2.4   The M/M/c/c Queueing System
(M/M/c Loss System)

This system is sometimes called the $M/M/c$ loss system because customers who arrive when all the servers are busy are not allowed to wait for

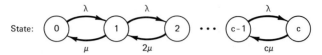

Fig. 5.2.5   State-transition diagram for M/M/c/c queueing system.

service and are lost. The state-transition diagram is given in Fig. 5.2.5. From the diagram we see that

$$C_n = u^n/n!, \qquad n = 1, 2, \ldots, c, \tag{5.2.72}$$

where, as usual, $u = \lambda E[s] = \lambda/\mu$, and

$$S = 1/p_0 = 1 + u + u^2/2! + \cdots + u^c/c!. \tag{5.2.73}$$

Thus

$$p_n = \frac{u^n/n!}{1 + u + u^2/2! + \cdots + u^c/c!}, \qquad n = 0, 1, 2, \ldots, c. \tag{5.2.74}$$

The distribution given by (5.2.74) is called the "truncated Poisson distribution," for obvious reasons. In particular, the probability that all the servers are busy, so that an arriving customer is lost, is

$$p_c = \frac{u^c/c!}{1 + u + u^2/2! + \cdots + u^c/c!} = B(c, u). \tag{5.2.75}$$

$B(c, u)$ is called "Erlang's B formula" or "Erlang's loss formula" in honor of its discoverer, A. K. Erlang. In Fig. 2 of Appendix C, graphs of $B(c, u)$ for selected values of c appear as a function of traffic intensity u. Just as with the $M/M/1/K$ model the actual average arrival rate into the system, λ_a, is less than λ because some arrivals are turned away. We must have

$$\lambda_a = \lambda(1 - B(c, u)). \tag{5.2.76}$$

Since no customers are allowed to wait, W_q and L_q are zero. However,

$$L = E[N] = \sum_{n=0}^{c} n p_n = p_0 \sum_{n=1}^{c} n \frac{u^n}{n!} = u p_0 \sum_{n=0}^{c-1} \frac{u^n}{n!} = u(1 - B(c, u)). \tag{5.2.77}$$

By Little's formula,

$$W = E[w] = L/\lambda_a = E[s]. \tag{5.2.78}$$

Of course (5.2.78) is obvious because there is no waiting. Thus w has the same distribution as s. This means that

$$W(t) = P[w \le t] = 1 - e^{-\mu t} = 1 - e^{-t/E[s]}. \tag{5.2.79}$$

It was conjectured by Erlang and later proven by others that all the formulas we have given for the $M/M/c/c$ queueing system (except (5.2.79), of course, which becomes $W(t) = P[w \le t] = P[s \le t]$) are also true for the M/G/c/c queueing system, that is, only the average value of the service time is important. (Such queueing systems are called "robust" systems.) For a proof see Gross and Harris [6].

Example 5.2.8 The Sad Sack Clothing Company has decided to install a tie-line telephone system between its east coast and west coast facilities. A caller receives a busy signal if the call is dialed when all the lines are in use. An average of 105 calls per hour with an average length of 4 minutes are expected. Enough lines are to be provided to ensure that the probability of getting a busy signal will not exceed 0.005. How many lines should be provided? With this number of lines, how many will be in use, on the average, during the peak period? How many lines are required if the probability of a busy signal is not to exceed 0.01? What would the performance be with 10 lines?

Solution The traffic intensity u is $(105/60) \times 4 = 7$ erlangs. By Fig. 2 of Appendix C, it appears that 15 tie lines are required. Using the APL function BCU we find that $B(15, 7) = 0.00332$ while $B(14, 7) = 0.00714$, so 15 lines are required. With 15 lines, the average number in use, L, is

$$7(1 - 0.00332) = 6.97675.$$

The smallest c such that $B(c, 7) \le 0.01$ is 14, so we save only one line if we double the allowed probability of a busy signal. With 14 lines the average number in use is 6.95. If only 10 tie-lines are provided, the probability of a busy signal is $B(10, 7) = 0.07874$ and the average number in use is $7(1 - 0.07874) = 6.4488$.

The formulas for the M/M/c/K queueing system can be derived much as they were for the M/M/c/c system and are given in Table 8 of Appendix C.

5.2.5 M/M/∞ Queueing System

No real life queueing system can have an infinite number of servers; what is meant, here, is that a server is immediately provided for each arriving customer. The state-transition diagram for this model is shown in Fig. 5.2.6. We can read off from the figure that

$$C_n = u^n/n!, \qquad n = 1, 2, 3, \ldots,$$

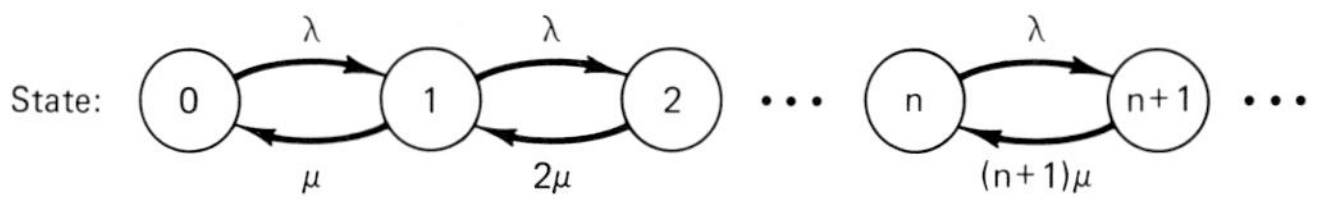

Fig. 5.2.6 State-transition diagram for M/M/∞ queueing system.

so that

$$S = 1/p_0 = \sum_{n=0}^{\infty} u^n/n! = e^u.$$

Hence,

$$p_n = e^{-u}(u^n/n!), \qquad n = 0, 1, 2, \ldots, \tag{5.2.80}$$

that is, N has a Poisson distribution! It can be shown (see Gross and Harris [6]) that (5.2.80) is also true for an $M/G/\infty$ queueing system. The fact that p_n has a Poisson distribution tells us that $L = E[N] = u$ is the average number of busy servers, with $\text{Var}[N] = u$. The $M/M/\infty$ queueing model can be used to estimate the number of lines in use in a large communication network or as a gross estimate of values in an $M/M/c$ or $M/M/c/c$ queueing system for large values of c. In Example 5.2.8, u was 7 erlangs which was close to the average number of servers in use for the $M/M/15/15$ queueing system. Also

$$p_{15} = 0.00332 \approx e^{-7}\frac{7^{15}}{15!} = 0.00331.$$

Example 5.2.9   Calls in a telephone system arrive randomly at an exchange at the rate of 140 per hour. If there are a very large number of lines available to handle the calls which last an average of 3 minutes, what is the average number of lines in use? Estimate the 90th and 95th percentile of number of lines in use.

Solution   The $M/M/\infty$ model can be used to estimate the requested values. For this example

$$u = \lambda E[s] = \frac{140}{60}\ \frac{\text{calls}}{\text{minute}} \times \frac{3}{\text{call}}\ \frac{\text{minutes}}{\text{call}} = 7\ \text{erlangs}.$$

Hence, the average number of lines in use is 7. We can use the normal approximation as a first estimate of percentile values. The 90th percentile value of the normal distribution is the mean plus 1.28 standard deviations; the 95th percentile value is the mean plus 1.645 standard deviations. Thus the 90th percentile value of number of lines is $7 + 1.28\sqrt{7} = 10.38$ or 11 lines; the 95th percentile value is $7 + 1.645\sqrt{7} = 11.35$ or 12 lines.

5.2.6   The M/M/1/K/K Queueing System
(*Machine Repair with One Repairman*)

This model, a limited source model in which there are only K customers, is variously called the machine repair model, the machine interference

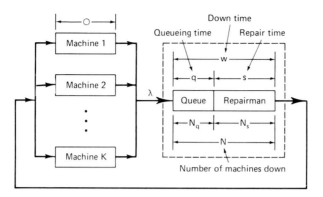

Fig. 5.2.7 Machine repair queueing system with one repairman. (M/M/1/K/K queueing system.)

model, or even the cyclic queue model. It is one of the most useful of all queueing theory models. One way to view this model is shown in Fig. 5.2.7. The population of potential customers for this queueing system consists of K identical devices, each of which has an operating time of O time units between breakdowns, O having an exponential distribution with average value $1/\alpha$. The one repairman repairs the machines at an exponential rate with an average repair time of $1/\mu$ time units. The operating machines are outside the queueing system (outlined by the dashed lines) and enter the system only when they break down; thus requiring repair. The queueing system always reaches a steady state because there can be no more than K customers in the system (one machine being repaired and $K - 1$ waiting for repairs).

When n of the machines are down (not operating) then $K - n$ of them are operating and the time until the next machine breaks down is the minimum of $K - n$ identical exponential distributions and thus, by Theorem 3.2.1h, is exponential with parameter $(K - n)\alpha$. Hence the state-transition diagram to describe the system is given by Fig. 5.2.8. From the figure we see that

$$p_n = \frac{K!}{(K - n)!}\left(\frac{\alpha}{\mu}\right)^n p_0, \qquad n = 0, 1, \ldots, K, \qquad (5.2.81)$$

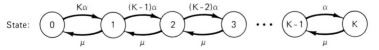

Fig. 5.2.8 State-transition diagram for M/M/1/K/K queueing system. (Machine repair).

where

$$p_0 = \left[\sum_{k=0}^{K} \frac{K!}{(K-k)!} \left(\frac{\alpha}{\mu}\right)^k \right]^{-1}. \tag{5.2.82}$$

The actual server (repairman) utilization is

$$\rho = 1 - p_0 = 1 - \frac{1}{\displaystyle\sum_{k=0}^{K} \frac{K!}{(K-k)!} \left(\frac{\alpha}{\mu}\right)^k}. \tag{5.2.83}$$

Thus ρ depends only upon K and the ratio $E[s]/E[O] = \alpha/\mu$. If we define $z = E[O]/E[s] = \mu/\alpha$, then (5.2.83) can be written as

$$\rho = 1 - \frac{z^K/K!}{\displaystyle\sum_{n=0}^{K} \frac{z^n}{n!}} = 1 - B(K, z), \tag{5.2.84}$$

where $B(K, z)$ is Erlang's loss formula. (To bring (5.2.83) into the form (5.2.84), multiply the numerator and denominator of the last expression in (5.2.83) by $(\mu/\alpha)^K/K!$.) ρ is given as a function of z for various values of K in Fig. 3 of Appendix C.

Since $\rho = \lambda E[s]$, we can calculate the actual arrival rate into the queueing system as

$$\lambda = \rho/E[s]. \tag{5.2.85}$$

To calculate $W_q = E[q]$, we reason as follows. For each machine a complete cycle consists of an operating period followed by a wait for service and a service time. Thus the average rate at which K machines break down (enter the queueing system), λ, is given by

$$\lambda = K/(E[O] + W_q + E[s]), \tag{5.2.86}$$

or

$$W_q = (K/\lambda) - E[O] - E[s]. \tag{5.2.87}$$

Little's result can then be used to calculate

$$L_q = \lambda W_q. \tag{5.2.88}$$

Then we have

$$W = W_q + E[s] = (K/\lambda) - E[O] \tag{5.2.89}$$

and

$$L = \lambda W. \tag{5.2.90}$$

Since W is the average time a machine spends in the down state we can calculate the probability that any machine n is down by

$$P[\text{machine} \quad n \quad \text{is down}] = W/(W + E[O]). \qquad (5.2.91)$$

The formulas for the $M/M/1/K/K$ queueing system are summarized in Table 10 of Appendix C. The calculations can be made using the APL function MACHΔREP.

The machine repair model can be used to study a system consisting of several disk storage devices attached to one computer channel, if it is assumed that a large data base is stored on K identical disk drives which correspond to the machines in the model, while the channel represents the repairman. We also assume that requests for information are distributed uniformly over the drives. The channel is released during a seek operation (while the read/write heads are positioned over the proper track). For a drive like the IBM 2314 or for any drive not utilizing RPS (rotational position sensing), we assume the channel service time (machine repair time), s, consists of the latency time (time for the requested sector to rotate under the read head) plus the read time and the control time. For any drive utilizing RPS, s consists of read time plus control time, since the channel is not in use during the latency time. In the former case, O consists of seek time; in the latter case it consists of seek time plus latency time. This is true because O represents the time between requests for channel service. We neglect the occasional extra disk rotation which occurs because the channel is not available when the read/write head is over the required record.

Algorithm M, below, finds the maximum inquiry rate, $\lambda_{\max}$, for a large data base stored as we have described. We assume the channel is being driven to capacity; that is, as soon as one access and read on a disk drive is completed, another is begun. In step 3 of the algorithm, ρ can be obtained by using Fig. 3 of Appendix C with $z = E[O]/E[s]$, or by calculation using (5.2.83) or (5.2.84). Although the hypotheses of the machine repair model such as exponential distribution of operating (seek) time and service (channel) time may be only a rough approximation, the results are sufficiently accurate for many design and performance analysis studies. For a more sophisticated model see Wilhelm [21].

Algorithm M This algorithm establishes an upper limit, $\lambda_{\max}$, for the average inquiry rate to a data base stored on K disk storage devices on one channel. The algorithm uses the machine repair $(M/M/1/K/K)$ queueing system as a model. It is assumed that a request for an access to each drive is always pending.

M1   Set $E[O] \leftarrow E[\text{seek time}] + C$ where C is the average latency time (the time for one-half of a rotation of the spindle) if the disk drive is used with RPS; otherwise set $E[O] \leftarrow E[\text{seek time}]$.

M2   Set $E[s] \leftarrow E[\text{channel time}]$ = average channel time to service one inquiry. (On an IBM 2314 type drive this is the read time (time to read one record) plus half of one rotation time. On an IBM 3330 or 3340 type drive, using RPS, it is the read time plus the control time.)

M3   Calculate the channel utilization, ρ, and set $\lambda_{\max} \leftarrow \rho/E[s]$.

Algorithm M is implemented by the APL function ALGΔM of Appendix B.

Example 5.2.10   An on-line information retrieval system is designed so that all queries retrieve records stored on 6 full disk storage devices attached to one block multiplexor channel. The average seek time for each drive is 30 milliseconds and the rotation time is 16.8 milliseconds. RPS is used in retrieving records and the average read plus control time is 6 milliseconds per record. Use Algorithm M to find the upper limit, $\lambda_{\max}$, for number of inquiries per second that the channel can handle.

Solution   Using Algorithm M, we set $E[O]$ to $30 + 8.4 = 38.4$ milliseconds = 0.0384 seconds and $E[s]$ to 0.006 seconds. Hence z is $0.0384/0.006 = 6.4$. Therefore, $\rho = 1 - B[6, 6.4] = 0.7076$, so that

$$\lambda_{\max} = \frac{0.7076}{0.006} = 117.93 \quad \text{queries/second.}$$

In Table 25 of Appendix C we give the formulas for the machine repair model $D/D/c/K/K$; that is, with constant operating time and repair time.

5.2.7   M/M/c/K/K Queueing System
(Machine Repair, Multiple Repairmen)

This queueing system is similar to the machine repair model considered in the last section except that we have c rather than one repairman, where $c \leq K$, as shown in Fig. 5.2.9. Thus we have a birth-and-death process queueing model with the state-transition diagram shown in Fig. 5.2.10. From this figure we can calculate the p_n's as

$$p_n = \begin{cases} \dbinom{K}{n}\left(\dfrac{\alpha}{\mu}\right)^n p_0, & n = 0, 1, \ldots, c \\[3mm] \dfrac{n!}{c!\,c^{n-c}}\dbinom{K}{n}\left(\dfrac{\alpha}{\mu}\right)^n p_0, & n = c + 1, \ldots, K, \end{cases} \qquad (5.2.92)$$

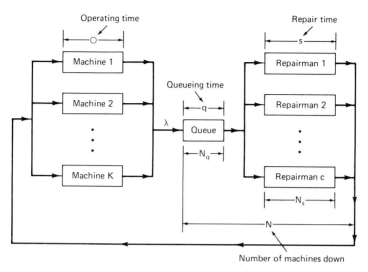

Fig. 5.2.9 Machine repair queueing system with c repairmen. (M/M/c/K/K queueing system).

where

$$p_0 = \left[\sum_{k=0}^{c} \binom{K}{k} \left(\frac{\alpha}{\mu} \right)^k + \sum_{k=c+1}^{K} \frac{k!}{c!\, c^{k-c}} \binom{K}{k} \left(\frac{\alpha}{\mu} \right)^k \right]^{-1}, \quad (5.2.93)$$

and, of course, $E[O] = 1/\alpha$, $E[s] = 1/\mu$.

We can now calculate

$$L_q = \sum_{n=c+1}^{K} (n - c)p_n. \quad (5.2.94)$$

Reasoning just as we did in (5.2.86) we see that

$$\lambda = K/(E[O] + W_q + E[s]). \quad (5.2.95)$$

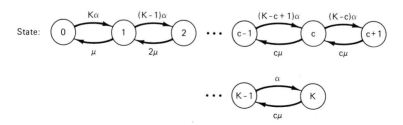

Fig. 5.2.10 State-transition diagram for M/M/c/K/K queueing system. (Machine repair with c repairmen.)

Then we apply Little's formula to obtain

$$W_q = L_q/\lambda = (E[O] + W_q + E[s])(L_q/K).\qquad(5.2.96)$$

Solving (5.2.96) for W_q yields

$$W_q = L_q(E[O] + E[s])/(K - L_q).\qquad(5.2.97)$$

The formulas for the $(M/M/c/K/K)$ queueing system are summarized in Table 11 of Appendix C. The formulas of this table apply to the single repairman model, also. In fact, they are arranged better for computation than those of Table 10 for the single repairman case. The calculations for the machine repair model using the formulas of Table 11 are implemented by the APL function MACHΔREP. In Table 25 of Appendix C we give the formulas for the machine repair model $D/D/c/K/K$; that is, with constant operating time and constant repair time.

Example 5.2.11   A company has 6 computer systems at each of its three computer centers for a total of 18 systems. The operating time of each computer is exponentially distributed with a mean time between failure of 30 hours. All computers are in continuous operation when not undergoing repairs. A customer engineering team is always on duty at each site and can repair a computer system in an average time of 30 minutes, the repair time being exponentially distributed. The company is considering consolidating all computers at one location with two customer engineering teams to maintain the equipment. The maintenance requirements are that the probability of no customer engineering team being free when a system goes down should not exceed 0.1 (10%), that the average down time per machine should not exceed 35 minutes, and the average waiting time for repairs to begin, for those systems with a delay, should not exceed 27 minutes. Will two customer engineering teams suffice?

Solution   The calculation of the probabilities p_n, $n = 0, 1, \ldots, 6$ are summarized in Table 5.2.3. The values displayed were calculated by the APL function MACHΔREP using the formulas from Table 11 and are shown accurate to five decimal places. As an example of how the calculations could be made by hand or with a desk calculator

$$\frac{p_2}{p_0} = 2! \binom{6}{2}\left(\frac{30}{1800}\right)^2 = \frac{2 \times 6!}{2!\,4!}\left(\frac{1}{60}\right)^2 = \frac{30}{3600} = \frac{1}{120} = 0.0083333.$$

The probabilities from Table 5.2.3 can be used to make other calculations using the formulas of Table 11. We summarize the statistics for each of the 3 $M/M/1/6/6$ systems and the one $M/M/2/18/18$ system in Table 5.2.4. It is clear that all three of the criteria are met by the proposed consol-

TABLE 5.2.3

Calculation of Probabilities for
Example 5.2.11 with One
Customer Engineering Team for
Six Computer Systems

n	p_n/p_0	p_n
0	1.0	0.90178
1	0.1	0.090178
2	0.0083333	0.0075148
3	0.00055556	0.00050099
4	2.7778×10^{-5}	2.5049×10^{-5}
5	9.2593×10^{-7}	8.3498×10^{-7}
6	1.5432×10^{-8}	1.3916×10^{-8}

TABLE 5.2.4

Summary of Calculations for Example 5.2.11

	Present system (each site)	Proposed consolidated system
L_q	0.0085954	0.005465
L	0.10682	0.30046
W_q	2.6253 minutes	0.55578 minutes
$E[q\|q > 0]$	26.729 minutes	15.311 minutes
W	32.625 minutes	30.556 minutes
$P[\text{machine } n \text{ down}]$	0.017803	0.016692
$P[\text{waiting for repair}]$	0.09822	0.0363

idated system. It is superior to the present system on all the statistics. For example, with the present system $3 \times 0.10682 \times 24 = 7.69$ hours of computer time are lost due to down time each 24 hours. With the proposed system $0.30046 \times 24 = 7.21$ hours of computer time are lost each 24 hours.

5.3 EMBEDDED MARKOV CHAIN QUEUEING SYSTEMS

In Section 5.2 we showed how, in some cases, a queueing system can be modeled as a birth-and-death process. This makes it relatively easy to calculate the steady state distribution of number of customers in the system and other performance measures; these include average queueing time, average waiting time in the system, etc. One fact makes analysis of such systems straightforward; it is that the stochastic process $\{N(t), t \geq 0\}$ is Markov. This is true because of the memoryless property of the exponential service time distribution; we need not account for the service time already expended by a customer receiving service. When more general service times are allowed, not only $\{N(t), t \geq 0\}$ but also $\{R(t), t \geq 0\}$, the remaining service time for the current customer, is needed to predict future values of $N(t)$. However, for a number of queueing systems, an embedded Markov chain can be constructed which makes it possible to compute many of the parameters of interest.

5.3.1 The M/G/1 Queueing System

We assume the queueing system has a Poisson input process with average value λ and a general service time distribution. Different customers have independent service times. The other restriction on the service time that must be made in order to calculate $L = E[N]$, $W = E[w]$, $L_q = E[N_q]$, and $W_q = E[q]$ is that the first and second moments of service time, $E[s]$ and $E[s^2]$, must exist. In order to calculate the standard deviations of each of the primary random values, $E[s^3]$ must exist and be known. Let $\{Z(t), t \geq 0\}$ denote the number of customers in the queueing system at time t. Let $0 < t_1 < t_2 < \cdots < t_n < \cdots$ denote the successive times at which a customer completes service. Then the sequence $\{Z(t_n)\}$ forms a discrete time process $\{X_n\}$ where $X_n = Z(t_n)$, $n = 1, 2, 3, \ldots$. X_n is thus the number of customers the nth departing customer leaves behind. We will show that $\{X_n\}$ is a Markov chain. First we note that

$$X_{n+1} = \begin{cases} X_n - 1 + A & \text{if } X_n \geq 1 \\ A & \text{if } X_n = 0, \end{cases} \tag{5.3.1}$$

where A is the number of customers who arrive during the service time of the $(n + 1)$st customer. The service time, s, of the $(n + 1)$st customer is independent of the service time of other customers and of the number of customers in the system. The arrival process is Poisson, which is stationary by Definition 4.2.1. Thus A depends only upon s and not upon when the service began or the length of the queue. Since X_{n+1} depends only upon the value of X_n, and the independent random variable A; and not upon X_{n-1}, X_{n-2}, etc., $\{X_n\}$ is a Markov chain.

The arrival process is Poisson but the length of the service time is a random variable, so A is not a Poisson random variable. However, we can write

$$P[A = n] = \int_0^\infty P[A = n \mid s = t] \, dW_s(t), \qquad n = 0, 1, \ldots, \tag{5.3.2}$$

by the law of total probability, where the integral is a Stieltjes integral with $W_s(t)$ the distribution function of the service time s. (Recall from Section 2.9

that the Stieltjes integral $\int_0^\infty h(x)\,dF(x)$ is evaluated as $\int_0^\infty h(x)f(x)\,dx$ or $\sum_{x_i} h(x_i)p(x_i)$ depending upon whether the random variable X with distribution function $F(\cdot)$ is continuous with density function $f(\cdot)$ or discrete with pmf $p(\cdot)$.) By Theorem 4.2.1,

$$P[A = n \mid s = t] = \frac{e^{-\lambda t}(\lambda t)^n}{n!}, \qquad n = 0, 1, 2, \ldots. \qquad (5.3.3)$$

Thus

$$P_{ij} = P[X_{n+1} = j \mid X_n = i] = P[A = j - i + 1]$$

$$= \int_0^\infty P[A = j - i + 1 \mid s = t]\,dW_s(t)$$

$$= \begin{cases} \int_0^\infty e^{-\lambda t}\dfrac{(\lambda t)^{j-i+1}}{(j-i+1)!}\,dW_s(t) & \text{if } j \geq i-1 \quad \text{and} \quad i \geq 1 \\ 0 & \text{if } j < i-1 \quad \text{and} \quad i \geq 1. \end{cases} \qquad (5.3.4)$$

Equation (5.3.4) allows us to calculate P_{ij} for $i \geq 1$. However, it does not tell us how to calculate the first row of the state-transition matrix P, that is, the probabilities of transition from the 0 state. We reason that, if the system is empty when a customer completes service and leaves the system, then no state transition can occur until a new customer arrives; the next transition occurs when this new customer departs. Thus the state transition probabilities are the same for $i = 0$ as for $i = 1$; that is, the first row of P is the same as the second row. It is convenient to represent P in terms of the probabilities $\{k_n\}$ where

$$k_n = P[n \quad \text{customers arrive during one service period}]$$

$$= P[A = n] = \int_0^\infty e^{-\lambda t}\frac{(\lambda t)^n}{n!}\,dW_s(t), \qquad n = 0, 1, 2, \ldots. \qquad (5.3.5)$$

We have

$$P = \begin{bmatrix} k_0 & k_1 & k_2 & k_3 & \cdots \\ k_0 & k_1 & k_2 & k_3 & \cdots \\ 0 & k_0 & k_1 & k_2 & k_3 & \cdots \\ 0 & 0 & k_0 & k_1 & k_2 & \cdots \\ \vdots & \vdots & \vdots & \vdots & \vdots \end{bmatrix}. \qquad (5.3.6)$$

It is intuitively clear that the queueing system should be stable, if and only if, the average number of customers who arrive during one service time, $E[A]$, is less than 1.

But

$$E[A] = \sum_{n=0}^{\infty} nk_n = \int_0^{\infty} e^{-\lambda t} \sum_{n=0}^{\infty} \frac{n(\lambda t)^n}{n!} \, dW_s(t)$$

$$= \int_0^{\infty} e^{-\lambda t}(\lambda t) \sum_{n=0}^{\infty} \frac{(\lambda t)^n}{n!} \, dW_s(t)$$

$$= \lambda \int_0^{\infty} t \, dW_s(t) = \lambda E[s] = u = \rho. \tag{5.3.7}$$

Hence ρ is the average number of customers to arrive during one service time. We will now use Theorem 4.4.4 to show that, if $\rho < 1$, then the embedded Markov chain $\{X_n\}$ is ergodic and thus, by Theorem 4.4.3, has a steady state probability distribution π. First we note that, since $k_n > 0$ for all n, the Markov chain is irreducible. Likewise, it is aperiodic, since $P_{ii} > 0$ for each i. So suppose $\rho < 1$ and $x_i = i/(1 - \rho)$, $i = 0, 1, 2, \ldots$. Then, if $i > 0$,

$$\sum_{j=0}^{\infty} P_{ij} x_j = \sum_{j=0}^{\infty} P_{ij} \frac{j}{1 - \rho} = \sum_{j=i-1}^{\infty} k_{j-i+1} \left(\frac{j}{1 - \rho} \right)$$

$$= \frac{1}{1 - \rho} \{ k_0(i - 1) + k_1 i + k_2(i + 1) + k_3(i + 2) + \cdots \}$$

$$= \frac{1}{1 - \rho} \{ k_0(i - 1) + k_1(i - 1) + k_2(i - 1) + \cdots \}$$

$$+ \frac{1}{1 - \rho} \{ k_1 + 2k_2 + 3k_3 + \cdots \}$$

$$= \frac{i - 1}{1 - \rho} \sum_{j=0}^{\infty} k_j + \frac{1}{1 - \rho} \sum_{j=1}^{\infty} jk_j$$

$$= \frac{i - 1}{1 - \rho} + \frac{\rho}{1 - \rho} = \frac{i}{1 - \rho} - 1 = x_i - 1.$$

Also

$$\sum_{j=0}^{\infty} P_{0j} x_j = \sum_{j=0}^{\infty} \frac{jk_j}{1 - \rho} = \frac{\rho}{1 - \rho} < \infty.$$

Hence, by Theorem 4.4.4, the embedded Markov chain $\{X_n\}$ is ergodic and thus, by Theorem 4.4.3, has a steady state probability distribution.

Karlin [9] shows that, if $\rho = 1$, then $\{X_n\}$ is recurrent null; while, if $\rho > 1$, $\{X_n\}$ is transient. In either case $\{X_n\}$ has no steady state probability distribution. Thus $\{X_n\}$ has a steady state probability distribution, if and only if, $\rho < 1$.

We will assume, henceforth, that $\rho < 1$ so the embedded Markov chain $\{X_n\}$ has a steady state distribution $\pi = (\pi_0, \pi_1, \ldots)$ with $\pi_i > 0$ for each i,

$$\sum_{i=0}^{\infty} \pi_i = 1, \quad \text{and} \quad \pi_j = \sum_{i=0}^{\infty} \pi_i P_{ij}, \quad j = 0, 1, 2, \ldots. \quad (5.3.8)$$

The above series of equations, (5.3.8), is the component by component statement of the matrix equation $\pi = \pi P$, expressing the fact that π is a stationary distribution. In terms of (5.3.6) the equations are

$$\pi_i = \pi_0 k_i + \sum_{j=1}^{i+1} \pi_j k_{i-j+1}, \quad i = 0, 1, 2, \ldots. \quad (5.3.9)$$

We define the generating functions (see Section 2.9) of the distributions π and $K = (k_0, k_1, k_2, \ldots)$ by

$$\pi(z) = \sum_{i=0}^{\infty} \pi_i z^i, \quad (5.3.10)$$

and

$$K(z) = \sum_{i=0}^{\infty} k_i z^i. \quad (5.3.11)$$

If we multiply (5.3.9) by z^i we get

$$\pi_i z^i = \pi_0 k_i z^i + \frac{1}{z} \sum_{j=0}^{i+1} \pi_j k_{i-j+1} z^{i+1} - \frac{\pi_0 k_{i+1} z^{i+1}}{z}, \quad i = 0, 1, 2, \ldots. \quad (5.3.12)$$

Summing (5.3.12) over i and recognizing $\sum_{j=0}^{i+1} \pi_j k_{i+1-j}$ as a convolution (see Theorem 2.9.2(c)), we see that

$$\sum_{i=0}^{\infty} \pi_i z^i = \pi(z) = \pi_0 K(z) + \frac{1}{z}[K(z)\pi(z) - \pi_0 k_0] - \frac{\pi_0}{z}[K(z) - k_0]. \quad (5.3.13)$$

This yields

$$\pi(z) = \frac{\pi_0(1-z)K(z)}{K(z) - z}. \quad (5.3.14)$$

Now

$$\pi(1) = \sum_{i=0}^{\infty} \pi_i = 1 = \sum_{i=0}^{\infty} k_i = K(1).$$

Also, by Theorem 2.9.2b,

$$K'(1) = E[A] = \sum_{n=0}^{\infty} n k_n = \rho.$$

Hence, applying L'Hôspital's rule to (5.3.14) yields

$$1 = \pi(1) = \lim_{z \to 1} \pi(z) = \lim_{z \to 1} \frac{\pi_0[(1-z)K'(z) - K(z)]}{K'(z) - 1} = \frac{\pi_0 K(1)}{1 - \rho} = \frac{\pi_0}{1 - \rho},$$

or

$$\pi_0 = 1 - \rho. \tag{5.3.15}$$

That is, the steady state probability that the system is empty is $1 - \rho$, just as it was for the M/M/1 queueing system!

We have shown that the queueing system has a steady state distribution of number of customers only at the times $\{t_n\}$ at which customers leave the system. However, it can be shown that what we have proven implies there exists a steady state probability distribution $P_n = \{p_0, p_1, p_n, \ldots\}$ of number of customers in the system at arbitrary points in time; furthermore $p_n = \pi_n$, $n = 0, 1, 2, \ldots$. These facts are proven in [6] and [10]. Thus

$$\pi(z) = \sum_{n=0}^{\infty} \pi_n z^n = \sum_{n=0}^{\infty} p_n z^n = P(z).$$

Therefore, we can write (5.3.14) as

$$\pi(z) = P(z) = \frac{(1 - \rho)(1 - z)K(z)}{K(z) - z}. \tag{5.3.16}$$

Since $\pi'(1) = E[N] = L$, by Theorem 2.9.2, we can calculate L from (5.3.16). In fact (see Exercise 28), if $\pi'(1)$ is calculated from (5.3.16) it yields

$$L = \pi'(1) = \rho + \frac{K''(1)}{2(1 - \rho)}. \tag{5.3.17}$$

The formula (5.3.17) is not very useful as it now stands because we do not know much about $K''(1)$. But, if we let $W_s^*(\theta)$ be the Laplace–Stieltjes transform of s (see formula (2.9.19)), then, by Theorem 2.9.3,

$$E[s] = -\frac{dW_s^*(\theta)}{d\theta}\bigg|_{\theta=0} = -W_s^{*(1)}(0), \tag{5.3.18}$$

and

$$E[s^2] = \frac{d^2 W_s^*(\theta)}{d\theta^2}\bigg|_{\theta=0} = W_s^{*(2)}(0). \tag{5.3.19}$$

By the proof of Theorem 2.9.2,

$$E[A^2] = K''(1) + K'(1). \tag{5.3.20}$$

But

$$K(z) = \sum_{n=0}^{\infty} k_n z^n = \int_0^{\infty} e^{-\lambda t} \sum_{j=0}^{\infty} \frac{(\lambda t z)^j}{j!} \, dW_s(t)$$

$$= \int_0^{\infty} e^{-\lambda t} e^{\lambda t z} \, dW_s(t) = \int_0^{\infty} e^{-(\lambda - \lambda z)t} \, dW_s(t) = W_s^*(\lambda - \lambda z). \quad (5.3.21)$$

Thus the generating function of $\{k_n\}$ can be represented as the Laplace–Stieltjes transform, $W_s^*(\theta)$, of the service time, evaluated at $\theta = \lambda(1 - z)$.

The formula (5.3.21) is a very useful relation. We can now use differentiation to solve for $E[A]$ (which we already know is ρ), $K''(1)$, and $E[A^2]$.

We calculate

$$K'(z) = -\lambda W_s^{*(1)}(\lambda - \lambda z) \qquad \text{and} \qquad K''(z) = \lambda^2 W_s^{*(2)}(\lambda - \lambda z).$$

Hence, by (5.3.18),

$$K'(1) = -\lambda W_s^{*(1)}(0) = \lambda E[s] = \rho,$$

and, by (5.3.19),

$$K''(1) = \lambda^2 W_s^{*(2)}(0) = \lambda^2 E[s^2]. \quad (5.3.22)$$

Substituting (5.3.22) into (5.3.17) yields

$$L = E[N] = \rho + \frac{\lambda^2 E[s^2]}{2(1 - \rho)} = \rho + \frac{\lambda^2 \operatorname{Var}[s] + \rho^2}{2(1 - \rho)}$$

$$= \rho + \frac{\rho^2(1 + C_s^2)}{2(1 - \rho)}. \quad (5.3.23)$$

The formula (5.3.23) in any of the forms shown is known as the *Pollaczek–Khintchine formula*.

Substituting (5.3.22) and $K'(1) = \rho$ into (5.3.20) yields

$$E[A^2] = K''(1) + K'(1) = \lambda^2 E[s^2] + \rho. \quad (5.3.24)$$

Now, by the law of total probability,

$$p_n = \int_0^{\infty} P[n \text{ arrivals during time } w \,|\, w = t] \, dW(t),$$

where $W(t)$ is the distribution function for waiting time in the queueing system, w. But, by Theorem 4.2.1,

$$P[n \text{ arrivals during time } w \,|\, w = t] = e^{-\lambda t}[(\lambda t)^n / n!].$$

Hence,

$$p_n = \pi_n = \int_0^{\infty} e^{-\lambda t} \frac{(\lambda t)^n}{n!} \, dW(t). \quad (5.3.25)$$

This means that

$$\pi(z) = P[z] = \sum_{n=0}^{\infty} p_n z^n = \int_0^{\infty} e^{-\lambda t} \sum_{n=0}^{\infty} \frac{(\lambda t z)^n}{n!} \, dW(t)$$

$$= \int_0^{\infty} e^{-\lambda t} e^{(\lambda t z)} \, dW(t) = \int_0^{\infty} e^{-\lambda t(1-z)} \, dW(t) = W^*[\lambda(1-z)], \quad (5.3.26)$$

where $W^*(\theta)$ is the Laplace–Stieltjes transform of the waiting time in the system, w.

We can differentiate (5.3.26) to get

$$\frac{dP(z)}{dz} = \frac{dW^*(v)}{dv} \frac{dv}{dz}\bigg|_{v=\lambda(1-z)} = -\lambda \frac{dW^*(v)}{dv}\bigg|_{v=\lambda(1-z)}$$

$$= \lambda \int_0^{\infty} t e^{-\lambda t(1-z)} \, dW(t). \quad (5.3.27)$$

Hence, by Theorem 2.9.2,

$$L = E[N] = \frac{dP(z)}{dz}\bigg|_{z=1} = \lambda \int_0^{\infty} t \, dW(t) = \lambda E[w] = \lambda W. \quad (5.3.28)$$

We have derived Little's formula for the special case of the M/G/1 queueing system. Proceeding as we did in deriving (5.3.28) we obtain

$$E[N(N-1)(N-2) \cdots (N-k+1)]$$

$$= \frac{d^k P(z)}{dz^k}\bigg|_{z=1} = \lambda^k E[w^k], \qquad k = 1, 2, 3, \dots . \quad (5.3.29)$$

This is a generalization of Little's formula.

Another important formula is obtained by substituting (5.3.21) into (5.3.16) yielding

$$\pi(z) = P(z) = \frac{(1-\rho)(1-z)W_s^*[\lambda(1-z)]}{W_s^*[\lambda(1-z)] - z}, \quad (5.3.30)$$

which is called the *Pollaczek–Khintchine transform equation* by many authors. (We shall see two other transform equations that are also called by this name.)

If we substitute (5.3.26) into (5.3.30) we obtain

$$W^*[\lambda(1-z)] = \frac{(1-\rho)(1-z)W_s^*[\lambda(1-z)]}{W_s^*[\lambda(1-z)] - z} \quad (5.3.31)$$

or

$$W^*(\theta) = \frac{(1-\rho)\theta W_s^*(\theta)}{\theta + \lambda[W_s^*(\theta) - 1]}. \quad (5.3.32)$$

But, since $w = q + s$, by Theorem 2.9.3d, we must have

$$W^*(\theta) = W_q^*(\theta)W_s^*(\theta).$$

Hence, by the uniqueness of the Laplace–Stieltjes transform,

$$W_q^*(\theta) = \frac{(1-\rho)\theta}{\theta + \lambda[W_s^*(\theta) - 1]}. \tag{5.3.33}$$

We now have three equations, each of which is sometimes called the Pollaczek–Khintchine transform equation; that is, (5.3.30), (5.3.32), and (5.3.33). From these equations we see that, in principle, if we know the Laplace–Stieltjes transform, $W_s^*(\theta)$, of the service time, we can calculate the steady state probability distribution, $\{p_n\}$, of number of customers in the system as well as the distribution functions, $W_q(t)$ and $W(t)$, of queueing time, q, and waiting time in the system, w, respectively. Of course we must be able to invert the Laplace–Stieltjes transforms in order to do this, that is, we must find the time dependent functions which have the transforms (5.3.30), (5.3.32), and (5.3.33). We demonstrate the procedure for the M/M/1 queueing system. If the service time has an exponential distribution with $E[s] = 1/\mu$, then, by Example 2.9.6,

$$W_s^*(\theta) = \frac{\mu}{\mu + \theta}. \tag{5.3.34}$$

Substituting (5.3.34) in (5.3.30) yields

$$P[z] = \frac{(1-\rho)(1-z)\mu/(\mu + \lambda(1-z))}{\{\mu/[\mu + \lambda(1-z)]\} - z} = \frac{\mu(1-\rho)(1-z)}{\mu - z(\mu + \lambda(1-z))}$$

$$= \frac{(1-\rho)(1-z)}{1 - z(1 + \rho(1-z))} = \frac{(1-\rho)(1-z)}{(1-z) - (1-z)\rho z}$$

$$= \frac{1-\rho}{1 - \rho z} = (1-\rho) \sum_{n=0}^{\infty} (\rho z)^n \quad \text{if} \quad |\rho z| < 1. \tag{5.3.35}$$

Hence, p_n can be read off as the coefficient of z^n in (5.3.35) or

$$p_n = (1-\rho)\rho^n, \quad n = 0, 1, 2, \ldots, \tag{5.3.36}$$

which agrees with (5.2.11).

Substituting (5.3.34) into (5.3.32) yields

$$W^*(\theta) = \frac{(1-\rho)\theta \left(\dfrac{\mu}{\mu + \theta}\right)}{\theta + \lambda \left[\dfrac{\mu}{\mu + \theta} - 1\right]} = \frac{(1-\rho)\theta\mu}{\theta(\mu + \theta) + \lambda(\mu - \mu - \theta)}$$

$$= \frac{(1-\rho)\theta}{(\theta/\mu)(\theta + \mu) - \rho\theta} = \frac{(1-\rho)\theta}{(1-\rho)\theta + (\theta^2/\mu)} = \frac{1}{1 + [\theta/\mu(1-\rho)]}$$

$$= \frac{\mu(1-\rho)}{\mu(1-\rho) + \theta}. \tag{5.3.37}$$

But, by Example 2.9.6, this is the Laplace–Stieltjes transform of an exponential random variable with parameter $\mu(1 - \rho)$, so we have

$$W(t) = P[w \le t] = 1 - e^{-\mu(1-\rho)t} = 1 - e^{-t/E[w]}. \qquad (5.3.38)$$

Similarly if, (5.3.34) is substituted into (5.3.33), the resulting transform can be inverted to yield

$$W_q(t) = P[q \le t] = 1 - \rho e^{-t/E[w]}. \qquad (5.3.39)$$

However, the inversion is less straightforward than for the above case. The interested reader can find the details in Kleinrock [10, pp. 202–203] (the reader is warned that Kleinrock's notation is much different from that of this author).

Takács [11] has generated a recurrence formula for calculating moments of queueing time in terms of moments of service time. We state one of his results without proof.

THEOREM 5.3.1 (*Takács Recurrence Theorem*) Consider an M/G/1 queueing system in which $E[s^{j+1}]$ exists. Then $E[q]$, $E[q^2]$, ..., $E[q^j]$ also exist and

$$E[q^k] = \frac{\lambda}{1-\rho} \sum_{i=1}^{k} \binom{k}{i} \frac{E[s^{i+1}]}{(i+1)} E[q^{k-i}], \qquad k = 1, 2, \ldots, j, \quad (5.3.40)$$

where $E[q^0] = 1$.

COROLLARY If the hypotheses of Theorem 5.3.1 are true then the moments $E[w]$, $E[w^2]$, ..., $E[w^j]$ exist and

$$E[w^k] = \sum_{i=0}^{k} \binom{k}{i} E[s^i]E[q^{k-i}], \qquad k = 1, 2, \ldots, j. \quad (5.3.41)$$

Proof of Corollary Since q and s are independent we can write

$$E[w^k] = E[(q+s)^k] = E\left[\sum_{i=0}^{k} \binom{k}{i} s^i q^{k-i} \right] = \sum_{i=0}^{k} \binom{k}{i} E[s^i]E[q^{k-i}].$$

Theorem 5.3.1 and the Corollary yield the formulas (see Exercise 19),

$$E[q] = \frac{\lambda E[s^2]}{2(1-\rho)} \qquad \text{(Pollaczek–Khintchine formula)}, \qquad (5.3.42)$$

$$E[q^2] = 2E[q]^2 + \frac{\lambda E[s^3]}{3(1-\rho)}, \qquad (5.3.43)$$

$$E[w] = E[q] + E[s], \qquad (5.3.44)$$

$$E[w^2] = E[q^2] + \frac{E[s^2]}{1-\rho}. \qquad (5.3.45)$$

We can now use Little's formula and (5.3.42) to show that

$$L_q = E[N_q] = \lambda E[q] = \lambda^2 E[s^2]/2(1 - \rho). \qquad (5.3.46)$$

Little's formula applied to (5.3.44) yields (5.3.23), as expected. Thus equations (5.3.23) and equations (5.3.42)–(5.3.46) together show that, if we know the first three moments of service time, we can calculate both the expected value and the standard deviation for the random variables q and w. (We can do the same for N_q and N, by Exercises 20 and 21.) However, if we know only the first and second moment of service time, we must be content with average values, only, for the random variables of interest.

In many cases knowledge of average values, only, will not enable us to make the kind of probability calculations we desire. It is especially valuable to be able to compute percentile values, such as we did for the random variables q and w in the M/M/1 queueing system. There is no such general formula for the M/G/1 queueing system but J. Martin [7] gives the estimates

$$\pi_w(90) = E[w] + 1.3\sigma_w, \qquad (5.3.47)$$

and

$$\pi_w(95) = E[w] + 2\sigma_w. \qquad (5.3.48)$$

(Actually Martin gives only the second estimate, (5.3.48), but the reasoning he gives to justify it yields (5.3.47), also.)

Another approach to estimating the percentile values of w and q is to calculate the C^2 values and use Table 5.3.1 or Table 5.3.2. This is equivalent to approximating the random variable of interest with a gamma random variable having the same mean and standard deviation.

Example 5.3.1 Four communication lines are connected to one central computer; each has an average transmission time per message of 2.4 seconds and operates at 80% line utilization. However, the message transmission time has a different distribution for each line. The transmission time for the

TABLE 5.3.1
Percentile Values of the Erlang-k Distribution
(Special Case of Gamma Distribution)

				$\pi_X(r)/E[X]$					
				$k = 1/C_X^2$					
$r/100$	1	2	3	4	5	10	20	40	100
0.90	2.30	1.94	1.77	1.67	1.60	1.42	1.30	1.21	1.13
0.95	3.00	2.37	2.09	1.94	1.83	1.57	1.39	1.27	1.17
0.99	4.61	3.32	2.80	2.51	2.32	1.88	1.59	1.40	1.25

TABLE 5.3.2
Percentile Values of the Gamma Distribution

					$\pi_X(r)/E[X]$						
						C_X^2					
$r/100$	1.25	1.5	1.75	2.0	2.5	3.0	4.0	6.0	8.0	10.0	20.0
0.90	2.43	2.54	2.63	2.71	2.82	2.91	3.00	3.00	2.86	2.66	1.53
0.95	3.24	3.46	3.66	3.84	4.16	4.42	4.84	5.38	5.68	5.80	5.25
0.99	5.16	5.68	6.17	6.63	7.50	8.30	9.74	12.16	14.17	15.88	11.09

first line is hyperexponential with $\alpha_1 = 0.4$, $\alpha_2 = 0.6$, $1/\mu_1 = 4.8$ seconds, and $1/\mu_2 = 0.8$ seconds; the distribution on the second line is exponential; it is Erlang-3 on the third line; and constant on the fourth line. Find W_q and W for each line. Then estimate $\pi_w(90)$ by Martin's estimate and by using Tables 5.3.1 and 5.3.2 (for the nonexponential cases).

Solution   The M/G/1 model applies for each line. For the hyperexponential service time

$$E[s] = 0.4 \times 4.8 + 0.6 \times 0.8 = 2.4 \quad \text{seconds.}$$

and

$$E[s^2] = 2 \times (0.4 \times 4.8^2 + 0.6 \times 0.8^2) = 19.2 \quad \text{seconds}^2$$

(by (3.2.59) and (3.2.60), respectively). Therefore

$$\text{Var}[s] = E[s^2] - E[s]^2 = 19.2 - 2.4^2 = 13.44,$$

and

$$C_s^2 = 13.44/2.4^2 = 2.33.$$

Using the formula

$$E[s^3] = 6 \sum_{i=1}^{k} \frac{\alpha_i}{\mu_i^3} \quad \text{(see Exercise 22)}$$

we calculate

$$E[s^3] = 6(0.4 \times 4.8^3 + 0.6 \times 0.8^3) = 267.264.$$

No special computations are necessary for the second line because it fits the M/M/1 model. For the third line with an Erlang-3 service time distribution we can use the formulas (see Exercise 25) valid for an Erlang-k distribution:

$$E[s^2] = \frac{(k+1)E[s]^2}{k}, \qquad E[s^3] = \frac{(k+1)(k+2)E[s]^3}{k^2}.$$

This gives

$$E[s^2] = 7.68, \qquad C_s^2 = 1/k = 1/3, \qquad \text{and} \qquad E[s^3] = 30.72.$$

(The formulas for the $M/E_k/1$ queueing system are summarized in Table 14 of Appendix C.)

Finally, for the fourth line, we have

$$E[s^2] = E[s]^2 = 5.76, \qquad C_s^2 = 0, \qquad \text{and} \qquad E[s^3] = E[s]^3 = 13.824.$$

(The formulas for the $M/D/1$ queueing system are given in Table 15 of Appendix C.) Substituting these results into the $M/G/1$ equations of Table 12 in Appendix C yields the number in Table 5.3.3. We illustrate the calculation of $\pi_w(90)$ by the "table" method for line 3. We have $C_w^2 = (\sigma_w/W)^2 = 0.7736$. Since this number does not appear in Table 5.3.2, we take the reciprocal $1/C_w^2$, which is 1.2927, so, by Table 5.3.1,

$$\pi_w(90)/E[w] = 2.195 \qquad \text{or} \qquad \pi_w(90) = 2.195 \times 8.8 = 19.32 \quad \text{seconds.}$$

This example dramatically demonstrates the inimical effect of "irregularity" in service time, as measured by C_s^2. (We have all been conditioned by television ads to recognize the deleterious effects of irregularity in our personal lives.) The average waiting time in the system, W, and the 90th percentile value of w is about one and a half times as large for the hyperexponential service time as for exponential service time; exponential service time yields significantly poorer performance than Erlang-3 or constant service time.

TABLE 5.3.3
Summary of Results in Example 5.3.1

Line	Distribution (line time)	$E[s^2]$	$E[s^3]$	W_q	W	σ_w	$\pi_w(90)$ Martin	$\pi_w(90)$ Tables
1	two-stage hyperexponential	19.2	267.264	16.00	18.40	20.44	44.98	44.56
2	exponential	11.52	82.944	9.60	12.00	12.00	27.60	27.60
3	Erlang-3	7.68	30.720	6.40	8.80	7.74	18.87	19.32
4	constant	5.760	13.824	4.80	7.20	5.54	14.40	14.77

Example 5.3.2 (*Example 5.2.2 Revisited*) We saw, in Example 5.2.2, that, if a large computer system could be modeled as an $M/M/1$ queueing model, then replacing this system by n $M/M/1$ systems, that is by n smaller computers, each with $1/n$ of the capacity and traffic, then the average time in the queue, W_q, and the average time in the system, W, would each increase n-fold. We now can show that the same holds true for an $M/G/1$ system, if

the service time provided by the smaller machines is ns, where s is the service time for the large computer. To see this, let λ be the present arrival rate. Then

$$E[q]_{\text{proposed}} = \frac{\lambda}{n} \frac{E[(ns)^2]}{2(1-\rho)} = \frac{\lambda n^2 E[s^2]}{2n(1-\rho)} = nE[q]_{\text{present system}}$$

and

$$E[w]_{\text{proposed}} = E[q]_{\text{proposed}} + E[s]_{\text{proposed}}$$

$$= nE[q]_{\text{present system}} + nE[s]_{\text{present system}}$$

$$= nE[w]_{\text{present system}}.$$

The $M/G/1$ queueing model is quite a useful one because random arrival patterns are quite common, although random service time is not. The model is often used as we used it in Example 5.3.1, that is, to calculate means and estimated percentile values rather than attempting to invert the Pollaczek–Khintchine transform equations. The interested readers can find the details of inverting these transforms for the $M/H_2/1$ queueing system in Kleinrock [10, Ch. 5]. The reader may find Tables 13–15 of Appendix C useful for models similar to those used in Example 5.3.1.

5.3.2 The GI/M/1 Queueing System

The $GI/M/1$ queueing system is another important model for which the embedded Markov chain technique enables us to obtain useful results. For this model we assume that the interarrival times are independent identically distributed random variables. (Such an arrival pattern is called a renewal process.) We represent the system state by the number of customers in the system at the instant of a customer arrival. This yields a stochastic process $\{X_n\}$, where X_n is the number of customers in the system when the nth customer arrives. By proceeding much as we did in Section 5.3.1, it can be shown that $\{X_n\}$ is a Markov chain, and that, if $\rho < 1$, then a steady state probability distribution $\{\pi_n\}$ exists where

$$\pi_n = P[\text{an arrival finds } n \text{ customers in the system}], \qquad n = 0, 1, 2, \ldots.$$

For the details see Kleinrock [10] or Gross and Harris [6]. It is also shown in the above references that

$$\pi_n = \pi_0(1 - \pi_0)^n, \qquad n = 0, 1, 2, \ldots, \tag{5.3.49}$$

where, of course, π_0 is the probability that an arriving customer finds the system empty. Furthermore, π_0 is the unique solution of the equation

$$1 - \pi_0 = A^*(\mu \pi_0), \tag{5.3.50}$$

such that

$$0 < \pi_0 < 1.$$

In equation (5.3.50), $A^*(\theta)$ is the Laplace–Stieltjes transform of the interarrival time, τ, and μ is the average service rate.

Since, by (5.3.49), the number of customers, X, an arriving customer finds in the system has a geometric distribution, we have

$$E[X] = \frac{(1 - \pi_0)}{\pi_0}, \qquad \text{Var}[X] = \frac{(1 - \pi_0)}{\pi_0^2}. \qquad (5.3.51)$$

Proceeding exactly as we did in (5.2.17)–(5.2.23) for the M/M/1 queueing system, we find that

$$W_q(t) = P[q \le t] = 1 - (1 - \pi_0)e^{-\pi_0 t/E[s]}, \qquad t \ge 0. \qquad (5.3.52)$$

It is not difficult to show from (5.3.52) (see Exercise 27) that

$$E[q] = (1 - \pi_0)E[s]/\pi_0, \qquad (5.3.53)$$

$$E[q^2] = 2(1 - \pi_0)(E[s]/\pi_0)^2, \qquad (5.3.54)$$

and

$$\text{Var}[q] = (1 - \pi_0^2)(E[s]/\pi_0)^2. \qquad (5.3.55)$$

The same argument we use for deriving (5.2.27) for the M/M/1 queueing system shows that

$$W(t) = P[w \le t] = 1 - e^{-\pi_0 t/E[s]} = 1 - e^{-t/W}, \qquad t \ge 0. \qquad (5.3.56)$$

Thus the waiting time in the system, w, has an exponential distribution, just as it did for the M/M/1 queueing system! This remarkable fact implies that

$$W = E[w] = E[s]/\pi_0, \qquad (5.3.57)$$

and

$$\text{Var}[W] = W^2. \qquad (5.3.58)$$

Also, since w is exponential,

$$\pi_w(r) = W \ln (100/(100 - r)), \qquad (5.3.59)$$

and

$$\pi_w(90) \approx 2.3W, \qquad \pi_w(95) \approx 3W. \qquad (5.3.60)$$

Differentiating (5.3.52) shows that the density function of q, f_q, which is defined only for $t > 0$, is given by

$$f_q(t) = \frac{(1 - \pi_0)\pi_0}{E[s]} e^{-\pi_0 t/E[s]}, \qquad t > 0. \qquad (5.3.61)$$

This means that the density function, f, of the queueing time of those who must queue, say q', is given by

$$f(t) = \frac{f_q(t)}{P[q > 0]} = \frac{f_q(t)}{1 - \pi_0} = \frac{e^{-t/W}}{W}, \qquad t > 0. \tag{5.3.62}$$

But this is the density function of W! Hence

$$P[q' \le t] = P[w \le t] = 1 - e^{-t/W}, \qquad t > 0. \tag{5.3.63}$$

We conclude that q' is exponential so

$$E[q'] = E[q \mid q > 0] = W,$$

and

$$\mathrm{Var}[q'] = \mathrm{Var}[q \mid q > 0] = W^2. \tag{5.3.64}$$

Kleinrock [10] shows that

$$p_0 = P[N = 0] = 1 - \rho, \tag{5.3.65}$$

and

$$p_n = P[N = n] = \rho \pi_0 (1 - \pi_0)^{n-1}, \qquad n = 1, 2, 3, \ldots. \tag{5.3.66}$$

Here we are dealing with the steady state probability that a *random* observer finds n customers in the system, rather than the probability that an arriving customer finds n customers in the system, as in (5.3.49) and (5.3.50). The two are the same, if and only if, the arrival pattern is random (Poisson).

We assure ourselves that the above equations really work, at least for a case we have examined before, by applying them to the M/M/1 queueing system, in the next example.

Example 5.3.3 Consider the M/M/1 queueing system. Here

$$A^*(\theta) = \lambda/(\lambda + \theta),$$

by Example 2.9.6. Hence, (5.3.50) becomes

$$1 - \pi_0 = \lambda/(\lambda + \mu \pi_0). \tag{5.3.67}$$

This equation yields

$$\pi_0 = 1 - \rho. \tag{5.3.68}$$

Since $p_n = \pi_n$ for all n, because the arrival pattern is random, we have

$$p_n = (1 - \rho)\rho^n, \qquad n = 0, 1, 2, \ldots. \tag{5.3.69}$$

All the other formulas for the M/M/1 now agree with those from the GI/M/1 model by making the substitution $\pi_0 = 1 - \rho$.

Example 5.3.4 Consider the $E_2/M/1$ queueing system. It is not difficult to show (see Exercise 31) that if the interarrival time distribution is Erlang-2 then

$$A^*(\theta) = \left(\frac{2\lambda}{2\lambda + \theta}\right)^2 \quad \text{if} \quad \theta < 2\lambda.$$

Hence, (5.3.50), becomes

$$1 - \pi_0 = A^*(\mu\pi_0) = \left(\frac{2\lambda}{2\lambda + \mu\pi_0}\right)^2 \quad \text{or} \quad (2\lambda + \mu\pi_0)^2(1 - \pi_0) = 4\lambda^2.$$

This yields

$$\pi_0 = \frac{-4\rho + 1 + \sqrt{8\rho + 1}}{2} = -2\rho + 0.5 + \sqrt{2\rho + 0.25}. \quad (5.3.70)$$

Hence, we can easily calculate π_0 as a function of ρ.

Example 5.3.5 Consider the $E_3/M/1$ queueing system. Then, since an Erlang-k distribution has the Laplace–Stieltjes transform $(k\lambda/(k\lambda + \theta))^k$, by Exercise 31, (5.3.50) is

$$1 - \pi_0 = \left(\frac{3\lambda}{3\lambda + \mu\pi_0}\right)^3,$$

which yields the cubic equation

$$\pi_0{}^3 + (9\rho - 1)\pi_0{}^2 + 9\rho(3\rho - 1)\pi_0 + 27\rho^2(\rho - 1) = 0. \quad (5.3.71)$$

This is a difficult equation to solve for π_0 as a function of ρ. By numerical methods we compiled the values of π_0 shown in the second column of Table 17 of Appendix C.

Example 5.3.6 Consider the $U/M/1$ queueing system where the interarrival times are independent random variables, each distributed uniformly on the interval 0 to $2/\lambda$. Then, by Exercise 32,

$$A^*(\theta) = \frac{\lambda}{2\theta}(1 - e^{-2\theta/\lambda}).$$

Thus, (5.3.50) becomes

$$1 - \pi_0 = \frac{\lambda}{2\mu\pi_0}(1 - e^{-2\mu\pi_0/\lambda}) = \frac{\rho}{2\pi_0}(1 - e^{-2\pi_0/\rho}). \quad (5.3.72)$$

This is a transcendental equation which must be solved numerically. We carried this out to construct the values in the "U" column of Table 17, Appendix C.

Example 5.3.7 Consider the D/M/1 queueing system in which the arrivals are equally spaced in time. For this interarrival time distribution

$$A^*(\theta) = e^{-\theta/\lambda},$$

so (5.3.50) is

$$1 - \pi_0 = e^{-\mu\pi_0/\lambda} = e^{-\pi_0/\rho}. \qquad (5.3.73)$$

This equation, too, must be solved numerically to yield the values in column "D" of Table 17, Appendix C.

Example 5.3.8 Consider the $H_2/M/1$ queueing system in which the interarrival time has a two-stage hyperexponential distribution (see Section 3.2.9). Then, by (3.2.64),

$$A^*(\theta) = \frac{\alpha_1 \lambda_1}{\lambda_1 + \theta} + \frac{\alpha_2 \lambda_2}{\lambda_2 + \theta},$$

where, of course,

$$E[\tau] = 1/\lambda = (\alpha_1/\lambda_1) + (\alpha_2/\lambda_2).$$

Equation (5.3.50) then is

$$1 - \pi_0 = A^*(\mu\pi_0) = \frac{\alpha_1 \lambda_1}{\lambda_1 + \mu\pi_0} + \frac{\alpha_2 \lambda_2}{\lambda_2 + \mu\pi_0}. \qquad (5.3.74)$$

Some tedious algebra applied to (5.3.74) yields the polynomial equation

$$\mu\pi_0^2 + \mu(\lambda_1 + \lambda_2 - \mu)\pi_0 + \mu(\alpha_1\lambda_1 + \alpha_2\lambda_2 - \lambda_1 - \lambda_2) + \lambda_1\lambda_2 = 0. \qquad (5.3.75)$$

The unique root of (5.3.75) which lies between 0 and 1 is the π_0 we seek. In the special case that Algorithm 6.2.1 of Chapter 6 is used to generate a two-stage hyperexponential distribution for the arrival pattern with a given λ and squared coefficient of variation $C^2 \geq 1$, then it is easy to show (see Exercise 42) that π_0 is given by

$$\pi_0 = 0.5 - \rho + 0.5\sqrt{(1 - 2\rho)^2 + 16\rho\alpha_1(1 - \alpha_1)(1 - \rho)}, \qquad (5.3.76)$$

where, of course,

$$\alpha_1 = \frac{1}{2}\left[1 - \left(\frac{C^2 - 1}{C^2 + 1}\right)^{1/2}\right].$$

The formula (5.3.76) is used to generate the last column of Table 17, Appendix C, when $C^2 = 20$.

The values of π_0 as a function of ρ for several GI/M/1 queueing systems are given in Table 17, Appendix C. The detrimental effects of "irregularity" are evident. The probability that an arriving customer will not have to wait for service is much higher for completely regular arrivals (constant interarrival time) than for somewhat regular arrivals, such as an Erlang-2 pattern, especially when the server utilization is low. The situation is even more dramatic when we consider a hyperexponential interarrival time distribution. For the particular hyperexponential distribution shown in the last column of the table, with $C_\tau^2 = 20$, the probability an arriving customer will find the server busy is almost 19% $(1 - \pi_0 = 0.189425)$ when the server utilization is only 0.1; that is, when the server is busy only one-tenth of the time, an arriving customer will have to wait for service almost two-tenths of the time! Although the Erlang-3 and the uniform distribution each have $C_\tau^2 = 1/3$, their π_0 values are different. The π_0 values are much higher for the Erlang-3 distribution than for the uniform distribution when ρ is small; however, π_0 for Erlang-3 arrivals approaches π_0 for uniformly distributed arrivals as ρ approaches 1. It is interesting to note, also, that the π_0 values for Erlang-2 interarrival times are larger than those for uniformly distributed interarrival times, at low server utilizations (but not for high server utilizations) although the C^2 value of the latter is smaller than that of the former.

We now give some examples of the use of Tables 16 and 17.

Example 5.3.9 Consider Example 5.3.1. Suppose that the message transmission time on each of the four lines is exponential with an average value of 2.4 seconds and that each line operates at 80% utilization. Suppose further that the arrival pattern of messages to the lines is different for each line. The interarrival time for the first line is two-stage hyperexponential with $\alpha_1 = 0.4$, $\alpha_2 = 0.6$, $1/\mu_1 = 6.0$ seconds, $1/\mu_2 = 1.0$ seconds, using the notation of Section 3.2.9; the interarrival time on the second line is exponential; it is Erlang-3 on the third line; and constant on the fourth line. Find W_q, W, $\pi_q(90)$, $\pi_w(90)$, L_q, and L for each line.

Solution The GI/M/1 system equations of Table 16 apply with π_0 values taken from Table 17. The results are summarized in Table 5.3.4, which should be compared to Table 5.3.3, the results for Example 5.3.1. We illustrate with the calculations for the first line. By Table 17, $\pi_0 = 0.124695$. Hence,

$$W_q = (1 - \pi_0)E[s]/\pi_0 = 16.85 \quad \text{seconds},$$

$$W = W_q + E[s] = 19.25 \quad \text{seconds},$$

$$\pi_q(90) = W \ln (10(1 - \pi_0)) = 41.76 \quad \text{seconds},$$

$$\pi_w(90) = 2.3W = 44.28 \quad \text{seconds},$$

$$L_q = (1 - \pi_0)\rho/\pi_0 = 5.62 \quad \text{messages},$$

$$L = \rho/\pi_0 = 6.42 \quad \text{messages}.$$

TABLE 5.3.4
Summary of Results of Example 5.3.9[a]

| | Line arrival pattern | | | |
	H_2	M	E_3	D
W_q	16.85	9.60	5.90	4.06
W	19.25	12.00	8.30	6.46
$\pi_q(90)$	41.76	24.95	16.28	11.88
$\pi_w(90)$	44.28	27.60	19.09	14.86
L_q	5.62	3.20	1.97	1.35
L	6.42	4.00	2.77	2.15

[a] All times in seconds.

It is evident from Table 5.3.4 that irregularity in the arrival process is inimical to the performance of a queueing system, just as Table 5.3.1 showed the harmful effects of lack of regularity in the service time distribution. In both Example 5.3.1 and Example 5.3.9 we have $E[\tau] = 3$ seconds and $E[s] = 2.4$ seconds for all queueing theory models considered. It is interesting to compare the $M/E_2/1$ queueing system to the $E_2/M/1$ system, the $M/E_3/1$ system to the $E_3/M/1$ system, and the $M/D/1$ to the $D/M/1$ system; however, the differences are not striking. The $E_3/M/1$ and the $D/M/1$ queueing systems have slightly lower W_q and W values than the corresponding queueing systems, $M/E_3/1$ and $M/D/1$; but the values of $\pi_w(90)$ are larger than those of the $M/G/1$ type systems. This is true because the distribution of w is more regular than exponential for these systems; the C_w^2 values are 0.7736 and 0.5920, respectively. It is somewhat surprising, therefore, to note that the $M/H_2/1$ system has slightly *lower* values of W_q and W than the $H_2/M/1$ queueing system; but $\pi_w(90)$ is slightly *larger* than that for the $H_2/M/1$ system. C_w^2 is 1.234 for the $M/H_2/1$ queueing system but, of course, only 1 for the $H_2/M/1$ system, since w is exponential for the latter.

Example 5.3.10   Consider an $E_2/E_2/1$ queue with $\rho = 0.95$ and $E[s] = 2$ seconds. Find an upper bound for W and $\pi_w(90)$.

Solution   We can get two conservative estimates by (1) approximating the queueing system by an $M/E_2/1$ model and (2) by an $E_2/M/1$ model. The smaller of these estimates will be an upper bound for W. Assuming an $M/E_2/1$ model yields, by Table 14 or the APL function MΔEKΔ1,

$$W = 30.5 \quad \text{seconds}, \qquad \pi_w(90) = 67.24 \quad \text{seconds}.$$

These are conservative estimates because the Erlang-2 distribution is more regular than the exponential distribution. Applying the $E_2/M/1$ approximation, with $\pi_0 = 0.066288$ from Table 17, yields

$$W = 30.17 \quad \text{seconds}, \qquad \pi_w(90) = 69.39 \quad \text{seconds}.$$

Thus we can safely use $W = 30.17$ seconds and $\pi_w(90) = 67.24$ seconds as upper bounds for these quantities. Later in the chapter we will see some other approximations that could be used here. In fact we shall see that W cannot exceed 22.03 seconds.

The GI/M/c queueing system for $c > 1$ can be solved, analytically, but the computational difficulty increases enormously with c. The interested reader should consult Gross and Harris [6] or Kleinrock [10].

5.4 M/G/1 PRIORITY QUEUEING SYSTEMS

As we mentioned in Section 5.1, all customers in a queueing system need not be treated equally. That is, just as in most organizations, some individuals may get preferential treatment. Queueing systems in which this is true are called *priority queueing systems*. The simplest queue discipline in which there are no priorities, which is tacitly assumed in the Kendall notation, unless something is said to the contrary, is the first-come, first-served system, abbreviated as FCFS or FIFO. Other nonpriority disciplines include last-come, first-served. (LCFS or LIFO), and random-selection-for-service (RSS or SIRO). There are some whimsical queue disciplines; they are part of the queueing theory folklore. These include BIFO, biggest-in-first-out (actually a priority system); FISH, first-in-still-here; and WINO, whenever-in-never-out. The reader can, no doubt, think of others to describe personal experiences with queueing systems.

In all the *priority* queueing systems we will study, customers are divided into priority classes, numbered from 1 to n. We assume that the lower the priority class number, the higher the priority; that is, customers in priority class i are given preference over customers in priority class j, if $i < j$. Customers within a given priority class are served, with respect to that class, by the FCFS queue discipline.

There are two basic control policies to resolve the situation wherein a customer of class i arrives to find a customer of class j in service, where $i < j$, called *preemptive* and *nonpreemptive* queues. In a *preemptive priority* queueing system, service is interrupted and the newly arrived customer begins service. The customer whose service was interrupted returns to the head of the jth class. As a further refinement, in a *preemptive-resume* priority queueing system, the customer whose service was interrupted begins service at the point of interruption upon the next access to the service facility. There are other variations, including *preemptive repeat*, in which the lower priority customer repeats the entire service from the beginning. In a *nonpreemptive* priority queueing system, the newly arrived customer waits until the customer in service completes service before gaining access to the service facility. This type of system is called a head-of-the-line system, abbreviated HOL.

We give the equations for the most common types of M/G/1 queueing systems, called nonpreemptive (HOL), and preemptive resume. In both cases each of the priority classes has a Poisson arrival pattern with average arrival rate λ_i, and a general independent service time distribution with average value $E[s_i] = 1/\mu_i$. Thus, by Section 3.1.4, the total arrival rate to the system has a Poisson distribution with average rate

$$\lambda = \lambda_1 + \lambda_2 + \cdots + \lambda_n. \tag{5.4.1}$$

By the law of total expectation, Theorem 2.8.1,

$$E[s] = \frac{\lambda_1}{\lambda} E[s_1] + \frac{\lambda_2}{\lambda} E[s_2] + \cdots + \frac{\lambda_n}{\lambda} E[s_n], \tag{5.4.2}$$

and, by the law of total moments,

$$E[s^2] = \frac{\lambda_1}{\lambda} E[s_1^2] + \frac{\lambda_2}{\lambda} E[s_2^2] + \cdots + \frac{\lambda_n}{\lambda} E[s_n^2], \tag{5.4.3}$$

for both queueing systems. The remainder of the equations for the HOL queueing system are given in Table 18, Appendix C, while those for the preemptive resume queueing system are given in Table 19.

We illustrate the effects of priority queueing by the following example.

Example 5.4.1 An on-line inquiry system receives two types of inquiries. Type 1 inquiries arrive in a Poisson pattern at an average arrival rate of 0.9 per second. The time required for the system to respond is nearly constant with an average value of 0.4 seconds. The Type 2 inquiries arrive at an average rate of 1 every 10 seconds. The system response time for Type 2 inquiries has a two-stage hyperexponential distribution with $\alpha_1 = 0.4$, $\alpha_2 = 0.6$, $1/\mu_1 = 10$ seconds, $1/\mu_2 = 5/3$ seconds; so the average system response time for Type 2 inquiries is 5 seconds, with a second moment of 83.33 seconds^2. Contrast the operation of the system with (a) no priorities, (b) with an HOL priority system that gives priority to Type 1 inquiries, and (c) with preemptive-resume priority given to Type 1 inquiries.

Solution (a) For a nonpriority system the average service time

$$E[s] = 0.9 \times 0.4 + 0.1 \times 5 = 0.36 + 0.5 = 0.86 \quad \text{seconds}$$

$$E[s^2] = 0.9 \times 0.4^2 + .1 \times 83.33 = 8.477 \quad \text{seconds}^2$$

$$\rho = \lambda E[s] = 0.86$$

$$W_q = E[q] = \frac{\lambda E[s^2]}{2(1 - \rho)} = 30.275 \quad \text{seconds}.$$

Average time in the system for Type 1 inquiries is

$$W_1 = 30.275 + 0.4 = 30.675 \quad \text{seconds.}$$

Average time in the system for Type 2 inquiries is

$$W_2 = 30.275 + 5 = 35.275 \quad \text{seconds.}$$

Overall average waiting time in the system is

$$W = W_q + E[s] = 31.135 \quad \text{seconds.}$$

(b) For an HOL queueing system with Type 1 inquiries having non-preemptive priority over Type 2 inquiries

$$u_1 = 0.9 \times 0.4 = 0.36 \quad \text{seconds,} \qquad u_2 = 0.36 + 0.1 \times 5 = 0.86 \quad \text{seconds.}$$

The average queueing times are

$$W_{q1} = \frac{\lambda E[s^2]}{2(1 - u_1)} = 6.6227 \quad \text{seconds}$$

$$W_{q2} = \frac{\lambda E[s^2]}{2(1 - u_1)(1 - u_2)} = 47.3047 \quad \text{seconds.}$$

The average times in the system are

$$W_1 = W_{q1} + E[s_1] = 7.0227 \quad \text{seconds,}$$
$$W_2 = W_{q2} + E[s_2] = 52.3047 \quad \text{seconds.}$$

The overall average queueing time

$$W_q = 0.9 \times 6.6227 + 0.1 \times 47.3047 = 10.6909 \quad \text{seconds.}$$

The overall average waiting time in the system is

$$W = W_q + E[s] = 11.5509 \quad \text{seconds.}$$

(c) For a priority queueing system with Type 1 inquiries receiving preemptive-resume priority over Type 2 inquiries, using u_1 and u_2 from (b), yields the waiting times in the system

$$W_1 = E[s_1] + \frac{\lambda_1 E[s_1^2]}{2(1 - u_1)} = 0.4 + \frac{0.9 \times 0.16}{2(1 - 0.36)} = 0.5125 \quad \text{seconds}$$

$$W_2 = \frac{1}{(1 - u_1)}\left[E[s_2] + \frac{\lambda_1 E[s_1^2] + \lambda_2 E[s_2^2]}{2(1 - u_2)}\right]$$

$$= \frac{1}{1 - 0.36}\left[5 + \frac{0.9 \times 0.16 + 0.1 \times 83.33}{2(1 - 0.86)}\right] = 55.1172 \quad \text{seconds.}$$

The corresponding average queueing times for the two inquiry types are

$$W_{q1} = W_1 - E[s_1] = 0.1125 \quad \text{seconds,}$$

$$W_{q2} = W_2 - E[s_2] = 50.1172 \quad \text{seconds.}$$

The overall average queueing time

$$W_q = 0.9 \times 0.1125 + 0.1 \times 50.1172 = 5.1130 \quad \text{seconds,}$$

and the overall average waiting time in the system

$$W = 0.9 \times 0.5125 + 0.1 \times 55.1172 = 5.9224 \quad \text{seconds.}$$

We summarize this data from this example in Table 5.4.1.

TABLE 5.4.1

Results of Example 5.4.1[a]

	No priority	HOL priority	Preemptive-resume priority
W_{q1} (type 1)	30.275	6.6227	0.1125
W_{q2} (type 2)	30.275	47.3047	50.1172
W_1 (type 1)	30.675	7.0227	0.5125
W_2 (type 2)	35.275	52.3047	55.1172
W_q	30.275	10.6909	5.1130
W	31.124	11.5509	5.9724

[a] All times in seconds.

The results shown in Table 5.4.1 illustrate how a priority system can dramatically improve the performance of a queueing system.

The average queueing time for a Type 1 inquiry drops from 30.275 seconds for a nonpriority system to 6.6227 seconds for an HOL queueing system; for a preemptive-resume system, it is only 0.1125 seconds! The overall average queueing time drops from 30.275 seconds to 10.6909 seconds, and then to 5.113 seconds; the improvement in average system time is similar. The performance of the system for Type 2 inquiries suffers, but not severely.

The reader is asked to show in Exercise 37 that, if the Type 2 inquiries were given priority over Type 1 inquiries, the overall average queueing and system times would be larger for the priority systems than for the original nonpriority system.

5.5 APPROXIMATIONS

For many queueing systems either (a) there is no analytic solution which enables one to calculate useful measures of performance such as W_q, W, and $\pi_w(90)$ or (b) analytic solutions are available but either require inversion of complicated Laplace–Stieltjes transforms or involve very complex calculations. In this kind of situation approximate methods can be very helpful.

5.5.1 Bounds on Queueing Systems

It is common to find a queueing system for which, during a peak period or periods, the server utilization may be close to 1, although the utilization is low much of the time. The following theorem can be helpful in anticipating the performance of the system during a peak period.

THEOREM 5.5.1 (*Heavy Traffic Approximation*) Consider a GI/G/1 queueing system. As ρ approaches 1, the distribution of queueing time, q, approaches that of an exponential distribution with

$$W_q = E[q] = \frac{\lambda(\text{Var}[\tau] + \text{Var}[s])}{2(1 - \rho)} = \frac{\rho E[s]}{2(1 - \rho)} \left\{ \frac{\text{Var}[\tau]}{E[s]^2} + C_s^2 \right\}. \quad (5.5.1)$$

The proof of this theorem is due to Kingman [13]; it can also be found in Kleinrock [12] and Gross and Harris [6], but will not be repeated here.

Example 5.5.1 A proposed on-line computer system has a communication line reserved for a special application. This line is expected to have a utilization of 95% for several peak periods per day; for some short periods the utilization may reach 99%. For the traffic expected on this line $E[s] = 2$ seconds with $\text{Var}[s] = 5$ seconds^2. During the 95% utilization periods $E[\tau] = 2.1053$ seconds with $\text{Var}[\tau] = 3$ seconds^2; when the utilization is 99% $E[\tau] = 2.02$ and $\text{Var}[\tau] = 2.8$. The users demand that, for 95% utilization periods, $\pi_q(90) \leq 90$ seconds. The second criterion is that, for the 99% utilization periods, the probability q exceeds 90 seconds should not exceed 0.5. Is one line enough for the special application?

Solution When $u = \rho = 0.95$, then $\lambda = \rho/E[s] = 0.475$ transactions per second. Hence, by (5.5.1),

$$W_q = \frac{0.475(3 + 5)}{2(1 - 0.95)} = 38 \quad \text{seconds.}$$

Thus $\pi_q(90) = 2.3W_q = 87.4$ seconds, and the first criterion is satisfied. When $u = \rho = 0.99$ and $\lambda = 0.495$, we have

$$W_q = \frac{0.495(2.8 + 5)}{2(1 - 0.99)} = 193.05 \quad \text{seconds.}$$

Thus

$$P[q > 90] = e^{-90/193.05} = 0.627,$$

and the proposed system fails the second criterion. Hence, at least two lines are needed for the special application. It seems reasonable, based on the numbers calculated above, that two lines will be enough.

The heavy traffic approximation was extended to the GI/G/c queueing system by Köllerström [14] and is stated without proof in Theorem 5.5.2 below.

THEOREM 5.5.2 (*Heavy Traffic Approximation*) Consider a GI/G/c queueing system. As $\rho = \lambda E[s]/c$ approaches 1, the distribution of queueing time, q, approaches that of an exponential distribution with

$$W_q = E[q] = \frac{\lambda(\text{Var}[\tau] + \text{Var}[s]/c^2)}{2(1 - \rho)} = \frac{uE[s]}{2(1 - \rho)}\left(\frac{\text{Var}[\tau]}{E[s]^2} + \frac{C_s^2}{c^2}\right). \quad (5.5.2)$$

Example 5.5.2 The 20 terminals in a student work area of a university computer center have 95% utilization during the peak period. The time each student spends at a terminal has a mean of 38 minutes and a standard deviation of 35 minutes. The student interarrival time has a mean of 2 minutes and a standard deviation of 4 minutes. Find the average queueing time that students experience during the busy period, and the 90th percentile value of this time.

Estimate the number of terminals needed to reduce the average queueing time to 20 minutes or less.

Solution By (5.5.2) with $\lambda = 0.5$ students per minute,

$$W_q = \frac{0.5[16 + (35/20)^2]}{2(1 - 0.95)} = 5[16 + (35/20)^2] = 95.3125 \quad \text{minutes,}$$

and

$$\pi_q(90) = 2.3W_q = 219.22 \quad \text{minutes.}$$

To get a gross approximation for the number of terminals needed, we assume the heavy traffic approximation holds, and solve (5.5.2) for c, using $\rho = 19/c$.

That is, set

$$20 = \frac{19 \times 38}{2(1 - 19/c)} \left| \frac{16}{38^2} + \left(\frac{35}{38} \right)^2 \times \frac{1}{c^2} \right|,$$

which yields $c = 24.53$. Thus 25 terminals would certainly lower the average queueing time below 20 minutes; in fact, since, if $c = 25$, $\rho = 19/25 = 0.76$, which is below the heavy traffic range, W_q would probably be much lower than 20 minutes.

It is not difficult to show (see Exercise 41) that if ρ, C_τ^2, C_s^2, and $E[s]$ are the same for two heavy-traffic queueing systems, one GI/G/1 and the other GI/G/c, then the average queueing time for the latter system is $1/c$ times the average queueing time for the first system, that is,

$$E[q]_{\mathrm{GI/G}/c} = \frac{1}{c} \times E[q]_{\mathrm{GI/G}/1}.$$

This is what intuition would suggest but is *not* true for low values of ρ. For example, the ratio of $E[q]$ for the M/M/2 queueing system, to that for the M/M/1 system is $\rho/(1 + \rho)$; this is approximately ρ for small values of ρ but approaches $1/2$ as ρ approaches 1 (the heavy traffic case). ($\rho/(1 + \rho)$ is 0.09091 for $\rho = 0.1$ but 0.4975 when $\rho = 0.99$.)

In the next theorem we see that both upper and lower bounds exist for W_q in a GI/G/1 queueing system; they do not require the heavy-traffic assumption, although the upper bound is the W_q of Theorem 5.5.1. The lower bound is due to Marchal [15]; the proof that it holds is also given in Kleinrock [12].

THEOREM 5.5.3 (*Bounds for* W_q) Consider a GI/G/1 queueing system. Then

$$\frac{\rho^2 C_s^2 + \rho(\rho - 2)}{2\lambda(1 - \rho)} \le W_q \le \frac{\lambda(\mathrm{Var}[\tau] + \mathrm{Var}[s])}{2(1 - \rho)}. \tag{5.5.3}$$

The lower bound in (5.5.3) is not sharp. In fact it is negative if $C_s^2 < (2 - \rho)/\rho$. Some sharper lower bounds are given in Kleinrock [12]. However, the sharper bounds require certain assumptions about the arrival pattern and are difficult to calculate. Recall that, for the M/G/1 queueing system,

$$W_q = \frac{\rho E[s]}{1 - \rho} \left| \frac{1 + C_s^2}{2} \right| = \frac{\rho^2 C_s^2 + \rho(\rho - 2)}{2\lambda(1 - \rho)} + \frac{E[s]}{1 - \rho}.$$

That is, the lower bound given in (5.5.3) is too small by the amount $E[s]/(1 - \rho)$, for the M/G/1 system.

Example 5.5.3 Consider Example 5.3.10. There we had an $E_2/E_2/1$ system with $\rho = 0.95$ and $E[s] = 2$ seconds. We can apply the heavy traffic approximation to yield, by (5.5.1)

$$W_q = \frac{0.95 \times 2}{2(1 - 0.95)} \left\{ \frac{2.216066 + 2}{4} \right\} = 20.026 \quad \text{seconds,}$$

since for an Erlang-2 distribution the variance is one half the square of the mean $(C_X^2 = \text{Var}[X]/E[X]^2 = 1/2)$. Thus $W = 22.026$ seconds. This is much smaller than the 30.17 seconds we found as an upper bound for W in Example 5.3.10. (This upper bound would be valid, by Theorem 5.5.3, even if we did not have heavy traffic.) We see, also, that $\pi_q(90) = 2.3W_q = 46.06$ seconds, but it is difficult to calculate $\pi_w(90)$ from $\pi_q(90)$. However, we reason that, since q exceeds $\pi_q(90)$ only 10% of the time, and, since s exceeds $\pi_s(90)$ only 10% of the time, we must have $\pi_w(90) \le \pi_q(90) + \pi_s(90)$. By Table 5.3.1, $\pi_s(90) = 2 \times 1.94 = 3.88$ seconds, so $\pi_w(90) \le 46.06 + 3.88 = 49.94$ seconds. This is smaller than the 67.24 seconds we estimated in Example 5.3.10. It is interesting to note that the lower bound given by Theorem 5.5.3 for W_q is negative, in this case.

The estimate of an upper bound for W_q in a GI/G/c queueing system is not as simple as for GI/G/1. Kingman has conjectured the W_q of (5.5.2) is an upper bound for W_q for a GI/G/c queueing system with $0 \le \rho < 1$. Brumelle [16] has shown that this is, indeed, true for the GI/M/c queueing system. Suzuki and Yoshida [17] have shown that (5.5.2) is an upper bound for W_q when $0 \le \rho < 1/c$. In the theorem below we give an upper bound for W_q due to Kingman [18] and a lower bound given in Kleinrock [12].

THEOREM 5.5.4 Consider a GI/G/c queueing system. Then the following inequalities are true for $0 \le \rho < 1$.

$$W_q = E[q] \le \frac{\text{Var}[\tau] + \text{Var}[s]/c + [(c-1)/c^2]E[s]^2}{2E[\tau](1 - \rho)} \qquad (5.5.4)$$

$$\frac{\rho^2 C_s^2 - \rho(2 - \rho)}{2\lambda(1 - \rho)} - \frac{[(c-1)/c]E[s^2]}{2E[s]} \le W_q. \qquad (5.5.5)$$

Example 5.5.4 Consider a multiprocessor computer system with 4 processing units, each with $E[s] = 0.04$ seconds and $C_s^2 = 10$. Suppose $E[\tau] = 0.05$ seconds with $C_\tau^2 = 8$. Thus $\rho = E[s]/4E[\tau] = 0.2$.

Find an upper bound for W_q. How will this change if $E[\tau]$ decreases to 0.0102 seconds?

Solution By (5.5.4), we have

$$W_q \le \frac{0.02 + 0.016/4 + 0.1875 \times 0.04^2}{2 \times 0.05 \times 0.8} = 0.30375 \quad \text{seconds,}$$

and, by (5.5.5), we have

$$W_q \geq \frac{0.2^2 \times 10 - 0.2 \times 1.8}{2 \times (1/0.05) \times 0.8} - \frac{0.75 \times 0.0176}{2 \times 0.04}$$

$$= 0.00125 - 0.165 = -0.16375.$$

The negative lower bound is, of course, meaningless. Since $\rho = 0.2 < 1/c = 0.25$, we can use (5.5.2) as an upper bound for W_q, by the result of Suzuki and Yoshida [17]. Hence

$$W_q \leq \frac{0.8 \times 0.04}{2 \times 0.8} \left| \frac{0.02}{0.04^2} + \frac{10}{4^2} \right| = 0.2625 \quad \text{seconds.}$$

This is a more satisfactory estimate.
    If $E[\tau] = 0.0102$, then

$$\rho = E[s]/4E[\tau] = 0.9804.$$

The heavy-traffic approximation applies and, by (5.5.2),

$$W_q = \frac{1/0.0102(8 \times 0.0102^2 + 0.016/16)}{2(1 - 0.9804)} = 4.5826 \quad \text{seconds.}$$

My colleague at the Los Angeles IBM Systems Science Institute, John T. Cunneen, and I have developed the following approximation for GI/G/c queues, which we modestly call the "Allen–Cunneen Approximation Formula."

Approximation 5.5.1 (*Allen–Cunneen Approximation Formula*) For any GI/G/c queueing system, it is approximately true that

$$W_q = E[q] = \frac{C(c, u)E[s]}{c(1 - \rho)} \left| \frac{C_\tau^2 + C_s^2}{2} \right|. \tag{5.5.6}$$

The reader should note that, for the M/M/c queueing system, (5.5.6) is exact. It is also exact for M/G/1 queueing systems. For M/G/c systems it gives Martin's estimate (see Martin [7, p. 461] and Exercise 11). We found, by comparing this approximation to the exact solution given by Hillier and Lo [19] for $E_m/E_k/c$ queueing systems, that the error was very small. In fact Hillier and Lo give an approximation for L_q which is algebraically equivalent to (5.5.6). The approximation gave slightly optimistic results for the $H_2/M/1$ system of Example 5.3.8. We have found no queueing system for which the error in (5.5.6) would exceed the error normally expected because of uncertainties in the values of λ, μ, etc. The approximation has the special virtue of being easy to calculate.

If we apply this approximation to Example 5.3.10, the $E_2/E_2/1$ system with $\rho = 0.95$ and $E[s] = 2$ seconds, we obtain

$$W_q = \frac{0.95 \times 2}{(1 - 0.95)}\left(\frac{1/2 + 1/2}{2}\right) = 19 \quad \text{seconds.}$$

This gives

$$W = 21 \quad \text{seconds.}$$

By the heavy-traffic approximation theorem (Theorem 5.5.1), q is nearly exponential, so we can write

$$\pi_w(90) \leq \pi_q(90) + \pi_s(90) \quad 2.3W_q + 1.94 \times 2 = 47.58 \quad \text{seconds.}$$

(See Example 5.5.3 for a solution of this problem using the heavy traffic approximation.)

In Example 5.5.4 the Allen–Cunneen approximation formula yields the values 0.001 seconds and 4.392 seconds, respectively, for mean queueing times.

There are a number of other approximate methods but they are too sophisticated to discuss here. The interested reader should consult Kleinrock [12, Ch. 2] and the references mentioned therein.

5.5.2 Graphical Methods

We have demonstrated graphical techniques to calculate $C(c, u)$ for the M/M/c queueing system and $B(c, u)$ for the M/M/c/c system. Graphical techniques are used primarily for easing the calculations required in a queueing model whose solution is known. Such techniques also help one visualize what happens to important system parameters, such as W_q and W, as the load changes. For example, Fig. 5.2.2 could be used for calculating the performance of most M/M/1 queueing systems, if $\rho < 0.9$, as demonstrated in the following example.

Example 5.5.5 Consider Example 5.2.3, an M/M/1 system with $E[s] = 30$ minutes and $\lambda = 1/48$ engineers per minute, so that $\rho = 5/8$. By Fig. 5.2.2, $W/E[s] = 2.7$, and thus

$$W = 2.7E[s] = 81 \quad \text{minutes,}$$

$$W_q = W - E[s] = 51 \quad \text{minutes,}$$

$$L = \lambda W = 81/48 = 1.69 \quad \text{engineers,}$$

$$L_q = \lambda W_q = 51/48 = 1.06 \quad \text{engineers, etc.}$$

The graph in Fig. 5.2.2 has been normalized by dividing W by $E[s]$ so that the same graph will suffice for all values of $E[s]$. However, there is a major problem with this graph for high values of ρ. If we scale the graph so that fairly large values of ρ can be represented, then it is difficult to read the values of $W/E[s]$ for small values of ρ. One way around this problem is to plot $E[s]/W$, as we have done in Fig. 5.5.1. Then it is easy to read the values of $E[s]/W$ for all values of ρ, since this ratio is always between 0 and 1. Such a graph has been called a *system map*. In Fig. 5.5.1 we have plotted the ratio $E[s]/W$ for several GI/M/1 queueing systems. We have done the same for some M/G/1 systems in Fig. 5.5.2.

Example 5.5.6 Consider Example 5.3.10 in which we approximated an $E_2/E_2/1$ queueing system first by an $M/E_2/1$ system and then by an $E_2/M/1$ system when $\rho = 0.95$ and $E[s] = 2$ seconds. For the $E_2/M/1$ system, we can

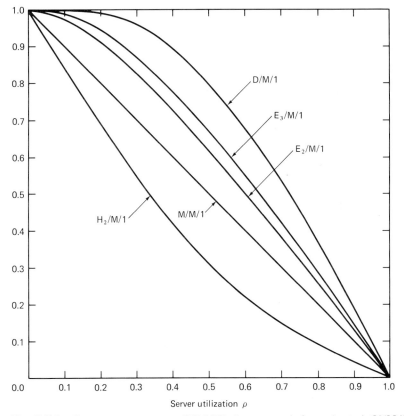

Fig. 5.5.1 Some system maps ($E[s]/E[w]$ versus ρ) for selected GI/M/1 queueing systems.

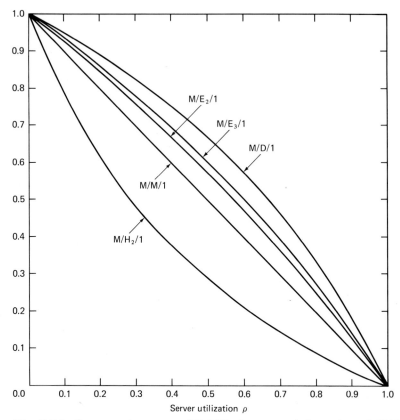

Fig. 5.5.2   Some system maps ($E[s]/E[w]$ versus ρ) for selected M/G/1 queueing systems.

read from Fig. 5.5.1 that $E[s]/W = 0.07$ or $W = 2/0.07 = 28.57$ seconds. Thus, since w is exponential, $\pi_w(90) = 2.3 \times 28.57 = 65.7$ seconds. For the $M/E_2/1$ system we read from Fig. 5.5.2 that $E[s]/W = 0.07$ or $W = 28.57$ seconds. It is purely coincidental that we got the same value of W for the $M/E_2/1$ queueing system and the $E_2/M/1$ queueing system; the ratio $E[s]/E[w]$ is 0.066 for both systems, but 0.95 is the only value of ρ for which this is true.

Graphical techniques can also be used for the calculation of such quantities as $\text{Var}[q]$, $\text{Var}[w]$, $\pi_q(90)$, etc., for many queueing models. However, scaling problems often exist and cannot always be obviated by simple tricks such as we used for graphically solving for $E[w]$ in M/G/1 and GI/M/1 queueing systems. The interested reader may wish to peruse Chapters 31–33 of Martin [7] for many examples of graphical methods of making queueing theory calculations. (The reader is warned that Martin's notation is very different from that used in this book.)

A comparison of Figs. 5.5.1 and 5.5.2 yields some interesting facts. (The reader may want to superimpose a foil copy of one upon the other.) Since $E[w] = E[q] + E[s]$, when the ratio of $E[s]$ to $E[w]$ is near one there is very little queueing time and $E[w]$ is approximately $E[s]$; conversely, when the ratio is small (close to zero), the queueing time is large relative to $E[s]$, and the performance is poor. Thus we see from a comparison of the figures that, for all values of ρ, an $E_3/M/1$ system performs better than a $M/E_3/1$ system, and an $E_2/M/1$ system better than a $M/E_2/1$ system. Surprisingly, an $M/H_2/1$ system performs better than an $H_2/M/1$ system for all values of ρ! We noticed this phenomenon in Example 5.3.8 with $\rho = 0.8$. The hyperexponential distribution used in Fig. 5.5.1 was generated by Algorithm 6.2.1 so that $C_\tau^2 = 4$. Thus $\alpha_1 = 0.1127$, $\alpha_2 = 0.8873$, $\lambda_1 = 2\alpha_1\lambda$, and $\lambda_2 = 2\alpha_2\lambda$; that for Fig. 5.5.2 was generated by the same algorithm and thus $\alpha_1 = 0.1127$, $\alpha_2 = 0.8873$, $\mu_1 = 2\alpha_1\mu$, and $\mu_2 = 2\alpha_2\mu$.

5.6 SUMMARY

In this chapter we have introduced the reader to the fundamental ideas of queueing theory and discussed some of the basic queueing systems which are especially useful in computer science. We have illustrated the use of these systems with a number of examples. In Chapter 6 we will show how some of these basic queueing theory models can be combined to study more complex systems.

Some of the material in this chapter is discussed in a more condensed format by Allen [20].

Student Sayings

Roses are red;
Violets are blue
If λ is big,
then ρ is too!

Did you say KAMAKAZY Airline or KAMAKAZY Erlang?

I have ρ-ed and ρ-ed until I'm c-sick.

MINO: "Meekest-in, never-out"; a queue discipline which describes the process of being snubbed by a snotty head waiter in an exclusive restaurant.

Exercises

1. [HM18] (a) Show that, for an M/M/1 queueing system, the probability there are n or more customers in the system is ρ^n.

(b) Use this result to find the value of μ such that, for given values of λ, n and α, with $0 < \alpha < 1$, the probability of n or more customers in the system is α. This value of μ should be found explicitly in terms of λ, n and α.

(c) In particular find μ if $\lambda = 10$, $n = 3$, and $\alpha = 0.05$.

2. [HM20] Show that for an M/M/1 queueing system the conditional density function for waiting time in the queue given that a wait occurs $(q > 0)$, say q', is given by $(\mu - \lambda)e^{-(\mu - \lambda)t}$, $t > 0$. Use this formula to calculate the conditional distribution function

$$P[q \le t \,|\, q > 0] = P[q' \le t] = 1 - e^{-\mu(1 - \rho)t}.$$

Thus q' has the same distribution as w.

[Hint: The conditional density function of q given that $q > 0$, that is, q', is the density function of q divided by the probability that $q > 0$.]

3. [18] Show that for an M/M/1 queueing system $E[N_q \,|\, N_q > 0] = 1/(1 - \rho)$.

4. [HM20] Show that, for an M/M/1 queueing system,

$$\text{Var}[N_q] = \rho^2(1 + \rho - \rho^2)/(1 - \rho)^2.$$

[Hint: Apply Theorem 2.9.2.]

5. [18] The BRITE LITE company has machines which break down in a Poisson pattern at an average rate of three per hour during the 8-hour working day. BRITE LITE is considering the repair services of I. M. Slow and I. M. Fast. Slow repairs machines with an exponential repair time distribution at an average rate of four machines per hour for a service charge of $10.00 per hour. Fast provides exponential repair time for $18.00 per hour but at an average rate of six machines per hour. Which person should be hired on a daily basis if the cost of an idle machine is $30.00 per hour?

6. [15] People arrive at a telephone booth in a random pattern, with an average interarrival time of 12 minutes. The length of phone calls from the booth, including the dialing time, wrong numbers, etc., is exponentially distributed with an average of 4 minutes.

(a) What is the probability that a person arriving at the booth will have to wait? Do not assume your mother-in-law is in the booth.

(b) What is the average length of the waiting lines that form from time to time; that is, the average of those that are not of zero length?

(c) What is the probability that an arrival will have to wait more than 10 minutes before the phone is available?

(d) The telephone company plans to install a second booth when convinced that an arriving customer would expect to have to wait at least five minutes to use the phone. At what average interarrival time will this occur?

7. [20] A clerk provides exponentially distributed service to customers who arrive randomly at the average rate of 15 per hour. What average service time must the clerk provide in order that 90% of all customers will not have to queue for service longer than 12 minutes.

[*Hint:* A graphical or iterative technique is necessary.]

8. [15] Consider an M/M/1/K queueing system. Let q_n be the probability that there are n customers in the system just before a customer arrival who actually enters the system. Assume that $q_n = kp_n$ for some constant k. Prove that $q_n = p_n/(1 - p_K)$, $n = 0, 1, \ldots, K - 1$.

9. [HM25] Carry out the details of (5.2.48), that is, show that

$$W(t) = P[w \leq t] = 1 - \sum_{n=0}^{K-1} q_n \left(\sum_{k=0}^{n} e^{-\mu t} \frac{(\mu t)^k}{k!} \right)$$

for the M/M/1/K queueing system, where $q_n = p_n/(1 - p_K)$.

[*Hint:* Write

$$W(t) = \sum_{n=0}^{K-1} \left[\int_0^t \frac{\mu(\mu x)^n e^{-\mu x}}{n!} dx \right] q_n = \sum_{n=0}^{K-1} \left[1 - \int_t^{\infty} \frac{\mu(\mu x)^n e^{-\mu x}}{n!} dx \right] q_n$$

$$= 1 - \sum_{n=0}^{K-1} q_n \int_t^{\infty} \frac{\mu(\mu x)^n}{n!} e^{-\mu x} dx.$$

Then make the change of variable $y = x - t$ in each of the integrals. By recognizing the integral form of the gamma function (see formula (3.2.35)) and the property of the gamma function expressed in (3.2.36), deduce that

$$\int_t^{\infty} \frac{\mu(\mu x)^n}{n!} e^{-\mu x} dx = \sum_{k=0}^{n} e^{-\mu t} \frac{(\mu t)^k}{k!}, \qquad n = 0, 1, \ldots, K - 1.]$$

10. [HM18] Prove that, for an M/M/c queueing system, the average number of busy servers is $u = \lambda E[s]$.

11. [15] Martin [7, p. 461] claims that, for an M/G/c system, one can approximate the average waiting time in queue, $W_q = E[q]$, by

$$W_q = \frac{C(c, u)E[s]}{c(1 - \rho)} \left\{ \frac{1 + C_s^2}{2} \right\}$$

where, of course, $C_s^2 = \text{Var}[s]/E[s]^2$. W can then be approximated by $W_q + E[s]$. Consider Example 5.2.7. Suppose KAMAKAZY Airlines installs the new reservations office with 8 agents and the system performs even better than expected. Each agent has an Erlang-3 service time distribution with average value 5 minutes. The OR department estimates that just before the holidays the peak calling rate may go up to 64.8 calls per hour. Use Martin's estimate to calculate W_q and W for the increased traffic. Note that Martin's estimate is a special case of the Allen–Cunneen approximation.

12. [12] The data processing manager at a certain company provides three consultants to help open-shop programmers debug their programs. Programmers with "buggy" programs arrive randomly, at an average rate of 20 per 8-hour day. The amount of time that a consultant spends with a programmer has an exponential distribution with an average value of 40 minutes. Programmers are assigned to consultants in the order of their arrival.

(a) What is the average number of hours, per 40 hour week, that each consultant spends with programmers seeking help?

(b) What is the average amount of time a programmer spends in the consulting facility?

13. [C20] The WEIRDOENGINEER Company of Examples 5.2.3 and 5.2.6 installs five terminals at customer locations about town and finds the distribution of users and average driving times as shown in Table E5.13. (The "turnpike effect" has caused the number of users to rise to an average of 30 per day.) Assuming that the average time at a terminal is 30 minutes calculate the following.

(a) W_q, W, and $\pi_q(90)$ for each terminal.

(b) The (weighted) average values of W_q, W, and $\pi_q(90)$ over all the terminals.

(c) The average value of time required for an engineer to drive to the assigned terminal, complete a work session, and return (the average total time, that is).

TABLE E5.13

Terminal number	Average number of users per day	Average driving time per user (minutes)
1	6	2
2	8	5
3	4	4
4	10	1
5	2	10

14. [25] Customers arrive randomly (during the evening hours) at the Kittenhouse, the local house of questionable services, at an average rate of five per hour. Service time is exponential with a mean of 20 minutes per customer. There are two servers on duty.

(a) What is the probability an arriving customer must queue?

(b) That one or both servers are idle?

(c) What is the average time a customer spends at the Kittenhouse?

(d) If the house is raided, how many customers will be caught, on the average?

(e) What is the probability that five or more customers will be captured in a raid?

(The data for this problem were conjectured by the author. Observed data from readers would be appreciated.)

15. [18] JETSET Airlines, a fierce competitor of KAMAKAZY Airlines (Example 5.2.7), also is planning a new telephone reservations office. Their agents provide customers who call with an exponential service time; the average time is 3 minutes. Like KAMAKAZY Airlines, calls that arrive when all agents are busy are held (with appropriate background music) until an agent is free. They expect a random pattern of customer calls with an average of 30 calls per hour during the peak period.

(a) If the two criteria are (1) that the average queueing time should not exceed 1 minute and (2) that 90% of all callers must wait less than 2 minutes for service to begin, how many agents are required?

(b) With the number of agents determined by part (a), what is the average queueing time, and the average number of customers waiting for service?

(c) What is the probability that all the agents are busy during the peak period? That all are idle?

16. [C15] YOUTOOLCOMPUTE has 10 portable computers available for rent. The average rental time is 2.5 days and is exponentially distributed. Customers arrive randomly at an average rate of two customers per day. If a computer is not available, a customer will go to another store.

(a) What fraction of arriving customers will be lost?

(b) What is the average number of computers in use?

(c) What fraction of customers will be lost if one of the computers is out of service for an extended period?

17. [15] The SUPERCOMPUTER Company offers computer service bureau services to drive-in customers. Twelve customer parking spaces are provided for customers who arrive randomly at the average rate of 14 per hour; those who arrive when all spaces are in use take their business across the street to the SUPER-DUPERCOMPUTER Company. Each parking space is occupied for an average of 30 minutes, occupancy time having an exponential distribution. Find the following.

(a) The effective average arrival rate.

(b) The fraction of arriving customers turned away.

(c) The average number of spaces in use.

18. [HM25] Use the Cauchy–Schwartz inequality (stated below) to show that, if X has a k-stage hyperexponential distribution, then $C_X^2 \geq 1$. The Cauchy–Schwartz inequality asserts that, if $a_1, a_2, \ldots, a_n$ and $b_1, b_2, \ldots, b_n$ are real, then

$$\left(\sum_{i=1}^{n} a_i b_i \right)^2 \leq \left(\sum_{i=1}^{n} a_i^2 \right) \left(\sum_{i=1}^{n} b_i^2 \right).$$

19. [18] Use Theorem 5.3.1 and the Corollary to show that formulas (5.3.42)–(5.3.45) are true.

20. [20] Prove that, for the M/G/1 queueing system,

$$\sigma_N^2 = \frac{\lambda^3 E[s^3]}{3(1-\rho)} + \left(\frac{\lambda^2 E[s^2]}{2(1-\rho)} \right)^2 + \frac{\lambda^2(3-2\rho)E[s^2]}{2(1-\rho)} + \rho(1-\rho).$$

[*Hint*: Use (5.3.29) to show that

$$E[N(N-1)] = E[N^2] - E[N] = \lambda^2 E[w^2].$$

Then calculate $E[N^2]$ from this formula, (5.3.45), and Little's formula.]

21. [HM20] Show that, for an M/M/c queueing system,

$$\sigma_{Nq}^2 = \frac{\rho C(c, u)[1 + \rho - \rho C(c, u)]}{(1-\rho)^2}.$$

[*Hint*: Apply Theorem 2.9.2.]

22. [HM15] Show that, if the service time, s, has a k-stage hyperexponential distribution, then

$$E[s^3] = 6 \sum_{i=1}^{k} \frac{\alpha_i}{\mu_i^3}.$$

23. [15] Repairing a small computer requires four steps in sequence. The time to complete each of these steps is exponentially distributed with a mean time of 3 minutes; the steps being independent of one another. If a facility has one person who can repair these computers and they break down in a Poisson pattern at an average rate of three per hour, what is the average down time of a computer?

24. [HM20] Suppose the service time has a gamma distribution with parameters α and β. (Since $E[s] = 1/\mu$, this means $\alpha/\beta = 1/\mu$ or $\alpha = \beta/\mu$.) Show that $E[s] = 1/\mu$, $E[s^2] = E[s]^2 + 1/\beta\mu$, $\mathrm{Var}[s] = 1/\beta\mu$,

$$C_s^{\,2} = \mu/\beta \qquad \text{and} \qquad E[s^3] = (\beta^2 + 3\beta\mu + 2\mu^2)/\beta^2\mu^3.$$

Then verify the equations of Table 13, Appendix C.

[*Hint:* By Table 2 of Appendix A, $\psi(\theta) = (\beta/\beta - \theta)^\alpha$, and, by Theorem 2.9.1,

$$E[s^k] = \left. \frac{d^k\psi}{d\theta^k} \right|_{\theta = 0}.]$$

25. [12] Apply the results of Exercise 24 to an Erlang-k service time, which is a special case of the gamma distribution with $\alpha = k$ and $\beta = k\mu$, to obtain

$$E[s] = \frac{1}{\mu}, \qquad E[s^2] = \frac{(k+1)}{k}\frac{1}{\mu^2} = \frac{(k+1)}{k}E[s]^2,$$

$$\mathrm{Var}[s] = 1/k\mu^2 = E[s]^2/k, \qquad C_s^{\,2} = 1/k,$$

$$E[s^3] = \frac{(k+1)(k+2)}{k^2}E[s]^3.$$

Then verify the equations of Table 14, Appendix C.

26. [C20] Suppose four communication lines are connected to one computer and that each line has an average incoming message transmission time of 2 seconds with a utilization of 0.6. The message transmission time is gamma with parameters $\alpha = 1/3$, $\beta = 1/6$ for the first line; it is exponential for the second line. The third line has an Erlang-3 service time, while it is constant for the fourth line. Calculate W_q, W, σ_w, L_q, and L for each line. Then estimate $\pi_w(90)$ by
(a) Martin's estimate and
(b) Tables 5.3.1 and 5.3.2.

27. [HM20] Show that, for the GI/M/1 queueing system, given that

$$P[q \le t] = 1 - (1 - \pi_0)e^{-t/W} = 1 - (1 - \pi_0)e^{-\pi_0 t/E[s]},$$

then
(a) $E[q] = (1 - \pi_0)E[s]/\pi_0$,
(b) $E[q^2] = 2(1 - \pi_0)(E[s]/\pi_0)^2$,
(c) $\mathrm{Var}[q] = (1 - \pi_0^2)(E[s]/\pi_0)^2$.

28. [HM20]   If $\pi(z)$ is given by (5.3.16), show that

$$L = \pi'(1) = \rho + \frac{K''(1)}{2(1 - \rho)}.$$

29. [12]   Consider Example 5.2.10. Suppose that all the parameters are the same as given except that RPS is not used. Find the maximum number of inquiries per second that the channel can handle. Use a read time of 6 milliseconds.

30. [C18]   Suppose 30 buffered terminals on one communication line are used for data entry to a computer system. (Buffered terminals are those in which entries are first keyed into a local memory and then transmitted as one message when the line is free to accept it.) The average time to key in an entry is 60 seconds; the keying time has an exponential distribution. The average system response time (line, both ways, plus computer time) is 2 seconds; the response time is also exponential.

(a)   Find the line utilization, the mean rate entries can be processed (λ), and the average queueing time for an entry (time spent waiting for the line to become free). If you have APL available you should compare your solution using MACHΔREP or direct calculation to that obtained by Fig. 3 of Appendix C.

(b)   Repeat the calculations for 40 terminals on the line.

(c)   For 50 terminals on the line.

31. [HM10]   Suppose the interarrival time has an Erlang-k distribution. Show that the Laplace–Stieltjes transform of τ, $A^*(\theta)$, is given by

$$A^*(\theta) = \left(\frac{k\lambda}{k\lambda + \theta}\right)^k, \qquad \theta < k\lambda.$$

[*Hint:* Recall that an Erlang-k random variable with parameters λ can be represented as the sum of k independent, exponential random variables, each with parameter $k\lambda$.]

32. [HM15]   Show that if the interarrival time in a GI/M/1 queueing system is uniformly distributed over the interval 0 to $2/\lambda$, then the Laplace–Stieltjes transform of τ, $A^*(\theta)$, is

$$A^*(\theta) = (\lambda/2\theta)(1 - e^{-2\theta/\lambda}).$$

33. [HM18]   Consider the interarrival time distribution for which the π_0 values are shown in the next to last column of Table 17, Appendix C. That is, a two-stage hyperexponential distribution with (in the notation of Section 3.2.9) $\alpha_1 = 0.4$, $\alpha_2 = 0.6$, $\mu_1 = 0.5\lambda$, and $\mu_2 = 3\lambda$. Show that $E[\tau] = 1/\lambda$ and (5.3.50) for π_0 reduces to

$$\pi_0{}^2 + (3.5\rho - 1)\pi_0 + 1.5\rho(\rho - 1) = 0.$$

This implies that the unique value of π_0 between 0 and 1 is given by

$$\pi_0 = 0.5 - 1.75\rho + (1.5625\rho^2 - 0.25\rho + 0.25)^{1/2}.$$

34. [HM20]   Show that, for a GI/M/1 queueing system

$$\text{Var}[N_q] = \frac{\rho(1 - \pi_0)\{2 - \pi_0 - \rho(1 - \pi_0)\}}{\pi_0{}^2}$$

[*Hint:* Use Theorem 2.9.2.]

35. [HM18]   Show that, for a GI/M/1 queue,

$$\text{Var}[N] = \frac{\rho(2 - \pi_0 - \rho)}{\pi_0{}^2}.$$

36. [15]   Suppose that in Exercise 6 the arrival pattern to the telephone booth is Erlang-2, rather than random, that the average interarrival time is 10 minutes, and that the length of phone calls is exponentially distributed with mean 3 minutes. Find (a) through (c) of Exercise 6.

37. [18]   Consider Example 5.4.1. Recalculate the values of Table 5.4.1 under the assumption that, for both the HOL queueing system and the preemptive-priority queueing system, Type 2 inquiries will be given preference over Type 1 inquiries.

38. [15]   Consider a communication line as a GI/G/1 queueing system in which $E[\tau] = 0.05$ seconds, $\text{Var}[\tau] = 0.003125$ seconds^2, $E[s] = 0.0475$ seconds, and $\text{Var}[s] = 0.001805$ seconds^2. Find $E[q]$, $E[w]$, $\pi_q(90)$, and the probability that the queueing time for a message will exceed 1.5 seconds. Check your results by approximating the system with a M/M/1 queueing system with the same average interarrival time and the same average service (transmission) time. Find the mean queueing time by the Allen–Cunneen approximation formula (5.5.6).

39. [12]   Consider a D/M/1 queueing system with $E[\tau] = 2$ seconds and $E[s] = 1.6$ seconds. Apply Theorem 5.5.3 to compute upper and lower bounds for $E[q]$. Then calculate the exact value using Tables 16 and 17 of Appendix C.

40. [12]   Consider an $H_2/H_2/1$ queueing system in which, for τ, $\alpha_1 = \alpha_2 = 0.5$, $1/\lambda_1 = 0.2$ seconds, and $1/\lambda_2 = 1.8$ seconds. Suppose, for s, $\alpha_1 = \alpha_2 = 0.5$ but $1/\mu_1 = 0.2$ seconds and $1/\mu_2 = 1.6$ seconds. Find upper and lower bounds for $E[q]$. Also compute the mean queueing time by the Allen–Cunneen approximation formula (5.5.6).

41. [10]   Let $E[q_1]$ be the average queueing time for GI/G/1 queueing system and $E[q_2]$ the average queueing time for a GI/G/c queueing system where ρ, $E[s]$, $C_s{}^2$, and $C_\tau{}^2$ are the same for both systems. Show that, if ρ is less than but close to 1, so that the heavy traffic approximation applies, then

$$E[q_2] = E[q_1]/c.$$

42. [HM20]   Suppose a two-stage hyperexponential distribution is generated using Algorithm 6.2.1 of Chapter 6. The distribution is to represent the interarrival time to a GI/M/1 queueing system with a given average interarrival time, $1/\lambda$, and a given squared coefficient of variation, $C^2 \geq 1$. Thus

$$\alpha_1 = \frac{1}{2}\left[1 - \left(\frac{C^2 - 1}{C^2 + 1}\right)^{1/2}\right],$$

$\alpha_2 = 1 - \alpha_1$, $\lambda_1 = 2\alpha_1\lambda$, and $\lambda_2 = 2\alpha_2\lambda$. Substitute these values into (5.3.74) and show that π_0 is given by the formula (5.3.76). Note that you must prove that, if $0 < \rho < 1$, then $0 < \pi_0 < 1$.

References

1. J. D. C. Little, A proof of the queueing formula: $L = \lambda W$, *Operations Res.* **9** (3), (1961), 383–387.
2. Lajos Takács, *Introduction to the Theory of Queues*. Oxford University Press, London and New York, 1962.
3. W. Feller, *An Introduction to Probability and Its Applications*, Vol. II, 2nd ed. Wiley, New York, 1971.
4. D. N. Streeter, Centralization or dispersion of computing facilities, *IBM Systems J.* **12** (3), (1973), 283–301.
5. E. Parzen, *Stochastic Processes*. Holden–Day, San Francisco, 1962.
6. D. Gross and C. M. Harris, *Fundamentals of Queueing Theory*. Wiley, New York, 1974.
7. J. Martin, *Systems Analysis for Data Transmission*. Prentice–Hall, Englewood Cliffs, New Jersey, 1972.
8. Analysis of some queueing models in real-time systems, IBM Report Number GF20-0007-1, IBM Data Processing Division, 1133 Westchester Avenue, White Plains, New York 10604.
9. S. Karlin, *A First Course in Stochastic Processes*. Academic Press, New York, 1969.
10. L. Kleinrock, *Queueing Systems, Volume I: Theory*. Wiley, New York, 1975.
11. L. Takács, A single-server queue with Poisson input, *Operations Res.* **10**, (1962), 388–397.
12. L. Kleinrock, *Queueing Systems, Volume II: Computer Applications*. Wiley, New York, 1976.
13. J. F. C. Kingman, On queues in heavy traffic, *J. Roy. Statist. Soc. Ser. B* **24**, (1962), 383–392.
14. J. Köllerström, Heavy traffic theory for queues with several servers. I, *J. Appl. Probability* **11**, (1974), 544–552.
15. W. G. Marchal, Some simple bounds and approximations in queueing, Technical Memorandom Serial T-294, Institute for Management Science and Engineering, The George Washington University, Washington, D.C., January 1974.
16. S. L. Brumelle, Bounds on the wait in a GI/M/k queue, *Management Sci.* **19** (7), (1973), 773–777.
17. T. Suzuki and Y. Yoshida, Inequalities for many-server queue and other queues, *J. Operations Res. Soc. Japan* **13**, (1970), 59–77.
18. J. F. C. Kingman, Inequalities in the theory of queues, *J. Roy. Statist. Soc. Ser. B* **32**, (1970), 102–110.
19. F. S. Hillier and F. D. Lo, Tables for multiserver queueing systems involving Erlang distributions, Tech. Rep. 31, December 28, 1971, Department of Operations Research, Stanford University, Stanford, California.
20. A. O. Allen, Elements of queueing theory for system design, *IBM Systems J.* **14** (2), (1975).
21. N. C. Wilhelm, A general model for the performance of disk systems, *J. ACM* **24** (1), (1977), 14–31.

All knowledge resolves itself into probability.
David Hume

Chapter Six

QUEUEING THEORY MODELS
OF COMPUTER SYSTEMS

6.1 INTRODUCTION

In Chapter 5 we examined basic queueing theory and applied the theory to selected parts of computer systems. The queueing systems we studied were essentially single resource systems; that is, there was but one service facility, although in some cases, there were multiple identical servers in the facility. Actual computer systems are multiple resource systems. Thus we may have on-line terminals, communication lines, line concentrators, and communication controllers as well as the computer itself. The computer has multiple resources, too, including main storage, channels, input/output (I/O) devices, etc. There may be a queue associated with each of these resources in a computer system. Thus a computer system is a network of queues. In this chapter most of the models studied are multiple resource models; that is, they consist of a network of the simple queueing systems we considered in Chapter 5. By "network of queues" we mean that the input(s) to one queueing system may be the output(s) from one or more other queueing system(s). Unfortunately very little can be done, analytically, with general queueing networks such as would be needed to model computer systems in such a way as to account for every resource (and every queue). Fortunately, however, a number of useful queueing network models do exist for modeling computer systems. We will study several of them in this chapter.

6.2   INFINITE MODELS

In most of the queueing models considered in Chapter 5 it was assumed the customer population was infinite (the exception was the machine repair models of Sections 5.2.6 and 5.2.7) and in most cases we also assumed that the arrival process was Poisson. The infinite population assumption merely asserts that the number of customers in the system has no effect on the arrival rate. In many of the models we study in this chapter this is not true; actually, there are no real world infinite queues. However, there are some infinite population models which do reasonably approximate computer systems. For an excellent discussion of the use of infinite source models to represent finite source computer systems see Buzen and Goldberg [1].

6.2.1   M/G/1 Processor-Sharing Queueing System

This model had its genesis in the analytical model of Kleinrock [2] for the first time-sharing computer systems, which used the round-robin algorithm for allocating the CPU resource to users. This model is shown in Fig. 6.2.1. Customers were assumed to arrive at the processor in a Poisson stream. Each arriving customer entered the single queue and waited in a FCFS fashion for a quantum Δs of CPU service (the size of the quantum was fixed for all customers). When the quantum of service was used up, if the

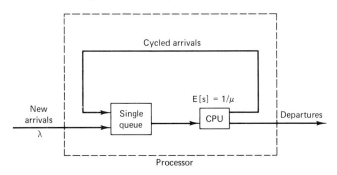

Fig. 6.2.1   The round-robin time-sharing system.

customer required more service, a return was made to the tail of the same queue to repeat the cycle. The cycle was repeated by the customer until the required service was received, whereupon the customer departed. Kleinrock [3] discovered that by letting the quantum Δs shrink to zero he got an analytical model with much simpler expressions for the performance measures but which was a good approximation to a round-robin system with a finite quantum. Such a system is called a *processor-sharing* system since, if there are k customers in the system, each is receiving the fraction $1/k$ of the processor capacity; that is the customers are sharing the processor equally. This model is described in great detail by Kleinrock [4]. The processor-sharing model is combined with other models in Section 6.3. We now give a formal description of the M/G/1 processor-sharing model.

The M/G/1 processor-sharing queueing system has a Poisson arrival pattern with average arrival rate λ. The service time distribution is general with average rate μ. The queue discipline is processor-sharing, which means each arriving customer immediately starts receiving his share of service, so there is no queue. Thus, if a customer arrives when there are already $n - 1$ customers in the system, the customer receives service at the average rate μ/n. The following steady state relations hold, when $\rho = \lambda/\mu < 1$:

$$p_n = P[N = n] = (1 - \rho)\rho^n, \qquad n = 0, 1, 2, \ldots, \qquad (6.2.1)$$

(The number of customers in the system is geometrically distributed!)

$$L = E[N] = \rho/(1 - \rho), \qquad (6.2.2)$$

$$E[w \mid s = t] = t/(1 - \rho), \qquad (6.2.3)$$

$$W = E[w] = E[s]/(1 - \rho). \qquad (6.2.4)$$

Although there is no queue (and thus no queueing time), a customer requiring t units of service time suffers a delay because the full capacity of the CPU is not available. The average of this delay, which we denote by $E[q \mid s = t]$, is the difference between $E[w \mid s = t]$ and t, the (full capacity) amount of CPU time needed. Thus we have the following:

$$E[q \mid s = t] = \rho t/(1 - \rho), \qquad (6.2.5)$$

$$E[q] = \rho E[s]/(1 - \rho). \qquad (6.2.6)$$

Here $E[q]$ is the average delay experienced by customers.

These equations are exactly the same as the corresponding equations for the M/M/1 queueing system! In addition the departure stream is Poisson. The distribution of w and q is not known in general. However, for the special case of exponential service time, Coffman et al. [5] have derived the Laplace transform of the conditional delay and its variance.

Equations (6.2.1)–(6.2.6) show that this model has some very nice properties. Equation (6.2.3) shows that the mean conditional response time is a linear function of required service time. A customer who requires twice as much service time as another will, on the average, spend twice as much time in the system. The mean response time W, by (6.2.4), is independent of the service time distribution and depends only upon its mean value. This contrasts with the M/G/1 queueing system in which, by Pollaczek's formula,

$$W = \frac{\lambda E[s^2]}{2(1 - \rho)} + E[s],$$

the average response time clearly depends upon the second moment of the service time. Other nice properties of the model are discussed by Kleinrock [4].

The M/G/1 processor-sharing queueing system can be used to approximate some time-shared computer systems.

Example 6.2.1 We consider a simplified form of an example given by Reiser [6, p. 317] (we will solve Reiser's original example in our Example 6.3.2). A number of active terminals feed a Poisson stream of requests to a computer which operates approximately in processor-sharing mode with an average processing rate of 500,000 instructions/second. The average arrival rate $\lambda = 4.77$ customers (requests)/second and the average number of instructions per customer is 100,000 (including overhead). Find the average response time W and the average number of customers being processed by the CPU, $L = E[N]$. Also find the probability that there are 5 or more customers being processed by the computer.

Solution $E[s] = 100,000/500,000 = 0.2$ second. Thus

$$\rho = \lambda E[s] = 4.77 \times 0.2 = 0.954.$$

Hence, by (6.2.4),

$$W = E[s]/(1 - \rho) = 4.348 \quad \text{seconds.}$$

By (6.2.2),

$$L = E[N] = \rho/(1 - \rho) = 20.74.$$

With the geometric distribution of (6.2.1) we have

$$P[N \geq n] = \rho^n. \quad \text{(See Problem 1, Chapter 5.)}$$

Hence, $P[N \geq 5] = (0.954)^5 = 0.79$.

In the next section we consider a profound theorem about networks of queues due to Jackson [7].

6.2.2 Jackson's Theorem

THEOREM 6.2.1 (*Jackson's Theorem*) Suppose a queueing network consists of m nodes satisfying the following three conditions.

(i) Each node consists of an M/M/c queueing system where node i has c_i servers each of whom has average service time $1/\mu_i$.

(ii) Customers arriving at node i from outside the system arrive in a Poisson pattern with average rate λ_i. (Customers also arrive at node i from other nodes within the network.)

(iii) Once served at node i, a customer goes (instantly) to node j $(j = 1, 2, \ldots, m)$ with probability p_{ij}; or leaves the network with probability $1 - \sum_{j=1}^{m} p_{ij}$.

For each node i $(i = 1, 2, \ldots, m)$ it can be seen that the average arrival rate to the node, Λ_i, is given by

$$\Lambda_i = \lambda_i + \sum_{j=1}^{m} p_{ji} \Lambda_j. \tag{6.2.7}$$

Then, if we let $p(k_1, k_2, \ldots, k_n)$ denote the steady state probability that there are k_i customers in the ith node $(i = 1, 2, \ldots, m)$, and, if $\Lambda_i < c_i \mu_i$ for all i, it is true that

$$p(k_1, k_2, \ldots, k_n) = p_1(k_1)p_2(k_2) \cdots p_m(k_m), \tag{6.2.8}$$

where $p_i(k_i)$ is the steady state probability there are k_i customers in the ith node if it is treated as an M/M/c_i queueing system with average arrival rate Λ_i and average service time $1/\mu_i$ for each of the c_i servers; the formula is given by (5.2.56). Furthermore, each node i behaves as if it were an independent M/M/c_i queueing system with average Poisson arrival rate Λ_i.

The proof of this theorem can be found in Jackson [7]. As Kleinrock [4] notes, the arrival stream to a node in a Jackson-type queueing network need not necessarily be Poisson but the theorem says we can operate as though it were.

Example 6.2.2 Consider the simple M/M/1 feedback queue of Fig. 6.2.2. Suppose the M/M/1 queueing system represents a message switching facility which transmits messages to the required destination, the time to transmit a message and receive an acknowledgment of correct receipt (assuming some type of error detecting code is used) being exponential. The probability a message is transmitted correctly is p; with probability $1 - p$ the message must be retransmitted. Then, by Jackson's theorem,

$$\Lambda = \lambda + (1 - p)\Lambda, \quad \text{or} \quad \Lambda = \lambda/p, \quad \text{and} \quad \rho = \Lambda E[s] = \lambda/p\mu.$$

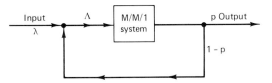

Fig. 6.2.2 Simple feedback queueing system of Example 6.2.2.

Since we have an $M/M/1$ system,

$$L = \rho/(1 - \rho) = \lambda/(p\mu - \lambda), \qquad \text{and} \qquad W = E[s]/(1 - \rho) = p/(p\mu - \lambda).$$

Consider the message switching center of Example 5.2.4 where $\lambda = 4$ messages/second and $E[s] = 0.22$ seconds. Suppose it is the basis for our present system and the probability p of correctly transmitting a message is 0.99. Then we have $\Lambda = \lambda/p = 4.0404$,

$$\rho = \Lambda E[s] = 0.8889,$$

$$L = \rho/(1 - \rho) = 8 \quad \text{messages},$$

$$W = E[s]/(1 - \rho) = 1.98 \quad \text{seconds}.$$

The numbers for the system without feedback were $L = 7.33$ messages and $W = 1.83$ seconds.

A number of queueing theory models have been used to study multiprogrammed computer systems. In such a system several programs are stored simultaneously in the main memory. Each program consists of a sequence of CPU and I/O instructions. While an I/O unit is processing some input or output for one program, which cannot execute any more CPU instructions until the I/O is completed, the CPU processes another program. The execution of a program in such a system is characterized by its cyclical movement between the CPU, and the I/O units until execution is completed; the program then leaves the system. Thus a program in main memory is in one of four states: waiting for the CPU, in CPU execution, waiting for I/O, or in I/O execution. Hence such a system can be modeled by the two-stage cyclic queueing network shown in Fig. 6.2.3.

In the preceeding paragraphs we described multiprogramming in a computer system without virtual memory. However, the model shown in Fig. 6.2.3 can also be used to study a computer having virtual memory operating under demand paging. In such a system both main memory and programs are partitioned into pieces of equal size called pages. Each program in main memory is assigned a fixed number of pages of memory (usually fewer than the number of pages of the program). If a program in CPU execution references a page not in main memory (we say a " page fault " has occurred) the missing page is brought in from secondary memory to replace a page in

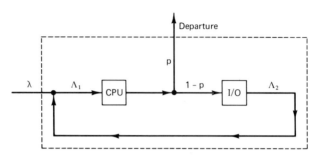

Fig. 6.2.3 **Two-stage cyclic queueing model of multiprogramming.**

main memory. While this paging activity occurs, the CPU processes another program. Eventually the original program returns to CPU execution until it experiences another page fault. Hence, the cyclic queueing network of Fig. 6.2.3 can be used to study this multiprogramming model, also, with paging corresponding to the I/O activity (which is what paging is).

In this cyclic queueing model we assume a Poisson arrival process with average arrival rate λ. It is also assumed that both the CPU and the I/O unit provide exponential service with average service times $1/\mu_1$ and $1/\mu_2$, respectively. An arriving program queues for service at the CPU. When a CPU service is completed, the program either leaves the system (with probability p) or queues for an I/O service (with probability $1 - p$). When the I/O service is over, the program rejoins the CPU queue. This cycle is repeated until the program completes execution. In this model the queues are assumed to have infinite capacity; there is no blocking due to lack of space in a queue.

We can calculate the steady state statistics of this model by using Jackson's theorem. Let Λ_1 and Λ_2 be the average arrival rates to the CPU and I/O unit, respectively. Then we have

$$\Lambda_1 = \lambda + \Lambda_2 = \lambda + (1 - p)\Lambda_1 \qquad (6.2.9)$$

or

$$\Lambda_1 = \lambda/p. \qquad (6.2.10)$$

Clearly

$$\Lambda_2 = (1 - p)\Lambda_1 = (1 - p)\lambda/p. \qquad (6.2.11)$$

The respective server utilizations for the CPU and I/O units are calculated by

$$\rho_1 = \Lambda_1/\mu_1 = \lambda/p\mu_1, \qquad (6.2.12)$$

and

$$\rho_2 = \Lambda_2/\mu_2 = (1 - p)\lambda/p\mu_2. \qquad (6.2.13)$$

The average thruput λ_T, measured in jobs per unit time, is given by

$$\lambda_T = p\Lambda_1 = \lambda. \tag{6.2.14}$$

It is clear that (6.2.14) must be true if no jobs are lost. By Jackson's theorem, the probability there are n_1 programs in the CPU and n_2 in the I/O unit (queueing or in service) is given by

$$p(n_1, n_2) = p(n_1)p(n_2) = (1 - \rho_1)\rho_1{}^{n_1}(1 - \rho_2)\rho_2{}^{n_2}. \tag{6.2.15}$$

The average number L in the system is given by

$$L = E[N] = E[N_1] + E[N_2] = \frac{\rho_1}{1 - \rho_1} + \frac{\rho_2}{1 - \rho_2}. \tag{6.2.16}$$

The average response time, by Little's formula (5.1.9), is

$$W = L/\lambda. \tag{6.2.17}$$

Example 6.2.3 Transylvania University has a batch computer system they model using the cyclic queueing system of Fig. 6.2.3. Each of their jobs, which are student generated, use, on the average, 4 seconds of CPU time with an I/O interrupt every 0.25 seconds of CPU time; each I/O service has a mean time of 0.2 seconds. Transylvania's computer is in operation 10 hours/day, processing an average of 8000 jobs. Find (a) the utilization of the CPU and of the I/O unit, (b) the average number of jobs in the CPU queue (including the job being executed), (c) the average number of jobs in the I/O queue (including the job receiving I/O service), (d) the average number of jobs in the queueing network, and (e) the average response time of the computer system.

Solution $\lambda = 8000/36,000 = 2/9$ jobs/second. The average number of I/O interrupts is $4/0.25 = 16$, so $p = \frac{1}{16}$. Hence, $\Lambda_1 = \lambda/p = 16 \times \frac{2}{9} = \frac{32}{9}$ jobs/second and $\Lambda_2 = \frac{15}{16}\Lambda_1 = \frac{10}{3}$ jobs/second.

(a) Thus $\rho_1 = \Lambda_1 E[s_1] = \frac{32}{9} \times \frac{1}{4} = \frac{8}{9}$

and

$$\rho_2 = \Lambda_2 E[s_2] = \frac{10}{3} \times \frac{1}{5} = \frac{2}{3}.$$

(b) The average number of jobs in the CPU queueing system is

$$E[N_1] = \rho_1/(1 - \rho_1) = 8 \quad \text{jobs.}$$

(c) The average number in the I/O queueing system is

$$E[N_2] = \rho_2/(1 - \rho_2) = 2 \quad \text{jobs.}$$

(d) $L = E[N] = E[N_1] + E[N_2] = 10 \quad \text{jobs.}$
(e) The average response time, by (6.2.17), is

$$L/\lambda = 10 \times \frac{9}{2} = 45 \quad \text{seconds!}$$

This response time seems huge for a job that requires only 4 seconds of CPU time and $15 \times 0.2 = 3$ seconds of I/O time. However, the job requires 16 passes through the CPU, each requiring $E[w_1] = E[s_1]/(1 - \rho_1) = 2.25$ seconds, for 36 seconds, plus 15 passes through the I/O unit, each requiring

$$E[w_1] = E[s_2]/(1 - \rho_2) = 0.6 \quad \text{seconds}$$

for 9 seconds of I/O time. This totals to the 45 seconds calculated above.

6.2.3   Other Two-Stage Cyclic Queueing Models

One characteristic of most multiprogramming computer systems is that the number of programs kept in main memory (the level of multiprogramming) is constant; whenever a program completes execution, another replaces it in main memory. The cyclic queueing model we used to study multiprogramming systems in the last section did not have this feature. For our next system we use the model shown in Fig. 6.2.3 but add the condition that the queueing network always contain exactly K customers (jobs).

This means that the incoming traffic must be sufficiently heavy that, whenever a job completes its service and departs, another is ready to take its place. The arrival pattern *into* the queueing network can no longer be regarded as Poisson; however it is an infinite arrival pattern.

The distribution of jobs in the system is completely specified by giving the number n in queue or service at the CPU, for then $K - n$ are in queue or service at the I/O unit. Let p_n be the probability there are n jobs queued for or in service at the CPU.

Mitrani [8] shows that formulas (6.2.18)–(6.2.21) are true:

$$p_n = \left(\frac{1 - r}{1 - r^{K+1}} \right) r^n, \qquad n = 0, 1, 2, \ldots, K, \qquad (6.2.18)$$

where $r = \mu_2/(1 - p)\mu_1$.

The server utilizations for the CPU and I/O unit are given by

$$\rho_1 = 1 - p_0 = (r - r^{K+1})/(1 - r^{K+1}), \qquad (6.2.19)$$

and

$$\rho_2 = 1 - p_K = (1 - r^K)/(1 - r^{K+1}), \qquad (6.2.20)$$

respectively.

The average response time is given by

$$K/\mu_1 p \rho_1. \qquad (6.2.21)$$

To get the throughput λ_T, which is the average rate at which jobs depart the system, we reason that the departure rate is $\mu_1 p$ when the CPU is busy and zero otherwise, so

$$\lambda_T = \mu_1 p(1 - p_0) + 0 p_0 = \mu_1 p \rho_1. \tag{6.2.22}$$

The equations of this model are shown in Table 22 of Appendix C.

Example 6.2.4   Consider Example 6.2.3. Suppose Transylvania University sets the multiprogramming level $K = 10$. Then

$$r = \mu_2/(1 - p)\mu_1 = 5/(\tfrac{15}{16} \times 4) = \tfrac{4}{3}.$$

Thus

$$\rho_1 = (r - r^{11})/(1 - r^{11}) = 0.9853, \qquad \rho_2 = (1 - r^{10})/(1 - r^{11}) = 0.738976,$$

and the average response time (time a job is in the system) is

$$K/\mu_1 p \rho_1 = 40.596 \quad \text{seconds.}$$

The average throughput $\lambda_T = \mu_1 p \rho_1 = 0.24632$ jobs/second.

The performance of the system is improved by fixing the level of multiprogramming at 10! However, to maintain this average throughput and thus satisfy the requirement (of this model) that a job is always waiting to enter main memory, Transylvania University must either (a) generate $0.24632 \times 36,000 = 8,868$ jobs/10-hour day or (b) push an average of 8,000 jobs through the system in $8,000/0.24632 \times 3600 = 9.02169$ hours.

The next cyclic queueing model we consider is shown in Fig. 6.2.4. As in the model just considered, there are a fixed number K of jobs in the queueing network; each job that completes service is immediately replaced by another one. The first server, which can be thought of as either the CPU or the I/O unit, has a general service time distribution with distribution function $F(\cdot)$ and mean rate μ. The second server provides exponentially distributed service with mean rate λ. Reiser and Kobayashi [9] have shown the following

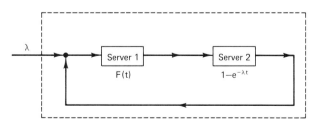

Fig. 6.2.4 Two-stage cyclic queueing model of multiprogramming with general service time distribution for first server; exponential for the second, fixed number of jobs (customers) in system.

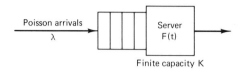

Poisson arrivals
λ

Server
F(t)

Finite capacity K

Fig. 6.2.5   M/G/1/K queueing system equivalent to queueing network of Fig. 6.2.4.

equivalence principle between the cyclic queueing model of Fig. 6.2.4 and the M/G/1/K (finite capacity) queueing system of Fig. 6.2.5: The joint probability distribution function for the queueing network of Fig. 6.2.4 is

$$p(n, K - n) = P[n \text{ jobs in queue 1 and } K - n \text{ jobs in queue 2}]$$

$$= p_K(n), \tag{6.2.23}$$

where $p_K(n)$ $(n = 0, 1, \ldots, K)$ is the probability there are n customers in the M/G/1/K queueing system shown in Fig. 6.2.5, which has the service time distribution $F(\cdot)$ and Poisson arrival rate λ.

Reiser and Kobayashi [9] also show the following:

$$p_K(n) = \begin{cases} C_K \hat{p}(n), & 0 \le n < K \\ 1 - [1 - C_K(1 - \rho)]/\rho, & n = K, \end{cases} \tag{6.2.24}$$

where $\rho = \lambda/\mu$, and C_K is defined to make the probabilities in (6.2.24) sum to 1 by

$$C_K = \left\{ 1 - \rho \left[1 - \sum_{n=0}^{K-1} \hat{p}(n) \right] \right\}^{-1}. \tag{6.2.25}$$

(The reader is warned that there are several troublesome missprints in [9], including several in the formulas we give as (6.2.24). Dr. Kobayashi has graciously verified that the formulas (6.2.24), (6.2.25), and (6.2.39)–(6.2.43) are correct.)

The values of $\hat{p}(n)$ can be computed by the formula

$$\hat{p}(n) = \frac{1}{n!} \frac{d^n P(z)}{dz^n} \bigg|_{z=0}. \tag{6.2.26}$$

The formula (6.2.26) is the same as the formula we developed in Theorem 2.9.2b when $P(z)$ was the generating function of the sequence $\{\hat{p}(n)\}$. $P(z)$ is then calculated by the formula

$$P(z) = \frac{(1 - \rho)(1 - z)W_s^*[\lambda(1 - z)]}{W_s^*[\lambda(1 - z)] - z}, \tag{6.2.27}$$

a formula we developed in Chapter 5 (formula (6.2.27) is our (5.2.30)), where $W_s^*(\theta)$ is the Laplace–Stieltjes transform of the service time distribution.

Thus, when $\rho < 1$, $\hat{p}(n)$ can be interpreted as the steady state probability of having n customers in the unrestricted M/G/1 queueing system with average arrival rate λ and the service time distribution $F(\cdot)$. Reiser and Kobayashi also demonstrate that the formula

$$\rho_1/\rho_2 = \lambda/\mu = \rho \tag{6.2.28}$$

relates the utilizations of the servers where

$$\rho_1 = 1 - p_K(0) \tag{6.2.29}$$

is the utilization of server 1 and

$$\rho_2 = 1 - p_K(K) \tag{6.2.30}$$

is the utilization of server 2. They derive (6.2.28) by showing that the average rate of jobs cycling through the system is $\rho_1 \mu = \rho_2 \lambda$. We can calculate the average number in each queueing subsystem by

$$E[N_1] = \sum_{n=1}^{K} np_K(n), \qquad E[N_2] = K - E[N_1]. \tag{6.2.31}$$

We could use (6.2.31) and Little's formula to calculate the average time a job spends in the CPU and in I/O for each cycle.

By Little's formula, the average time it takes a job to make a complete cycle is

$$K/\rho_1\mu = K/\rho_2\lambda. \tag{6.2.32}$$

If we know the average number of times job cycles through the system before completing execution, say m, we can calculate the average response time W by

$$W = mK/\rho_1\mu = mK/\rho_2\lambda, \tag{6.2.33}$$

and the average throughput (in jobs/unit time) λ_T by

$$\lambda_T = \text{average cycling rate}/m = \rho_1\mu/m = \rho_2\lambda/m. \tag{6.2.34}$$

Reiser and Kobayashi [9] give straightforward techniques for computing the values of $\hat{p}(n)$ if the service time of server 1 is either Erlang-2 or two-stage hyperexponential. Before we discuss their methods let us consider the case where $F(t) = 1 - e^{-\mu t}$ (server 1 has an exponential service time distribution). Then, if $\rho = \lambda/\mu \neq 1$, it is easy to show, using (6.2.24)–(6.2.26) (see Exercise 4) that

$$p_K(n) = (1 - \rho)\rho^n/(1 - \rho^{K+1}), \qquad n = 0, 1, \ldots, K, \tag{6.2.35}$$

$$\rho_1 = (\rho - \rho^{K+1})/(1 - \rho^{K+1}), \tag{6.2.36}$$

and

$$\rho_2 = (1 - \rho^K)/(1 - \rho^{K+1}). \tag{6.2.37}$$

(We developed (6.2.35) as (5.2.38) in Chapter 5, using another technique.)

Example 6.2.5   Let us apply this model to the system of Transylvania University (Example 6.2.4), assuming $K = 10$, $\mu = \mu_1 = 4$, and $\lambda = \mu_2 = 5$. Thus $\rho = 1.25$, so

$$\rho_1 = (\rho - \rho^{11})/(1 - \rho^{11}) = 0.97651,$$

and

$$\rho_2 = (1 - \rho^{10})/(1 - \rho^{11}) = 0.78121.$$

Since each job cycles through the system an average of 16 times, the average response time

$$W = mK/\rho_1 \mu = 32.71 \quad \text{seconds},$$

and

$$\lambda_T = \rho_1 \mu/m = 0.24413 \quad \text{jobs/second}.$$

The reader should compare these statistics to those obtained in Example 6.2.4.

The reader may recall from Chapter 3 that an Erlang-k distribution can often be used to approximate an empirical distribution which has a squared coefficient of variation less than or equal to one. That is, one can find an Erlang-k distribution with the same mean as the given distribution and a variance that is fairly close to the given one. To get an exact match of mean and variance a gamma distribution can be used, although gamma distributions are more difficult to use for probability calculations. In Algorithm H, which follows, we show how to construct a two-stage hyperexponential distribution with a given mean and squared coefficient of variation, provided the latter is greater than or equal to one.

Algorithm 6.2.1   (*Algorithm H*)   Given $C^2 \geq 1$ and $\mu > 0$, this algorithm will produce the parameters for a two-stage hyperexponential random variable X with squared coefficient of variation $C_X^2 = C^2$ and mean $E[X] = 1/\mu$. The distribution function of X will be given by

$$F(x) = 1 - \alpha_1 e^{-\mu_1 x} - \alpha_2 e^{-\mu_2 x}, \qquad x > 0. \tag{6.2.38}$$

(We use the notation of Section 3.2.9.)

Step 1   [*Calculate α_1 and α_2*]   Set $\alpha_1 = \frac{1}{2}\{1 - [(C^2 - 1)/(C^2 + 1)]^{1/2}\}$ and $\alpha_2 = 1 - \alpha_1$.

Step 2 [*Calculate* μ_1 *and* μ_2] Set $\mu_1 = 2\alpha_1\mu$ and $\mu_2 = 2\alpha_2\mu$.

Step 3 [*Produce F*] The distribution $F(\cdot)$, defined by (6.2.38) with parameters calculated in Step 1 and Step 2, is the distribution function of a two-stage hyperexponential random variable having the required properties.

Proof The proof is a simple exercise using (3.2.59), (3.2.60) and some algebra (see Exercise 5).

We now give the Reiser and Kobayashi solution for $\hat{p}(\cdot)$ when the service time of server 1 is Erlang-2 or two-stage hyperexponential. In these two cases we have

$$\hat{p}(n) = C_1 z_1^{-n} + C_2 z_2^{-n} \qquad \text{for all} \quad n, \tag{6.2.39}$$

where

$$C_1 = (1 - \rho z_2)(1 - z_1)/(z_2 - z_1) \tag{6.2.40}$$

$$C_2 = (1 - \rho z_1)(1 - z_2)/(z_1 - z_2) \tag{6.2.41}$$

and where z_1 and z_2 are the roots of the following equations:

$$\rho^2 z^2 - \rho(\rho + 4)z + 4 = 0 \tag{6.2.42}$$

for Erlang-2 service, and

$$\beta_1\beta_2 z^2 - (\beta_1 + \beta_2 + \beta_1\beta_2)z + 1 + \beta_1 + \beta_2 - \rho = 0 \tag{6.2.43}$$

for two-stage hyperexponential service.

In (6.2.43), $\beta_1 = \lambda/\mu_1$ and $\beta_2 = \lambda/\mu_2$, where μ_1 and μ_2 are the parameters of the distribution and satisfy the equation $(\alpha_1/\mu_1) + (\alpha_2/\mu_2) = 1/\mu$.

We illustrate the use of this cyclic queueing model of multiprogramming by the following example.

Example 6.2.6 Consider the cyclic queueing model of Fig. 6.2.4. Suppose $\rho = \lambda/\mu = 0.75$, with $\mu = 1$ customer/unit time and $\lambda = 0.75$ customers/unit time. (We will allow the reader to choose an agreeable unit of time.) Suppose, on the average, jobs require 10 cycles through the system and the level of multiprogramming $K = 4$. We illustrate the effect of the service time distribution of server 1 (which we usually think of as being the CPU but could represent I/O) on the performance by calculating the statistics of the model when the service time is (a) two-stage hyperexponential with squared coefficient of variation 5, (b) exponential, (c) Erlang-2, and (d) constant. The summary of the calculations for the $p_4(n)$ values are shown in Tables 6.2.1–6.2.4, while the performance measures for all the service times are summarized in Table 6.2.5.

TABLE 6.2.1

Summary of Probability Calculations for Two-Stage Hyperexponential CPU Service Time (Example 6.2.6)

n	$\hat{p}(n)$	$p_4(n)$
0	0.25000	0.39385
1	0.12829	0.20211
2	0.07830	0.12335
3	0.05642	0.08888
4	0.04567	0.19180

$\hat{p}(n) = C_1 z_1^{-n} + C_2 z_2^{-n}$ where
$C_1 = 0.19077$, $C_2 = 0.05923$,
$z_1 = 2.57245$ and $z_2 = 1.09421$,
$C_4 = 1.5754$.

TABLE 6.2.2

Summary of Probability Calculations for Exponential CPU Service Time (Example 6.2.6)

n	$\hat{p}(n)$	$p_4(n)$
0	0.25000	0.32778
1	0.18750	0.24584
2	0.14063	0.18438
3	0.10547	0.13829
4	—	0.10371

$C_4 = 1.3113$.

TABLE 6.2.3

Summary of Probability Calculations for Erlang-2 CPU Service Time (Example 6.2.6)

n	$\hat{p}(n)$	$p_4(n)$
0	0.25000	0.30775
1	0.22266	0.27409
2	0.16315	0.20084
3	0.11399	0.14032
4	—	0.07700

$\hat{p}(n) = C_1 z_1^{-n} + C_2 z_2^{-n}$ where
$C_1 = -0.106776$, $C_2 = 0.356776$,
$z_1 = 4.87449$ and $z_2 = 1.45884$,
$C_4 = 1.231$.

TABLE 6.2.4

Summary of Probability Calculations for Constant CPU Service
Time (Example 6.2.6)

n	$\hat{p}(n)$	$p_4(n)$
0	0.25000	0.28405
1	0.27925	0.31729
2	0.19424	0.22070
3	0.11667	0.13256
4	—	0.04540

$C_4 = 1.13621.$

TABLE 6.2.5

Summary of Performance of Multiprogramming
System with Various CPU Service Time Distributions
(Example 6.2.6)

CPU service	ρ_1	ρ_2	W	λ_T
H_2	0.60615	0.80820	65.99	0.060615
Exponential	0.67222	0.89629	59.50	0.067220
Erlang-2	0.69225	0.92300	57.78	0.069230
Constant	0.71595	0.95460	55.87	0.071600

To construct the two-stage hyperexponential distribution with $C_s^2 = 5$ we use Algorithm H as follows:

$$\alpha_1 = \tfrac{1}{2}(1 - \sqrt{\tfrac{2}{3}}) = 0.09175, \qquad \alpha_2 = 1 - \alpha_1 = 0.90825,$$

$$\mu_1 = 2\alpha_1\mu = 0.18350, \qquad \mu_2 = 2\alpha_2\mu = 1.81650,$$

$$\beta_1 = \lambda/\mu_1 = 0.75/\mu_1 = 4.08719, \qquad \beta_2 = \lambda/\mu_2 = 0.75/\mu_2 = 0.41288.$$

Substituting these values into (6.2.43) yields

$$1.6875z^2 - 6.1875z + 4.75 = 0. \qquad (6.2.44)$$

The roots of (6.2.44) are

$$z_1 = 2.57245 \qquad \text{and} \qquad z_2 = 1.09421.$$

Then, by (6.2.40) and (6.2.41),

$$C_1 = 0.19077 \qquad \text{and} \qquad C_2 = 0.05923.$$

The formula for $\hat{p}(n)$ is

$$\hat{p}(n) = C_1 z_1^{-n} + C_2 z_2^{-n}.$$

The results of the calculations for $\hat{p}(n)$ and $p_4(n)$ are shown in Table 6.2.1. The calculation of the probabilities for exponential CPU service are straightforward by (6.2.35), although we also show the values of $\hat{p}(n)$ in Table 6.2.2.

For Erlang-2 CPU service, (6.2.42) yields

$$0.5625z^2 - 3.5625z + 4 = 0, \tag{6.2.45}$$

which has roots

$$z_1 = 4.87449 \quad \text{and} \quad z_2 = 1.45884.$$

By (6.2.40) and (6.2.41) we have

$$C_1 = -0.106776 \quad \text{and} \quad C_2 = 0.356776.$$

The values of $\hat{p}(n)$ shown in Table 6.2.3 were calculated by the formula

$$\hat{p}(n) = C_1 z_1^{-n} + C_2 z_2^{-n}.$$

The values of $\hat{p}(n)$ for the constant service time distribution were the most onerous to compute. We used formulas (6.2.26) and (6.2.27) and the fact that, for a constant service time, $W_s^*(\theta) = e^{-\theta/\mu}$. The results are shown in Table 6.2.4. Table 6.2.5 shows that irregularity in service time is inimical to good performance. The throughput is 18.123% higher for constant service time than it is for two-stage hyperexponential.

Exponential CPU service and Erlang-2 CPU service provide performance between that for hyperexponential and constant, as expected.

Franta [10] discusses several other two-stage cyclic queueing models of computer systems.

6.3   FINITE MODELS

In Section 6.2 we considered infinite population queueing models of computer systems. This means that the number of customers already in the system has no affect on the arrival rate. The models we discussed in Section 6.2.3 could also be considered finite population models because there are a fixed number of customers in such a system. However, Fig. 6.3.1 portrays what is commonly known as a finite population queueing model of an interactive computer system (see Muntz [11]). The central processor system consists of one or more CPUs plus the associated queues. This central processor system could be represented by one of the queueing models of Section 6.2.3 if there is a queue for the central processor system and this queue is never empty. (This is the assumption in the "straightforward model" of Boyse and Warn [12], which we consider in Section 6.3.3.) The customers (users) are interacting with the system through the N terminals.

Terminals

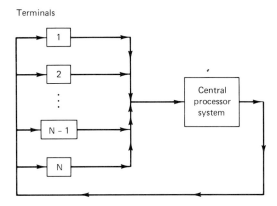

Fig. 6.3.1 Finite population queueing model of interactive computer system.

Each customer (user) is assumed to be in exactly one of three states at any instant of time: (1) "thinking" at the terminal (this time is called "think time t"), (2) queueing for CPU service, or (3) receiving CPU service. Thus a user at a terminal cannot submit a new request for CPU service until the previous request has been satisfied. In Fig. 6.3.1 the customer (user) can be represented as a token which circulates around the system and which at any instant is either at a terminal, in a CPU queue, or at a CPU. The model shown in Fig. 6.3.1 is sometimes called a *closed* model since no customer enters or leaves it. This contrasts with the *open* queueing models such as the ones shown in Figs. 6.2.1–6.2.3, in which customers enter the system and leave it. The particular queueing network we consider in Section 6.3 is determined by the model we select to enter in Fig. 6.3.1 for the central processor system.

6.3.1   Machine Repair Model

For this model the think time is assumed to be exponentially distributed with average value $1/\alpha$ and the single CPU central processor system is represented as a simple exponentially distributed service time; there is one queue for the central processor system. Thus the machine repair model $(M/M/1/K/K)$ of Section 5.2.6 applies. (The equations are also given in Table 10 of Appendix C.) (For a multiple CPU central processor system the multiple repairman machine repair model $(M/M/c/K/K)$ of Section 5.2.7 can be applied.) It seems incredible that the central processor system of a time-sharing system could be modeled as a simple exponential distribution, but Scherr [13] successfully used this model to analyze the Compatible Time-Sharing System (CTSS) at MIT. This was an early time-sharing system in which user programs were swapped in and out of main memory with only one complete program in memory at a time. Since there was no overlap of program execution and swapping, Scherr used the sum of program execution time and swapping time as the CPU service time. The machine repair analytic model gave results that were very close to those for a simulation model and to actual measured values.

For this model, since the operating time for machines corresponds to think time, with average think time $1/\alpha$, we have, by (5.2.89), the mean response time

$$W = (N/\lambda) - (1/\alpha). \tag{6.3.1}$$

But $\lambda = \rho/E[s]$, and, by (5.2.83),

$$\rho = 1 - p_0. \tag{6.3.2}$$

Therefore, the mean response time can be written as

$$W = \frac{NE[s]}{1 - p_0} - \frac{1}{\alpha}, \tag{6.3.3}$$

where, by (5.2.82),

$$p_0 = \left[\sum_{n=0}^{N} \frac{N!}{(N-n)!} \left(\frac{\alpha}{\mu} \right)^n \right]^{-1}. \tag{6.3.4}$$

Example 6.3.1 SLOBOVIAN SCIENTIFIC has an interactive time-sharing system of 20 active terminals which can be studied by the machine repair model. The average CPU service time, including swapping, is 2 seconds, while the mean think time is 20 seconds. Find p_0, ρ, λ, and the average response time W. Note that λ is the average throughput in interactions per second. What would be the effect of adding five terminals?

Solution For 20 terminals

$$p_0 = \left[\sum_{n=0}^{20} \frac{20!}{(20-n)!} \left(\frac{2}{20} \right)^n \right]^{-1} = 0.001869.$$

Then,

$$\rho = 1 - p_0 = 0.998131, \qquad \lambda = 0.49907 \quad \text{interactions/second},$$

and

$$W = (20/0.49907) - 20 = 20.075 \quad \text{seconds}.$$

With 25 terminals

$$p_0 = 0.00002927, \qquad \rho = 1 - p_0 = 0.99997073,$$

$$\lambda = \rho/E[s] = 0.499985365, \qquad \text{and} \qquad W = (25/\lambda) - 20 = 30 \quad \text{seconds}.$$

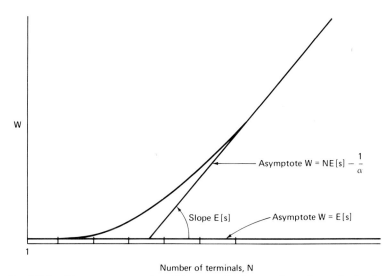

Fig. 6.3.2 Mean response time *W* versus N, the number of terminals, for the model of Fig. 6.3.1.

Thus the addition of 5 terminals has increased the average throughput by only 0.18% while increasing the mean response time by 49.44%.

This example illustrates the concept of system saturation. Consider Fig. 6.3.2, the graph of the mean response time $W = \{NE[s]/(1 - p_0)\} - (1/\alpha)$, versus N. For $N = 1$ there is no queueing so $W = E[s]$. For small values of N, the customers interfere with each other very little; that is, when one person wants a CPU interaction the others are usually in think mode so little queueing occurs. Thus the curve is asymptotic at $N = 1$ to the line $W = E[s]$. As $N \to \infty$, $p_0 \to 0$ since the likelihood of the CPU being idle must go to 0. Hence the curve is asymptotic to the line $NE[s] - (1/\alpha)$ as $N \to \infty$. Clearly the two asymptotes intersect where

$$N = N^* = \frac{E[s] + (1/\alpha)}{E[s]} = \frac{E[s] + E[t]}{E[s]}.$$

Kleinrock [4] calls N^* the *system saturation point*. He points out that, if each interaction required exactly $E[s]$ units of CPU service time and exactly $E[t]$ units of think time, then N^* is the maximum number of terminals that could be scheduled in such a way as to cause no mutual interference. For $N \ll N^*$ there is almost no mutual interference, and W is approximately $E[s]$. For $N \gg N^*$ users " totally interfere" with each other; that is the addition of a terminal raises everyone's average response time by $E[s]$. In Example 6.3.1, $N^* = 22/2 = 11$ terminals, and the increase in W due to the change from 20 terminals to 25 terminals was close to $5 \times 2 = 10$ seconds (actually it was 9.963 seconds).

Lassettre and Scherr [14] have also successfully used the machine repair model to develop the OS/360 time-sharing option (TSO).

Kobayashi [25] shows equations (6.3.1)–(6.3.3) hold for any system like that shown in Fig. 6.3.1 providing (a) the equilibrium or steady state of the system exists and (b) the queue discipline is "work conserving" in the sense that the service times of the individual requests are not affected by it. Thus these equations have a simple form independent of the distributional form of t and s. Of course the value of p_0 may depend upon the distributional form of t and s as well as on the queue discipline; we shall see this in two special cases of this model described in Sections 6.3.2 and 6.3.3.

6.3.2   Finite Processor-Sharing Model

This model is Fig. 6.3.1 with a single CPU operating with the processor-sharing queue discipline; that is, the CPU operates as an M/G/1 processor-sharing system but with the finite input shown in Fig. 6.3.1. Kleinrock [4] shows that if the CPU service time is only restricted to the extent that the Laplace–Stieltjes transform $W_s^*(\theta)$ is rational (the ratio of two polynomials in θ), and similarly for the think time, then exactly the same equations hold as we found in the last section for the exponential CPU service time with FCFS queue discipline. That is

$$W = \frac{NE[s]}{1 - p_0} - \frac{1}{\alpha}, \tag{6.3.5}$$

where $1/\alpha$ is the average think time and

$$p_0 = \left[\sum_{n=0}^{N} \frac{N!}{(N-n)!} \left(\frac{\alpha}{\mu}\right)^n \right]^{-1}. \tag{6.3.6}$$

Also

$$\rho = 1 - p_0, \tag{6.3.7}$$

and

$$\lambda = \rho/E[s]. \tag{6.3.8}$$

Of course all the other equations for the machine repair queueing system (M/M/1/K/K) hold as well, with K replaced by N and O by think time t.

Example 6.3.2   Let us consider the example on page 317 of Reiser [6], which he solves using a very sophisticated APL program called QNET4. He considers a finite processor-sharing model with 20 active terminals, $1/\alpha = 3$ seconds, a CPU average service rate of 500,000 instructions/second and an average interaction requirement of 100,000 instructions. Thus $E[s] = 100,000/500,000 = 0.2$ seconds. Hence $E[s]/E[t] = 0.2/3 = 1/15$, and

$$p_0 = \left[\sum_{n=0}^{20} \frac{20!}{(20-n)!}\left(\frac{1}{15}\right)^n\right]^{-1} = \left[20!\sum_{n=0}^{20} \frac{1}{(20-n)!}\left(\frac{1}{15}\right)^n\right]^{-1}$$

$$= 0.045593216.$$

The mean response time

$$W = \frac{NE[s]}{1-p_0} - \frac{1}{\alpha} = 1.191 \quad \text{seconds}$$

agrees with Reiser's solution as does the average throughput

$$\lambda_T = \rho/E[s] = (1-p_0)/E[s] = 4.772 \quad \text{interactions/second.}$$

The average number in the central processor system

$$\lambda_T W = 5.6835, \quad \text{and} \quad \rho = 1 - p_0 = 0.954407$$

also agrees with Reiser's solution.

Note that in the simplified form of Reiser's example considered in Example 6.2.1 we assumed the value of λ_T, whereas in this example we actually had to calculate it. Note also that in the infinite processor-sharing model used in Example 6.2.1 we computed $W = 4.348$ seconds and the average number in the CPU system as 20.74! This agrees with the results of Buzen and Goldberg [1] who show the infinite source approximation is poor for high server utilization.

6.3.3 The Straightforward Model of Boyse and Warn

Boyse and Warn [12] have developed a computer performance prediction model that is very useful for certain kinds of computer systems. The model, which is shown in Fig. 6.3.3, is essentially that of Fig. 6.3.1 with the "central processor system" box replaced by the system inside the dashed lines. The assumptions made in the Boyse and Warn model are as follows.

(1) A fixed multiprogramming level K, which implies the system is heavily loaded so the queue for main memory, is never empty.
(2) Multiple CPUs or a single CPU with each treated as a single server.
(3) There are K parallel I/O servers, which implies there is no queueing for I/O service.

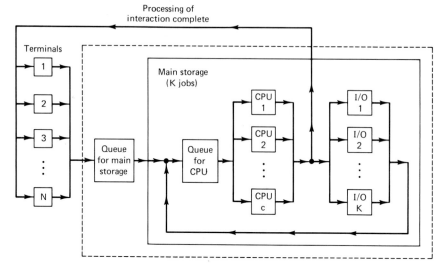

Fig. 6.3.3   The "straightforward model" of Boyse and Warn [12].

(4)   There is either (a) exponential service time at both the CPU(s) and the *I/O* stations or (b) constant service time at both the CPU(s) and the *I/O* stations.

(5)   There is no CPU and *I/O* overlap, which means the only *I/O* modeled is page reads.

(6)   Think time has a general distribution with mean $E[t] = 1/\alpha$.

Of these assumptions, (3) and (5) seem most subject to question. In the actual computer system studied by Boyse and Warn these conditions were close to reality because of the number of data paths available for *I/O*. If the assumption of no queueing for *I/O* is not quite true, the average *I/O* time can be adjusted to correct for this fact. If there is a great deal of queueing for *I/O*, this model would not apply but the central server model of Section 6.3.4 can be used. The only *I/O* explicitly considered in the model is that due to paging; an *I/O* operation begins when a page fault occurs and the job in CPU execution must be terminated until the required page is available in main memory. All other *I/O* is assumed to be overlapped with CPU activity.

The parameters necessary to specify the model are $E[s]$ the average CPU usage interval between page faults, m the average number of CPU intervals required per job (interaction) (thus $mE[s]$ is the average CPU time per customer interaction), and $E[O]$ the average service time of an *I/O* request. K is both the multiprogramming level (assumed to be constant) and the number of parallel *I/O* servers. N is the number of active terminals, and $E[t] = 1/\alpha$ is the average think time.

The principal output statistics are ρ, the average CPU utilization; λ_T, the average throughput in interactions per unit time; and the average response time, W. W is, of course, the average time from the submission of a request for a CPU interaction until the interaction is completed.

The first step in the analysis is to calculate the CPU utilization, ρ. The assumptions made ensure that the subsystem consisting of the I/O and the CPU(s) form a machine repair model with the I/O units playing the part of the machine and the CPU(s) playing the part of the repairman or repairmen. Boyse and Warn solve for ρ in the $D/D/c/K/K$ machine repair queueing system yielding

$$\rho = \min\left\{\frac{K}{c(1 + E[O]/E[s])}, 1\right\}. \tag{6.3.9}$$

(For completeness we give all of the equations for the $D/D/c/K/K$ machine repair problem in Table 25 of Appendix C.)

For the exponential case (exponential I/O service and exponential CPU service) we can use the $M/M/c/K/K$ machine repair equations of Table 11, Appendix C, to conclude that

$$\rho = \lambda E[s]/c, \tag{6.3.10}$$

where to get λ we must compute, successively,

$$\frac{p_n}{p_0} = \begin{cases} \binom{K}{n}\left(\dfrac{E[s]}{E[O]}\right)^n, & n = 1, 2, \ldots, c \\ \dfrac{n!}{c!\, c^{n-c}}\binom{K}{n}\left(\dfrac{E[s]}{E[O]}\right)^n, & n = c + 1, \ldots, K, \end{cases} \tag{6.3.11}$$

$$p_0 = \left[1 + \sum_{n=1}^{K}\left(\frac{p_n}{p_0}\right)\right]^{-1}, \tag{6.3.12}$$

$$p_n = \left(\frac{p_n}{p_0}\right)p_0, \qquad n = 1, 2, \ldots, c, \tag{6.3.13}$$

$$L_q = \sum_{n=c+1}^{K}(n - c)p_n, \tag{6.3.14}$$

$$W_q = \frac{L_q(E[O] + E[s])}{K - L_q}, \tag{6.3.15}$$

and, finally,

$$\lambda = \frac{K}{E[O] + W_q + E[s]}. \tag{6.3.16}$$

If there is just one CPU ($c = 1$) then (6.3.10)–(6.3.16) can be reduced to the two equations

$$p_0 = \left[\sum_{n=0}^{K} \frac{K!}{(K-n)!} \left(\frac{E[s]}{E[O]}\right)^n\right]^{-1}, \tag{6.3.17}$$

and

$$\rho = 1 - p_0. \tag{6.3.18}$$

Boyse and Warn give a simple argument to show the average throughput λ_T is given by

$$\lambda_T = c\rho/mE[s]. \tag{6.3.19}$$

This formula also follows from the "throughput law" of Buzen [15]. Boyse and Warn also show the average response time W is given by

$$W = (NmE[s]/c\rho) - E[t]. \tag{6.3.20}$$

When $c = 1$ and (6.3.19) is used for lambda, then (6.3.20) agrees exactly with (6.3.1). If $mE[s]$ is identified with the mean service time provided by the CPU per interaction, then (6.3.20) is the same as (6.3.3). This agrees with the results of Kobayashi [25] mentioned at the end of Section 6.3.1.

Example 6.3.3 Boyse and Warn [12] collected some system monitor data on their own computer system (a single CPU demand paging system) during a 30 minute period of high system load. They found the number of active terminals $N = 10$, average user think time $E[t] = 4$ seconds, degree of multiprogramming $K = 3$, total CPU job time of 613 seconds, total system CPU time of 827 seconds, total elapsed time of 1800 seconds (30 minutes), the number of interactions processed was 720, the total page reads was 75,600, and the average I/O service time $E[O] = 0.038$ seconds. The total CPU time per interaction was $(613 + 827)/720 = 2$ seconds $= mE[s]$. Since there were $75,600/720 = 105$ page reads per interaction, $m = 105$ and $E[s] = 2/105 = 0.019$ seconds. We have all the parameters we need to compute the model output.

First we note that the actual CPU utilization $\rho = (613 + 827)/1800 = 0.80$. Thus, by (6.3.19),

$$\lambda_T = c\rho/mE[s] = 0.8/(105 \times 0.019) = 0.40 \quad \text{interactions/second.}$$

Also, by (6.3.20),

$$W = (NmE[s]/c\rho) - E[t] = (10 \times 2/0.8) - 4 = 21 \quad \text{seconds.}$$

For the exponential model of CPU and I/O service we have, by (6.3.17),

$$p_0 = \left[3! \sum_{n=0}^{3} \frac{1}{(3-n)!} \left(\frac{0.019}{0.038}\right)^n\right]^{-1} = 0.2105,$$

so

$$\rho = 1 - p_0 = 0.7895.$$

This yields

$$\lambda_T = \rho/2 = 0.3948 \quad \text{interactions/second.}$$

Finally,

$$W = (20/\rho) - 4 = 21.33 \quad \text{seconds.}$$

This is in very close agreement to the actual system.

For the constant model of CPU and I/O service we have

$$\rho = \min\left\{\frac{K}{1 + E[O]/E[s]}, 1\right\} = \min\{1, 1\} = 1.$$

This gives

$$\lambda_T = \rho/2 = 0.5 \quad \text{interactions/second,}$$

and

$$W = (20/\rho) - 4 = 16 \quad \text{seconds.}$$

The constant model does not match the actual system as well as the exponential model does.

The constant model does, however, give a lower bound to the average response time that can be obtained with *any* service times for the CPU and I/O servers as well as an upper bound for CPU utilization and for throughput.

Boyse and Warn consider the following possible changes in hardware configuration: (a) add 0.5 megabytes of memory and (b) procure faster I/O devices which will decrease $E[O]$ to 0.0155 seconds.

They estimate that with the extra memory the multiprogramming level $K = 5$ can be utilized. For alternative (b), K will remain 3. In addition to evaluating the improvements in throughput and response time, they estimate the number of active terminals that could be supported with the *present* actual average response time $W = 21$ seconds. This can be done by solving (6.3.20) for N (with $c = 1$), yielding

$$N = \rho(W + E[t])/mE[s].$$

If we now let W_1 be the average response time for the initial configuration while all the other parameters are calculated for the new configuration, we get N^{**} the number of active terminals that can be supported at the old level of average response time W_1. Thus

$$N^{**} = \rho(W_1 + E[t])/mE[s]. \tag{6.3.21}$$

(Note that N^{**} is *not* the N^* of Section 6.3.2.) Then we have for alternative (a), using the exponential model with $K = 5$,

$$p_0 = \left[5! \sum_{n=0}^{5} \frac{1}{(5-n)!} \left(\frac{1}{2} \right)^n \right]^{-1} = 0.0367.$$

Hence

$$\rho = 1 - p_0 = 0.9633, \qquad \lambda_T = \rho/2 = 0.4817 \quad \text{jobs/interaction},$$

$$W = (NmE[s]/\rho) - E[t] = 16.76 \quad \text{seconds},$$

and

$$N^{**} = 0.9633(21 + 4)/2 = 12 \quad \text{terminals}.$$

For alternative (b) and $E[O] = 0.0155$ seconds but $K = 3$, using the exponential model,

$$p_0 = \left[3! \sum_{n=0}^{3} \frac{1}{(3-n)!} \left(\frac{0.0190}{0.0155} \right)^n \right]^{-1} = 0.0404, \qquad \rho = 1 - p_0 = 0.9596,$$

$$\lambda_T = \rho/2 = 0.4798 \quad \text{jobs/interaction}, W = (20/\rho) - 4 = 16.84 \quad \text{seconds},$$

and

$$N^{**} = 0.9596(21 + 4)/2 = 12 \quad \text{terminals}.$$

There is no significant difference in the improvement in performance for alternatives (a) and (b). Note that we used only the exponential model in our evaluation since the constant model showed a utilization of 1 for the initial system. Note also that making either of the changes brings W close to the 16 seconds predicted by the constant model for the initial configuration. Boyse and Warn [12] consider a number of alternative configurations using their model. The fact that they were able to do this illustrates the value of an analytic model. Any model is valuable only if it enables one to predict what would happen under various alternative strategies for change.

6.3.4   The Central Server Model
of Multiprogramming

   This model, shown in Fig. 6.3.4, is a specialization of the Jackson-type models we considered in Section 6.2.2. It is a closed model since it contains a fixed number of programs which can be thought of as markers that cycle around the system interminably. However, each time a marker (program) makes the cycle from the CPU directly back to the CPU we assume a program execution has been completed and a new program enters the

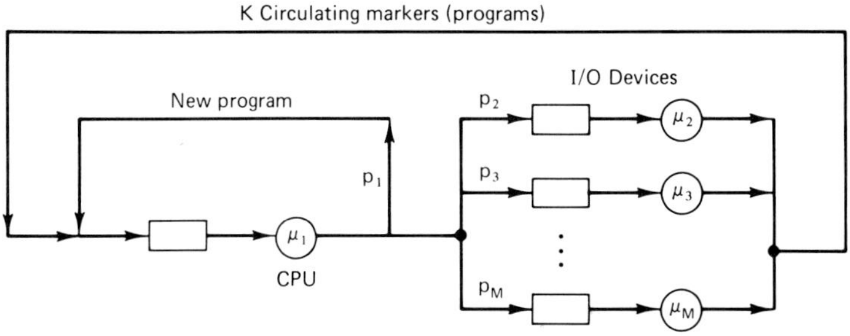

Fig. 6.3.4   Central server model of multiprogramming.

system. Thus, if we consider the model of Fig. 6.3.4 to represent the central processor system of Fig. 6.3.1, then we must assume the users at the terminals are sufficiently active to guarantee there is always an interaction pending. (Thus we could also classify the central server model as an infinite source model!) There are M-1 I/O devices, each with its own queue, and each exponentially distributed with average service rate μ_i $(i = 2, 3, \ldots, M)$. The CPU also is assumed to provide exponential service (with average rate μ_1). Upon completion of a CPU interval the job returns to the CPU (completes execution) with probability p_1 or requires service at I/O device i with probability p_i, $i = 2, 3, \ldots, M$. Upon completion of I/O service the job returns to the CPU queue for another cycle. If we let $\mathbf{k} = (k_1, k_2, \ldots, k_M)$ represent the state of the system, where k_i is the number of jobs (markers) at the ith queue (queueing or in service) then Buzen [16, 17] shows the probability $p(k_1, k_2, \ldots, k_M)$ that the system is in state $\mathbf{k}$ is given by

$$p(k_1, k_2, \ldots, k_M) = \frac{1}{G(K)} \prod_{i=2}^{M} \left(\frac{\mu_1 p_i}{\mu_i} \right)^{k_i}, \qquad (6.3.22)$$

where $G(K)$ is defined so as to make the probabilities sum to 1. Thus

$$G(K) = \sum_{k \in S(K,M)} \prod_{i=2}^{M} \left(\frac{\mu_1 p_i}{\mu_i} \right)^{k_i}, \qquad (6.3.23)$$

where

$$S(K, M) = \left\{ (k_1, k_2, \ldots, k_M) \,\middle|\, \sum_{i=1}^{M} k_i = K \text{ and each } k_i \geq 0 \right\}. \quad (6.3.24)$$

It can be shown (see Feller [28, p. 38]) that there are

$$\binom{M + K - 1}{M - 1} = \binom{M + K - 1}{K} \quad \text{elements in} \quad S(K, M).$$

We will show in Algorithm 6.3.1, below, the technique developed by Buzen for calculating $G(0) = 1, G(1), G(2), ..., G(K)$. Buzen [17] also shows that the server utilizations can be calculated by

$$\rho_i = \begin{cases} G(K-1)/G(K), & i = 1 \\ \mu_1 \rho_1 p_i / \mu_i, & i = 2, 3, ..., M. \end{cases} \tag{6.3.25}$$

He also shows that the throughput λ_T, in jobs per unit time, is given by

$$\lambda_T = \mu_1 p_1 \rho_1. \tag{6.3.26}$$

It follows from Buzen's general response time law (see Buzen [15]) that the average response time W (assuming N terminals in Fig. 6.3.1) is given by

$$W = (N/\lambda_T) - E[t] = (N/\mu_1 \rho_1 p_1) - E[t]. \tag{6.3.27}$$

Equation (6.3.27) also follows from the results of Kobayashi [25] quoted at the end of Section 6.3.1.

Buzen [16] shows that if we let $x_1 = 1$ and define

$$x_i = \mu_1 p_i / \mu_i, \qquad i = 2, 3, ..., M, \tag{6.3.28}$$

then the expected number of jobs at each queue $E[k_i]$ is given by

$$E[k_i] = \sum_{k=1}^{K} (x_i)^k \frac{G(K-k)}{G(K)}, \qquad i = 1, 2, ..., M. \tag{6.3.29}$$

All that remains to be done so that we may reap the benefits of the central server model is to give Buzen's efficient algorithm for computing

$$G(k), \qquad k = 0, 1, 2, ..., M.$$

Buzen defines the function $g(\cdot, \cdot)$ by

$$g(k, m) = \sum_{k \in S(k,m)} \prod_{i=1}^{m} (x_i)^{k_i}, \tag{6.3.30}$$

where $S(k, m)$ is defined by (6.2.24), $x_1 = 1$, and x_i for $i = 2, ..., M$ is defined by (6.3.28). Then we note that $G(K) = g(K, M)$ by (6.3.23) and, actually, $g(k, M) = G(k)$ for $k = 0, 1, ..., K$.

Buzen also shows that

$$g(k, m) = g(k, m-1) + x_m g(k-1, m), \tag{6.3.31}$$

if $k > 0$ and $m > 1$.

Furthermore,

$$g(k, 1) = 1 \qquad \text{for} \quad k = 0, 1, ..., K, \tag{6.3.32}$$

and

$$g(0, m) = 1 \qquad \text{for} \quad m = 1, 2, ..., M. \tag{6.3.33}$$

The initial conditions (6.3.32) and (6.3.33) plus the recursive relationship (6.3.31) define Buzen's algorithm. The algorithm can be visualized in Table 6.3.1. The table has $K + 1$ rows and M columns. The 0th row and the 1st column contain only ones. The algorithm proceeds row by row. For example, to calculate $g(k, m)$, the item to its left $g(k, m - 1)$ is added to x_m times the item above it $g(k - 1, m)$ to yield $g(k, m) = g(k, m - 1) + x_m g(k - 1, m)$.

TABLE 6.3.1
Buzen's Algorithm for Computing $G(K)$

	x_1	x_2	x_3	$\cdots$	x_m	$\cdots$	x_M
0	1	1	1		1		1
1	1	$g(1, 2)$	$\cdots$				$g(1, M) = G(1)$
2	1	$g(2, 2)$	$\cdots$				$g(2, M) = G(2)$
3	1	$g(3, 2)$	$\cdots$				$g(3, M) = G(3)$
.					$g(k - 1, m)$		.
.					$\downarrow \times$		.
.					x_m		.
k	1	$g(k, 2)$	$g(k, m - 1) \rightarrow g(k, m)$				$g(k, M) = G(k)$
.							.
.							.
K	1	$g(K, 2)$	$\cdots$				$g(K, M) = G(K)$

Algorithm 6.3.1 (*Buzen's Algorithm*) Given the parameters of the central server model of Fig. 6.3.4 (that is, μ_i, p_i for $i = 1, 2, \ldots, M$), this algorithm will generate $G(K)$ defined by (6.2.23) as well as $G(K - 1)$, $G(K - 2), \ldots, G(1), G(0) = 1$.

Step 1 [*Assign values to the* x_i] Set $x_1 = 1$ and then set $x_i = \mu_1 p_i / \mu_i$ for $i = 2, 3, \ldots, M$.

Step 2 [*Set initial values*] Set $g(k, 1) = 1$ for $k = 0, 1, \ldots, K$ and set $g(0, m) = 1$ for $m = 1, 2, \ldots, M$.

Step 3 [*Initialize k*] Set k to 1.

Step 4 [*Calculate kth row*] Set

$$g(k, m) = g(k, m - 1) + x_m g(k - 1, m), \qquad m = 2, 3, \ldots, M.$$

Step 5 [*Increase k*] Set k to $k + 1$.

Step 6 [*Algorithm complete?*] If $k \leq K$ return to Step 4. Otherwise terminate the algorithm. Then $g(n, M) = G(n)$ for $n = 0, 1, \ldots, K$.

Example 6.3.4 Consider the example on page 531 of Buzen [16]. Then $K = 4$, $M = 3$; $\mu_1 = \frac{1}{28}$, $\mu_2 = \frac{1}{40}$, $\mu_3 = \frac{1}{280}$ millisecond^{-1}, respectively; $p_1 = 0.1$, $p_2 = 0.7$, $p_3 = 0.2$. Applying Algorithm 6.3.1 in Step 1, we set

$$x_1 = 1, \qquad x_2 = \mu_1 p_2/\mu_2 = (\tfrac{1}{28} \times 0.7)/\tfrac{1}{40} = 1,$$

and

$$x_3 = \mu_1 p_3/\mu_3 = (\tfrac{1}{28} \times 0.2)/\tfrac{1}{280} = 10 \times 0.2 = 2.$$

In Step 2 we set the 0th row and the 1st column to 1's in Table 6.3.2, which corresponds to Table 6.3.1 for this data. Then we set k to 1 in Step 3. For the first execution of Step 4 we set $g(1, 2) = 1 + 1 = 2$, $g(1, 3) = 2 + 2 \times 1 = 4$. (Thus $G(1) = 4$.) In Step 5 we set k to 2, and in Step 6 we branch back to Step 4, since $2 \leq K = 4$. In the second execution of Step 4 we set $g(2, 2) = 1 + 1 \times 2 = 3$ and $g(2, 3) = 3 + 2 \times 4 = 11$. On the third execution of Step 4 we set $g(3, 2) = 1 + 3 = 4$ and $g(3, 3) = 4 + 2 \times 11 = 26$. On the fourth and last execution of Step 4 we set $g(4, 2) = 1 + 4 = 5$ and $g(4, 3) = 5 + 2 \times 26 = 57$. This terminates the algorithm. Thus $G(K) = G(4) = 57$, $G(3) = 26$, $G(2) = 11$, $G(1) = 4$, and $G(0) = 1$. Thus, as Buzen notes, the CPU utilization ρ_1, for 1, 2, 3, and 4 levels of multiprogramming, is $\frac{1}{4}$, $\frac{4}{11}$, $\frac{11}{26}$, and $\frac{26}{57}$, respectively. In particular, for $K = 4$ we have $\rho_1 = \frac{26}{57}$. In this case we have, by (6.3.25),

$$\rho_2 = \mu_1 \rho_1 p_2/\mu_2 = \tfrac{26}{57}, \qquad \text{and} \qquad \rho_3 = \mu_1 \rho_1 p_3/\mu_3 = \tfrac{52}{57}.$$

By (6.3.26), the average throughput λ_T is given by

$$\lambda_T = \mu_1 p_1 \rho_1 = 0.001629 \quad \text{jobs/millisecond} = 1.629 \quad \text{jobs/second}.$$

If the number of active terminals $N = 20$ with $E[t] = 10$ seconds, then, by (6.3.27), the average response time W is

$$(N/\lambda_T) - E[t] = 2.277 \quad \text{seconds}.$$

TABLE 6.3.2
Buzen's Algorithm for Example 6.3.4

	1	1	2
0	1	1	1
1	1	2	4
2	1	3	11
3	1	4	26
4	1	5	57

Example 6.3.5   Consider the central server model on pages 322–323 of Reiser [6]. Reiser measures all times in units of millions of CPU instructions where the CPU executes 0.75 million = 750,000 instructions/second. The multiprogramming level K is 10. There are two I/O devices: a drum and a disk. The mean path length between I/O operations to the drum $m_2 = 0.005$ million instructions and between I/O operations to the disk $m_3 = 0.01$ million instructions. The mean path length of a job $m_1 = 0.1$ million instructions. Thus the mean length of a CPU burst m_{CPU} is given by

$$m_{\text{CPU}} = \frac{1}{(1/m_1) + (1/m_2) + (1/m_3)} = \frac{1}{310} \quad \text{million instructions.}$$

This gives the probability of job completion, $p_1 = m_{\text{CPU}}/m_1 = \frac{10}{310}$,

$$p_2 = m_{\text{CPU}}/m_2 = \tfrac{200}{310}, \quad \text{and} \quad p_3 = m_{\text{CPU}}/m_3 = \tfrac{100}{310}.$$

The average access time of the drum is 0.03 seconds and of the disk it is 0.1 seconds. Thus

$$\mu_2 = 1/0.03 = 33.33 \quad \text{jobs/second,}$$

$$\mu_3 = 1/0.1 = 10 \quad \text{jobs/second,}$$

and

$$\mu_1 = 310 \times 0.75 = 232.5 \quad \text{jobs/second.}$$

These values are shown in Table 6.3.3. The result of applying Buzen's Algorithm is shown in Table 6.3.4. Therefore,

$$\rho_1 = G(9)/G(10) = 0.132972533.$$

Hence,

$$\lambda_{\text{T}} = \mu_1 \rho_1 p_1 = 0.997293997 \quad \text{jobs/second,}$$

which agrees with Reiser [6].

We also find, by (6.3.25), that

$$\rho_2 = \mu_1 \rho_1 p_2/\mu_2 = 0.598376 = \text{drum utilization,}$$

TABLE 6.3.3

Data for Example 6.3.5

i	μ_i (jobs/second)	p_i
1	232.50	10/310
2	33.33	200/310
3	10.00	100/310

TABLE 6.3.4

Buzen's Algorithm Applied to Example 6.3.5

	x_1 1	x_2 4.5	x_3 7.5
0	1	1	1
1	1	5.5	13.0
2	1	25.75	123.25
3	1	116.875	1041.25
4	1	526.9375	8,336.3125
5	1	2,372.21875	64,894.5625
6	1	10,675.98438	497,385.2032
7	1	48,042.92971	3,778,431.954
8	1	216,194.1837	28,554,433.84
9	1	972,874.8267	215,131,128.6
10	1	4,377,937.720	1,617,861,403

and

$$\rho_3 = \mu_1 \rho_1 p_3 / \mu_3 = 0.997294 = \text{disk utilization}.$$

It appears that the disk is the bottleneck in the system. If there were 20 terminals with average think time $E[t] = 10$ seconds, then by (6.3.27), the average response time W would be

$$(N/\lambda_T) - E[t] = 10.054 \quad \text{seconds}.$$

Moore [18] shows that, in the above example, the disk *is* a bottleneck in the sense that if K is allowed to increase without limit, an infinite queue will form at the disk while all the other nodes will have finite queues. What Moore showed was that, in a closed network of the Jackson type, the node with the highest utilization is a bottleneck node in the sense just described.

Muntz and Wong [19] have generalized Moore's results. They consider a general finite population interactive computing queueing model such as that in Fig. 6.3.1. Suppose resource j of the central processor system has the highest relative utilization for this system. They call this resource the "system bottleneck." Let $E[N_j]$ be the mean number of times this resource is utilized per interaction and $E[s_j]$ its mean service time. Muntz and Wong [19] show that for large N the asymptote for the mean response time is given by

$$W_2 = N \times E[N_j] \times E[s_j] - E[t].$$

If we let W_1 be the mean response time for one customer, and thus the asymptote for small N, then the asymptotes intersect at the system saturation point N^* given by

$$N^* = \frac{W_1 + E[t]}{E[N_j] \times E[s_j]}. \tag{6.3.34}$$

Note that (6.3.34) is very similar to the formula for the system saturation point for the machine repair model of Section 6.3.1, and has the same interpretation. It is the maximum number of terminals that could be scheduled with no mutual interference if each interaction required exactly the average response time. For $N \gg N^*$, the mean response time increases by $E[N_j] \times E[s_j]$ for each additional customer, which implies that the mean queue length of the bottleneck resource j tends to increase by one for each additional customer. Thus, from the point of view of mean response time, the best system is a balanced system in which each resource of the central processor system has the same utilization. Kleinrock [4, Section 4.12] discusses how one can investigate the feasibility of removing successive system bottlenecks on the way to a balanced system. (For example, if the mean response time asymptotes for the first and second system bottlenecks are close to each other, then little system performance improvement will result from removing only the first bottleneck; both must yield to get real improvement.) However, Buzen has shown that for the central server model of a batch computer system, where throughput rather than response time is the principal performance criterion, a balanced system is *not* optimal. Buzen [16] shows that the faster resources (those with large values of μ) should be loaded heavier than slower resources for optimal throughput, even though this may make the faster resource a system bottleneck. To achieve this effect files can be reallocated to the various processors; this policy means the drums should be more heavily loaded than disks, for example.

6.4 SUMMARY

We have considered a number of queueing models that are widely used to study the performance of computer systems. It has been found that some queueing models which are gross simplifications of the modeled system are often very useful for performance prediction. The trend now seems to be to build more complex queueing models which more accurately portray the system under study; then approximation techniques are applied to solve the equations of these models, which may have no simple closed form solutions. For example Herzog *et al.* [20] have developed a recursive technique for the approximate solution of rather general queueing networks. Sauer and Chandy [21] have an approximation technique for handling central server models with general rather than exponential service times. Another promising approximation technique is the diffusion approximation to queueing networks discussed by Kobayashi in [22] and [23] with a report on the accuracy by Reiser and Kobayashi [24]. The final approximation technique that should be mentioned is that of decomposition or the analysis of a large complex system by analyzing its subsystems separately and using the separate results to analyze the total system. This is essentially the approach we took in Section 6.3. Courtois [26, 27] investigates the conditions under which queueing and computer systems can be decomposed.

Student Sayings

Mary had a little λ.
From it she can get ρ.
And everywhere her λ goes,
Her ρ is sure to follow.

Exercises

1. [15]   Consider the customer loss ratio R for a queueing system discussed in Section 5.2 and defined by $R = E[q]/E[s]$. Compare this ratio in an M/G/1 queueing system to that in an M/G/1 processor-sharing queueing system; that is, let R_1 be the loss ratio for the first system, R_2 that for the latter, and calculate R_1/R_2. What is the value of R_1/R_2 for two identical systems (except for queue disciplines) if $C_s^2 = 3$? What do you conclude about the relative desirability of the two systems if the physical environment permits you to select the one you want?

2. [18]   Suppose the Highpockretease Medical Academy has a computer system very similar to that at Transylvania University, as described in Example 6.2.3; the system runs an average of 7,200 student jobs/12-hour day. The average job requires 5 seconds of CPU time with the CPU experiencing an I/O interrupt, on the average, every 0.2 seconds; an average I/O service time is 0.24 seconds. Find the values of the computer system parameters calculated in Example 6.2.3.

3. [15]   Suppose the Highpockretease Medical Academy of Exercise 2 decides to imitate Transylvania University (Example 6.2.4) by setting their level of multiprogramming at 10, so the model considered in Example 6.2.4 applies. Find ρ_1, ρ_2, the average response time, and λ_T.

4. [20]   Suppose that in the cyclic queueing model of Fig. 6.2.4, $F(t) = 1 - e^{-\mu t}$. Then the Laplace–Stieltjes transform of the service time of server 1 is $W_s^*(\theta) = \mu/(\mu + \theta)$. Use this fact in (6.2.27) to calculate the values of $\hat{p}(n)$ by (6.2.26) and thereby show that

$$p_K(n) = \frac{(1 - \rho)\rho^n}{1 - \rho^{K+1}}, \qquad n = 0, 1, \ldots, K,$$

if $\rho \neq 1$. Then show that

$$\rho_1 = \frac{\rho - \rho^{K+1}}{1 - \rho^{K+1}} \qquad \text{and} \qquad \rho_2 = \frac{1 - \rho^K}{1 - \rho^{K+1}}.$$

5. [18]   Prove that if $F(\cdot)$ is given by (6.2.38), where the parameters are generated by Algorithm 6.2.1 and X is a random variable with distribution function F, then $E[X] = 1/\mu$ and $C_X^2 = C^2$.

6. [C20]   Construct the cyclic queueing model shown in Fig. 6.2.4 for the computer system at Transylvania University (Example 6.2.3) using $K = 5$. Assume that the CPU (server 1) has a two-stage hyperexponential service time distribution s with $C_s^2 = 5$ and that each job needs, on the average, 16 CPU service periods for completion. (Use Algorithm 6.2.1 to calculate the parameters of the CPU service time.) Use $\mu = 4$ and $\lambda = 5$ in the equations of Table 23 of Appendix C. Calculate ρ_1, ρ_2, W, and λ_T. Compare the results to those obtained in Examples 6.2.3–6.2.5.

7. [C12]   Consider Example 6.3.3 in which two alternative hardware configurations were considered. Suppose both alternatives are adopted; that is, enough memory is added to allow multiprogramming level $K = 5$, and faster I/O devices are obtained so that $E[O] = 0.0155$ seconds. Calculate ρ, λ_T, W, and N^{**} for the new configuration.

8. [C18]   Consider Exercise 7. Suppose a new CPU is procured which is twice as fast as the old.

(a)   Find ρ, λ_T, W, and N^{**} for this system with $W_1 = 21$ seconds.

(b)   Calculate the statistics of (a) if, in addition to the faster CPU, more memory is obtained so that $K = 5$ can be used.

(c)   Calculate the statistics of (a) if the faster CPU is used with the faster I/O devices so $E[O] = 0.0155$ second.

(d)   Calculate the statistics of (a) if the faster CPU is obtained together with more memory and faster I/O units.

Use the exponential model in all cases.

9. [C18]   Consider Example 6.3.3. Suppose it is decided to proceed to a multiprocessor system by securing another CPU of the same speed plus enough additional memory to maintain a multiprocessing level $K = 5$. If all the other parameters are the same as in the original system, calculate p_0, ρ, λ_T, W, and N^{**} (using $W_1 = 21$ seconds). Use the exponential model. Note that N^{**} cannot be calculated by (6.3.21) because its derivation assumed $c = 1$.

References

1.   J. P. Buzen and P. S. Goldberg, Guidelines for the use of infinite source queueing models in the analysis of computer system performance, *Proc. Nat. Computer Conference* (1974), 371–374.

2.   L. Kleinrock, Analysis of a time-shared processor, *Naval Res. Logist. Quart.* **11** (1964), 59–73.

3.   L. Kleinrock, Time-shared systems: A theoretical treatment, *J. Assoc. Comput. Mach.* **14** (1967), 242–261.

4.   L. Kleinrock, *Queueing Systems, Volume II: Computer Applications.* Wiley, New York, 1976.

5.   E. G. Coffman, Jr., R. R. Muntz, and H. Trotter, Waiting time distributions for processor-sharing systems, *J. ACM* **17** (1), (Jan. 1970).

6.   M. Reiser, Interactive modeling of computer systems, *IBM Systems J.* **15** (4), (1976).

7.   J. R. Jackson, Networks of waiting lines, *Operations Res.* **5** (4), (August 1957).

8.   I. Mitrani, Nonpriority multiprogramming systems under heavy demand conditions—Customers' viewpoint, *J. ACM* **20** (3), (July 1973).

9. M. Reiser and H. Kobayashi, The effects of service time distributions on system perfor-mance, *Inform. Process. 74*, North-Holland Publishing Company, 1974.

10. W. R. Franta, The mathematical analysis of the computer system modeled as a two-stage cyclic queue, *Acta Informat.* **6** (1976), 187–209.

11. R. R. Muntz, Analytic modeling of interactive systems, *Proc. IEEE* **63** (June 1975), 946–953.

12. J. W. Boyse and D. R. Warn, A straightforward model for computer performance predic-tion, *ACM Comput. Surveys* **7** (2), (June 1975), 73–93.

13. A. L. Scherr, *An analysis of time-shared computer systems.* MIT Press, Cambridge, Mas-sachusetts, 1967.

14. E. R. Lassettre and A. L. Scherr, Modelling the performance of the OS/360 time sharing option (TSO), in *Statistical Computer Performance Evaluation* (W. Freiberger, ed.), pp. 57–72. Academic Press, New York, 1972.

15. J. P. Buzen, Fundamental laws of computer system performance, in *Proc. Internat. Symp. on Computer Performance Modeling, Measurement, and Evaluation*, March 29–31, 1976, ACM, New York, pp. 200–210.

16. J. P. Buzen, Computational algorithms for closed queueing networks with exponential servers, *Commun. ACM* **16** (9), (September 1973), 527–531.

17. J. P. Buzen, Queueing network models of multiprogramming, Ph.D. dissertation, Division of Engineering and Applied Physics (NTIS AD 731 575 August 1971) Harvard University, Cambridge, Massachusetts, May 1971.

18. C. G. Moore, III, Network models for large-scale time-sharing systems, Tech. Rep. No. 71-1, Department of Industrial Engineering, University of Michigan, Ann Arbor, Michi-gan, April 1971.

19. R. R. Muntz and J. Wong, Asymptotic properties of closed queueing networks, in *Proc. 8th Ann. Princeton Conference on Information Sciences and Systems*, March 1974.

20. U. Herzog, L. S. Woo, and K. M. Chandy, Solution of queueing problems by a recursive technique, *IBM J. Res. Develop.* **19** (3), (May 1975).

21. C. H. Sauer and K. M. Chandy, Approximate analysis of central server models, *IBM J. Res. Develop.* **19** (3), (May 1975).

22. H. Kobayashi, Application of the diffusion approximation to queueing networks I: Equilibrium queue distributions, *J. Assoc. Comput. Mach.* **21** (2), (1974), 316–328.

23. H. Kobayashi, Application of the diffusion approximation to queueing networks II: Nonequilibrium distributions and applications to computer modelling, *J. Assoc. Comput. Mach.* **21** (3), (1974), 459–469.

24. M. Reiser and H. Kobayashi, Accuracy of the diffusion approximation for some queueing systems, *IBM J. Res. Develop.* **18** (2), (March 1974).

25. H. Kobayashi, *Modeling and Analysis: An Introduction to System Performance Evaluation Methodology.* Addison–Wesley, Reading, Massachusetts, 1978.

26. P. J. Courtois, *Decomposability Queueing and Computer System Applications.* Academic Press, New York, 1977.

27. P. J. Courtois, Decomposability, instabilities, and saturation in multiprogramming systems, *Commun. ACM* **18** (7), (July 1975).

28. W. Feller, *An introduction to probability theory and its applications, Volume I*, 3rd ed. Wiley, New York, 1968.

There are lies, damned lies, and statistics.
Benjamin Disraeli

PART THREE

STATISTICAL INFERENCE

Probability must atone for the want of truth.
Matthew Prior

Chapter Seven

ESTIMATION

INTRODUCTION

Heretofore we have assumed in all our probability models that we knew the exact probability distributions of all random variables under consideration. That is, we assumed a knowledge of both the form of the probability laws and the values of the parameters for these laws. In practice, of course, we are not sure of either. (I am, of course, excepting the rare individual who is in direct communication with the Supreme Being. If you are one of these, you need not read the remainder of this book.) For most of us our information about a particular random variable must be based on a sampling of observed values. Nearly everyone uses this technique to make judgments concerning such entities as the quality of food and service at a restaurant, the entertainment value of a TV series, the talent of an actor or actress, etc.

Chapters 7 and 8 are part of a subject area called *statistical inference.* All of statistical inference is based upon a sample from the population of all items under consideration.

We will usually be concerned with obtaining a sample $x_1, x_2, \ldots, x_n$ of values from the *population* of all possible values of a random variable X. For the sample to have desirable mathematical properties it should be what is called a *random sample*. We can visualize the process of obtaining a random sample as a step-by-step experiment in which a sequence of observations are obtained in such a way that (a) each observed or selected value is independent of the others, and (b) at each step the selected value has the same probability of being chosen as any other element in the population. This can be conceptualized as a sequence $X_1, X_2, \ldots, X_n$ of independent random variables, each with the same distribution as X. We will therefore define a *random sample of size n* to be a sequence of independent, identically distributed random variables $X_1, X_2, \ldots, X_n$. Once a random sample has been taken (the random variables have assumed values), we indicate the sample by $x_1, x_2, \ldots, x_n$. Thus we take note of the fact that the values of two different random samples of size n from the same population are usually different; one random sample of five response times may be 5.2, 6.1, 5.8, 5.95, 5.7 seconds, while another random sample may yield 6.2, 5.75, 6.15, 5.1, 5.92 seconds. There may be some observed "randomness" between random samples.

A number of questions concerning the use of a random sample have probably occurred to the reader. We list some of the most common concerns as a sequence of "cosmic questions" below. We will not be able to fully answer all of these questions in this book but we shall attack each of them vigorously, if not rigorously.

(The reader is warned that, in Section 7.3 of the chapter, μ means "average service rate" while elsewhere in the chapter it may be used to denote the mean of any random variable.)

Some Cosmic Questions

Given a random sample $X_1, X_2, \ldots, X_n$ from a population determined by the random variable X (thus the X_i's are independent with the same distribution as X), the following cosmic questions occur.

1. How do we estimate the values of the parameters of X (such as $E[X]$ and $\text{Var}[X]$) and how do we make probability judgements about the quality of these estimates?
2. How do we make a probability judgement as to the type of random variable X is (such as normal, exponential, Erlang-k, etc.)?
3. How do we test whether or not a parameter of X, such as $E[X]$ or $\text{Var}[X]$, has a certain numerical value?
4. Assuming that the technique for estimating a parameter θ of X is to use a random variable $\hat{\theta}$ ($\hat{\theta}$ is pronounced "theta hat"), called an *estimator* of

θ, which depends upon the random sample, thus $\hat{\theta} = \hat{\theta}(X_1, X_2, \ldots, X_n)$; what are some desirable properties of estimators?

We shall consider the first cosmic question in the next two sections.

7.1 ESTIMATORS

An *estimator* $\hat{\theta}$ of a parameter θ of a random variable X is a random variable which depends upon a random sample $X_1, X_2, \ldots, X_n$. The two most common estimators are the *sample mean* $\bar{X}$ defined by

$$\bar{X} = (X_1 + X_2 + \cdots + X_n)/n, \tag{7.1.1}$$

and the *sample variance* S^2 defined by

$$S^2 = \sum_{i=1}^{n} \frac{(X_i - \bar{X})^2}{n-1}. \tag{7.1.2}$$

We define the *sample standard deviation* S, of course, by

$$S = \sqrt{S^2} = \left[\frac{\sum_{i=1}^{n}(X_i - \bar{X})^2}{n-1} \right]^{1/2}. \tag{7.1.3}$$

As the names suggest, $\bar{X}$ is an estimator of $E[X]$, S^2 of $\mathrm{Var}[X]$, and S of σ_X.

We must now consider a certain awkwardness of notation which occurs frequently in statistics, but is not unknown in other areas of mathematics. This awkwardness concerns the distinction between a function, itself, which is a mapping from one set into another, and a value of the function. For example, in Chapter 3, we tried to be consistent about indicating the distribution function of a random variable X by either the symbol F or $F(\cdot)$ (the latter to indicate that F was a function of one variable), but reserved the notation $F(x)$ to represent the *value* of F at the point x, such as in the formula

$$F(x) = 1 - e^{-x/E[X]}, \qquad x > 0,$$

for the distribution function of an exponential random variable. When we indicate a random variable, which is a function, we usually use a capital letter such as X; we use a small letter x to indicate a particular value of X. Thus we indicate a random sample, which is a collection of functions, by $X_1, X_2, \ldots, X_n$, while we indicate a particular random sample that has been selected by $x_1, x_2, \ldots, x_n$. Similarly, when we talk about the sample mean, as in (7.1.1), we use a capital letter; we do the same for the sample variance in (7.1.2). An actual calculated value of the sample mean would be written as

$$\bar{x} = (x_1 + x_2 + \cdots + x_n)/n, \tag{7.1.4}$$

and an actual calculated value of the sample variance would be written as

$$s^2 = [(x_1 - \bar{x})^2 + (x_2 - \bar{x})^2 + \cdots + (x_n - \bar{x})^2]/(n-1), \qquad (7.1.5)$$

where $\bar{x}$ was calculated in (7.1.4).

The next theorem which, if theorems were rated like restaurants, would have a "five star" rating, will help us answer part of "cosmic question 1" from the introduction to this chapter.

THEOREM 7.1.1 (*The Sampling Theorem*)   Let $X_1, X_2, \ldots, X_n$ be a random sample of size n from a population determined by the random variable X which has finite mean $E[X]$ and finite variance $\text{Var}[X] = \sigma^2$. Let $\bar{X}$ be defined by (7.1.1) and S^2 by (7.1.2). Then

(a) $E[\bar{X}] = E[X]$,
(b) $E[S^2] = \text{Var}[X] = \sigma^2$,
(c) $\text{Var}[\bar{X}] = (\text{Var}[X])/n = \sigma^2/n$,
(d)   for large n the random variable

$$Z = \frac{\bar{X} - E[X]}{\sigma/\sqrt{n}}$$

has approximately the standard normal distribution.

Furthermore, if X has a normal distribution, then the random variable

$$Y = \frac{\bar{X} - E[X]}{S/\sqrt{n}},$$

has a Student-t distribution with $n - 1$ degrees of freedom.

We omit the proof of this theorem. It can be found in Kreyszig [1].

Before applying this theorem we will discuss some desirable properties of estimators. Recall that we write $\hat{\theta} = \hat{\theta}(X_1, X_2, \ldots, X_n)$ for an estimator of the parameter θ. An estimator $\hat{\theta}$ of θ is *unbiased* if $E[\hat{\theta}] = \theta$. Intuitively this means that the distribution of the estimated values of θ centers about θ. (a) and (b) of Theorem 7.1.1 tell us that both the sample mean and the sample variance are unbiased estimators (how reassuring!). (The reason we divided by $n - 1$ in (7.1.2), rather than n, was to make S^2 an unbiased estimator.)

An estimator $\hat{\theta}$ having the property that, for each $\varepsilon > 0$,

$$\lim_{n \to \infty} P[|\hat{\theta} - \theta| < \varepsilon] = 1,$$

is called a *consistent estimator*. $\hat{\theta}$ is said to *converge in probability to* θ. Cramér [2] shows that both the sample mean and the sample variance are consistent estimators.

Let us consider what "unbiased" and "consistent" mean in terms of the sample mean $\bar{X}$; as an estimator of the population mean $E[X]$, it is both unbiased and consistent. If we took k random samples, each of size n, and calculated the corresponding k values of the sample mean, say $\bar{x}_1, \bar{x}_2, \bar{x}_3, \ldots, \bar{x}_k$, then these values would cluster about the value $E[X]$ because $\bar{X}$ is unbiased. The variance of $\bar{X}$ ($\text{Var}[\bar{X}] = \text{Var}[X]/n$) is a measure of the tightness of the clustering about $E[X]$; by Chebychev's inequality (Theorem 2.10.2), at least three-fourths of the values $\bar{x}_1, \bar{x}_2, \ldots, \bar{x}_k$ are within two standard deviations of $E[X]$, that is, not farther than $2\sigma/\sqrt{n}$ from $E[X]$. (Actually, as we shall see in Section 7.2, about 95% of the sample means are this close to $E[X]$, if n is larger than 30.) Because $\bar{X}$ is consistent we can make the probability that the estimate $\bar{x}$ is not in error by more than ε (for an arbitrary $\varepsilon > 0$) as close to one as we please, by choosing the sample size n sufficiently large. (Formally, given $\varepsilon > 0$, $\delta > 0$, there exists an integer N such that, if $n \geq N$, then $P[\,|\bar{x} - E[X]| < \varepsilon] > 1 - \delta$.) We illustrate these ideas in the following example.

Example 7.1.1 Consider the 8 random samples, each of size 10, shown in Table 7.1.1. These samples were taken from a normally distributed population with mean 50 and standard deviation 10. The first 10 items in each row of the table are the 10 values of one sample. The sample mean and the sample standard deviation of the sample are the last two items in a row. It is evident that the 8 sample means cluster about 50, which is $E[X]$, and the 8 sample standard deviations cluster about 10, which is σ. In fact,

$$\bar{\bar{x}} = (51.08 + 49.6 + 49.73 + \cdots + 53.34)/8 = 50.6625,$$

$$\bar{s} = (10.48 + 10.61 + \cdots + 10.41)/8 = 10.29625,$$

and

$$s_{\bar{X}}^2 = [(51.08 - 50.6625)^2 + \cdots + (53.34 - 50.6625)^2]/7$$

$$= 8.602 \quad \text{or} \quad s_{\bar{X}} = 2.93.$$

TABLE 7.1.1
Data for Example 7.1.1

				Sample values							
x_1	x_2	x_3	x_4	x_5	x_6	x_7	x_8	x_9	x_{10}	$\bar{x}$	s
37.51	39.69	51.44	49.29	54.42	73.48	53.83	52.97	40.74	57.46	51.08	10.48
49.53	64.07	60.40	46.97	39.01	28.61	49.06	46.05	59.98	52.30	49.60	10.61
69.25	35.18	52.50	42.11	64.13	53.97	54.06	38.15	36.07	51.89	49.73	11.68
41.76	25.96	53.80	47.78	55.79	52.39	47.02	53.40	50.15	36.16	46.42	9.36
59.69	56.01	47.00	48.24	40.94	44.23	38.33	28.96	57.16	57.64	47.82	9.99
48.30	34.84	51.34	54.56	45.81	42.16	51.35	55.74	65.45	68.46	51.80	10.09
44.49	64.48	55.10	42.90	62.28	45.84	52.98	74.14	56.43	56.50	55.51	9.75
44.71	57.99	51.69	61.19	48.81	37.70	62.39	64.45	39.15	65.32	53.34	10.41

$\bar{x}$, the average of the sample means, is, of course, the estimate of $E[X]$ we would have gotten by considering the 80 observed values as one big sample of size 80. It is reasonably close to 50. Likewise $\bar{s}$, the average of the calculated values of the sample standard deviation, is close to $\sigma = 10$.

We know by the sample theorem (Theorem 7.1.1) that

$$\text{Var}[\bar{X}] = \text{Var}[X]/n = 10.$$

Thus the estimate of 8.602 for $\text{Var}[\bar{X}]$, calculated on the basis of the sample of size 8 from the 11th column of Table 7.1.1, seems reasonable. If we merged all 80 observed values of X from the 8 samples, each of size 10, we would calculate

$$\bar{x} = 50.6635 \qquad \text{and} \qquad s = 10.23.$$

(The value of 50.6635 for $\bar{x}$ differs slightly from the $\bar{\bar{x}}$ value of 50.6625 because the numbers in column 11 of Table 7.1.1 were rounded to two decimal places; s based on a sample of size 80 is not the same as the average of 8 s's, each based on a sample of size 10, because of the value " $n - 1$ " in the denominator of formula (7.1.5).)

There are two more desirable properties of estimators that should be mentioned. The first of these concerns efficiency. Clearly, if we compare two unbiased estimators, $\hat{\theta}_1$ and $\hat{\theta}_2$, the one with the smallest variance would tend to be more efficient in terms of giving a better estimate of θ for a given sample size. We say that an unbiased estimator θ is the *minimum variance unbiased estimator* of θ if $\text{Var}[\hat{\theta}] < \text{Var}[\hat{\theta}_1]$ when $\hat{\theta}_1$ is any other unbiased estimator of θ. Some authors say the minimum variance unbiased estimator is *efficient*.

All the claims in the following theorem are proven by Hogg and Craig [3].

THEOREM 7.1.2   Let $X_1, X_2, \ldots, X_n$ be a random sample from a population determined by X. Then the following hold.

(a)   If X is normally distributed with mean μ and variance σ^2, then the estimators $\bar{X}$ and S^2 are unbiased, consistent, minimum variance estimators of the parameters μ and σ^2, respectively.

(b)   If X is Poisson with parameter (expected value) λ, then $\bar{X}$ is an unbiased, consistent, minimum variance estimator of λ.

(c)   If X is Bernoulli with parameter p, then k/n, where k is the number of successes observed in n independent trials, is the maximum likelihood estimator of p. It is also unbiased, consistent, and the minimum variance estimator of p. (Maximum likelihood estimation is discussed in Section 7.1.2.)

While unbiased estimators are desirable in many respects, such as by the fact that the estimates from several unbiased estimators can be averaged together to give a new (and hopefully better) unbiased estimate, unbiased estimators are not always available. In such cases we may consider a consistent estimator $\hat{\theta}$ with minimum mean-squared error where mean-squared error of $\hat{\theta} = E[(\hat{\theta} - \theta)^2] = \text{Var}[\hat{\theta}] + (E[\hat{\theta}] - \theta)^2$. The term $E[\hat{\theta}] - \theta$ is called the *bias* of $\hat{\theta}$ and, of course, is zero for unbiased estimators. It is possible for a biased estimator $\hat{\theta}$ to have a smaller mean-squared error than any unbiased estimator if $\text{Var}[\hat{\theta}]$ is small.

We have discussed some desirable properties of estimators and have shown that, for some special populations, the sample mean $\bar{X}$ and the sample variance S^2 have many of these properties. However, we have not given any general methods for constructing estimators. In the next two subsections we consider the two most popular techniques for estimation.

7.1.1 Method of Moments Estimation

Suppose we are given the values $x_1, x_2, \ldots, x_n$ of a random sample taken from a population determined by a random variable X; X is characterized by k parameters $\theta_1, \theta_2, \ldots, \theta_k$, which we wish to estimate. We define the jth sample moment by

$$M_j = \frac{\sum_{i=1}^{n} x_i^j}{n}, \qquad j = 1, 2, \ldots, k. \tag{7.1.6}$$

We then equate the k sample moments and population moments $E[X^j]$ (defined in Chapter 2), giving

$$M_j = E[X^j], \qquad j = 1, 2, \ldots, k. \tag{7.1.7}$$

The values $\hat{\theta}_1, \hat{\theta}_2, \ldots, \hat{\theta}_k$ obtained by solving the k simultaneous equations (7.1.7) are the method of moment estimates of the parameters. We illustrate this method with some examples.

Example 7.1.2 Suppose the processing time X of an inquiry for the on-line system developed at Bigbug Computing has a gamma distribution with parameters α and λ (see Section 3.2.4), and that n random values of processing time $x_1, x_2, \ldots, x_n$ have been observed. We calculate M_1 and M_2 by (7.1.6). Then we set

$$M_1 = E[X] = \alpha/\lambda, \qquad \text{and} \qquad M_2 - M_1^2 = \text{Var}[X] = \alpha/\lambda^2.$$

If we let $\hat{\alpha}$, $\hat{\lambda}$ denote the solution to these equations, then

$$\hat{\alpha} = M_1^2/(M_2 - M_1^2), \qquad \text{and} \qquad \hat{\lambda} = M_1/(M_2 - M_1^2).$$

Example 7.1.3 The arrival pattern to the main office of Bigbucks Financial is Poisson. The number of arrivals for each of n randomly selected

10-minute time intervals has been observed yielding the values $x_1, x_2, \ldots, x_n$. We want to estimate $\lambda = E[X]$, where X is the number of arrivals per 10-minute interval. Then, obviously, by the method of moments, we have

$$\hat{\lambda} = \bar{x} = M_1.$$

We used the method of moments in Example 3.2.7 to fit an Erlang-k distribution to an observed message length distribution.

The method of moments has the twin virtues of being intuitively satisfying, as well as easy to apply, in most cases. However, the method discussed in the next subsection, the maximum likelihood method, is even more intuitively appealing; it also has a deeper theoretical foundation. Thus anyone using it can feel smug about using a highly sophisticated technique which is also easy to understand and use. In many cases the two methods yield the same estimators.

7.1.2 Maximum Likelihood Estimation

The idea of maximum likelihood estimation of the parameters $\theta_1, \theta_2, \ldots, \theta_k$, which characterize a random variable X, is to choose the value(s) which make(s) the observed sample values $x_1, x_2, \ldots, x_n$ most probable. We illustrate the idea with an example before we set up the formal machinery.

Example 7.1.4 A computer installation has a large number of application programs which have been collected from a number of sources. The program librarian carefully tabulated the programs and found that a certain proportion P of the programs was written in FORTRAN. The librarian lost the paper containing the number P and can only remember that either $P = 0.6$ or $P = 0.8$. Fifteen of the programs are selected randomly and eight of the fifteen are found to be FORTRAN programs. What is the maximum likelihood estimate of P? Since X, the number of the programs written in FORTRAN has a binomial distribution with parameters 15 and P, we can calculate the probability of the observed result if $P = 0.6$, and if $P = 0.8$. If $P = 0.6$, the probability of 8 FORTRAN programs is

$$\binom{15}{8}(0.6)^8(0.4)^7 = 0.177,$$

while, if $P = 0.8$, the probability is

$$\binom{15}{8}(0.8)^8(0.2)^7 = 0.0138.$$

Hence we would estimate that $P = 0.6$, since this value has the greatest probability of yielding the observed sample.

Suppose now that X is a random variable, discrete or continuous, whose distribution depends upon a single parameter θ. Let $x_1, x_2, \ldots, x_n$ be an observed random sample. If X is discrete, the probability that a random sample consists of exactly these values is given by

$$l(\theta) = p(x_1)p(x_2) \cdots p(x_n), \tag{7.1.8}$$

where $p(\cdot)$ is the probability mass function of X. l is called the *likelihood function* and is a function of θ; that is, the value of (7.1.8) depends both upon the selected sample values and the choice of θ. If X is continuous with density function $f(\cdot)$, then the likelihood function $l(\cdot)$ is defined by

$$l(\theta) = f(x_1)f(x_2) \cdots f(x_n). \tag{7.1.9}$$

The *maximum likelihood estimate of* θ is the value of θ that maximizes the value of the likelihood function (7.1.8) or (7.1.9). If l is a differentiable function of θ, then a necessary condition for l to have a maximum value is that

$$\partial l/\partial \theta = 0. \tag{7.1.10}$$

We indicate a partial derivative in (7.1.10) because l depends both on θ and the sample values $x_1, x_2, \ldots, x_n$. Thus to find the maximum likelihood estimate of θ we solve (7.1.10) to find the value of θ which maximizes l. If we replace the values $x_1, x_2, \ldots, x_n$ in the solution by the random sample $X_1, X_2, \ldots, X_n$, we obtain the random variable $\hat{\theta}$, which is the *maximum likelihood estimator* of θ.

If the distribution of X involves several parameters $\theta_1, \theta_2, \ldots, \theta_k$, then to find the values which maximize the likelihood function we can solve the system of equations

$$\partial l/\partial \theta_1 = 0, \ \partial l/\partial \theta_2 = 0, \ \ldots, \ \partial l/\partial \theta_k = 0, \tag{7.1.11}$$

to determine maximum likelihood estimates for the parameters. Some care must be taken to ensure that the solution of the k simultaneous equations (7.1.11) maximizes l; a minimum point is also characterized by zero partial derivatives.

In many cases it is more convenient to work with $L = \ln l$, the logarithmic likelihood function. Since the logarithm function $\ln$ is a monotonically increasing function, a maximum of L is a maximum of l and vice versa. In this case we replace the equations (7.1.11) by

$$\partial L/\partial \theta_1 = 0, \ \partial L/\partial \theta_2 = 0, \ \ldots, \ \partial L/\partial \theta_k = 0. \tag{7.1.12}$$

Example 7.1.5 Consider Example 7.1.4. Suppose the librarian now reports that perhaps the proportion p of FORTRAN programs was not 0.6 or

0.8. We want to make a maximum likelihood estimate on the basis of the data. If the sample is of size n, then X is Bernoulli with probability of success p. Thus, if the observed number of FORTRAN programs is k, then the likelihood function l is, by (7.1.8),

$$l(p) = p^k (1 - p)^{n-k}.$$

Thus

$$L = \ln l = k \ln p + (n - k) \ln(1 - p),$$

and (7.1.12) becomes

$$\partial L / \partial p = (k/p) + (n - k)/(1 - p) = 0.$$

Solving for p yields

$$\hat{p} = k/n.$$

In Example 7.1.4, $n = 15$ and $k = 8$, so

$$\hat{p} = \tfrac{8}{15} = 0.533.$$

Example 7.1.6 Pourtnoy's Complaint Service receives requests for service in a Poisson pattern and wants to estimate the average arrival rate λ from the random sample $k_1, k_2, \ldots, k_n$ of arrivals per one-minute interval. Thus

$$l(\lambda) = \left(e^{-\lambda} \frac{\lambda^{k_1}}{k_1!} \right) \left(e^{-\lambda} \frac{\lambda^{k_2}}{k_2!} \right) \cdots \left(e^{-\lambda} \frac{\lambda^{k_n}}{k_n!} \right) = \frac{1}{k_1! \cdots k_n!} e^{-n\lambda} \lambda^{n\bar{k}},$$

where

$$\bar{k} = (k_1 + k_2 + \cdots + k_n)/n.$$

Thus

$$L = \ln l = -\ln(k_1! \cdots k_n!) - n\lambda + n\bar{k} \ln \lambda,$$

and (7.1.12) is

$$\partial L/\partial \lambda = -n + (n\bar{k}/\lambda) = 0.$$

Solving for λ yields

$$\hat{\lambda} = \bar{k} = (k_1 + k_2 + \cdots + k_n)/n.$$

Thus the sample mean is the maximum likelihood estimate for λ! This is the same solution we got by the method of moments in Example 7.1.3.

Example 7.1.7 Suppose the random variable X has a normal distribution with parameters μ and σ^2. If we have obtained a random sample $x_1, x_2, \ldots, x_n$, then the likelihood function is

$$l(\mu, \sigma^2) = \{(1/\sigma\sqrt{2\pi}) \exp[-(x_1 - \mu)^2/2\sigma^2]\} \cdots$$
$$\times \{(1/\sigma\sqrt{2\pi}) \exp[-(x_n - \mu)^2/2\sigma^2]\},$$
$$= \left(\frac{1}{2\pi\sigma^2} \right)^{n/2} \exp\left[-\sum_{i=1}^{n} \frac{(x_i - \mu)^2}{2\sigma^2} \right].$$

Thus

$$L(\mu, \sigma^2) = -\frac{n}{2}\ln 2\pi - \frac{n}{2}\ln \sigma^2 - \sum_{i=1}^{n} \frac{(x_i - \mu)^2}{2\sigma^2}.$$

Setting the partial derivatives equal to zero yields

$$\frac{\partial L}{\partial \mu} = \sum_{i=1}^{n} \frac{(x_i - \mu)}{\sigma^2} = 0,$$

and

$$\frac{\partial L}{\partial \sigma^2} = \frac{-n}{2\sigma^2} + \sum_{i=1}^{n} \frac{(x_i - \mu)^2}{2\sigma^4} = 0.$$

Solving the first of these equations, we obtain

$$\hat{\mu} = (x_1 + x_2 + \cdots + x_n)/n = M_1 = \bar{x}.$$

Substituting this solution into the second equation yields

$$\widehat{\sigma^2} = \sum_{i=1}^{n} \frac{(x_i - \bar{x})^2}{n} = M_2 - M_1^2.$$

These estimates are the same as those we would get using the method of moments. Although $\hat{\mu}$ is unbiased, $\widehat{\sigma^2}$ is a biased estimator.

So far we have nibbled a little on the " cosmic questions " of the introduction. We have completely ignored questions two and three. However, we have fairly well cleared up question four about desirable properties of estimators and have even exhibited a few estimators with these nice properties. We have answered the first part of question one in that we have given some estimators for $E[X]$ and $Var[X]$. In the next section we will attack the problem of making probability judgements about the quality of our estimates.

7.2 CONFIDENCE INTERVALS

In Section 7.1 we talked about some nice properties of estimators such as being unbiased, consistent, the minimum variance unbiased estimator, etc. (The reader may feel like adding " trustworthy, loyal,..... reverent.") However, not one of these desirable properties is of any help in making a probability judgement about the quality or accuracy of the estimate delivered. The confidence interval, as you probably suspected from the title of this section, is what enables us to do that.

The idea of a confidence interval is very similar to that of an error limit in numerical analysis. If we calculate a value x and know that the error in the calculation does not exceed δ (where $\delta > 0$), then we know the true value lies between $x - \delta$ and $x + \delta$. In the case of an estimator, we are dealing with a random variable, so we cannot predict with certainty that the true value θ of the parameter is within any finite interval. However, we can choose a high probability, such as 0.95 (95%), and then construct an interval, called a *confidence interval*, such that the probability the true value of θ lies in the interval is 0.95 (the probability it doesn't is then 0.05, so that 19 times out of 20 the true value is in the interval; and 1 time out of 20 it isn't). This is usually stated in the form of a $100(1 - \alpha)\%$ confidence interval where α, sometimes called the "level of significance," is the probability of error. Thus $1 - \alpha$ is a measure of confidence and α of nonconfidence.

The sequence of theorems that follow will indicate how to calculate confidence intervals for the parameters of some random variables of interest. The theorems will be interspersed with partial proofs and examples.

THEOREM 7.2.1   Let $x_1, x_2, \ldots, x_n$ be the values of a random sample from a population determined by the random variable X which has finite mean $\mu = E[X]$ and variance σ^2. Suppose further that either (a) X is normally distributed, or (b) n is large enough that, by Theorem 7.1.1, $\bar{X}$ can be considered normally distributed. Then, if we assume σ is known, the $100(1 - \alpha)\%$ confidence interval is given by

$$\bar{x} \pm E, \tag{7.2.1}$$

where

$$E = z_{\alpha/2}\sigma/\sqrt{n}. \tag{7.2.2}$$

(Recall that by the definition in Chapter 3, z_α is defined to be the largest value of z such that $P[Z > z] = \alpha$, where Z is the standard normal random variable. It can be picked out of Table 3 of Appendix A or chosen from Table 7.2.1. The reader is cautioned that to use Table 3 one sets $z_{\alpha/2}$ to the value of z which yields the value $1 - \alpha/2$ in the table. Thus, for a 95% confidence interval, $\alpha = 0.05$ and $z_{\alpha/2} = z_{0.025} = 1.96$, since this is the z value corresponding to the Table 3 value of $1 - 0.025 = 0.975$.)

Proof If the stated conditions are true $(\bar{X} - \mu)/\sigma/\sqrt{n}$ has a standard normal distribution so

$$P\left[-z_{\alpha/2} \leq \frac{\bar{x} - \mu}{\sigma/\sqrt{n}} \leq z_{\alpha/2}\right] = 1 - \alpha, \tag{7.2.3}$$

by the symmetry of the normal distribution. A little manipulation of (7.2.3) yields

$$P[\bar{x} - z_{\alpha/2}\sigma/\sqrt{n} \leq \mu \leq \bar{x} + z_{\alpha/2}\sigma/\sqrt{n}] = 1 - \alpha, \tag{7.2.4}$$

which completes the proof.

TABLE 7.2.1

Values of $z_{\alpha/2}$ for
calculating confidence intervals

$1 - \alpha$:	0.90	0.95	0.99	0.999
$z_{\alpha/2}$:	1.645	1.960	2.576	3.291

The major problem with Theorem 7.2.1 is that sometimes we know *neither* the mean μ nor the standard deviation σ. In these cases, when n is fairly large, say $n \geq 30$, we can use s for σ in (7.2.2). When n is small, but X is normally distributed, we can use Theorem 7.2.2.

Example 7.2.1 Suppose a random sample of 225 response times of an on-line terminal at Bonanza Banana yields $\bar{x} = 7$ jerks (a jerk is a time unit known only to the chief systems analyst at Bonanza) with a sample standard deviation of 3 jerks. Find a 95% confidence interval for the average response time at the terminal.

Solution Since n is relatively large, $\bar{X}$ is approximately normally distributed, and s is a good approximation of σ. For a 95% confidence interval $\alpha = 0.05$ and, by Table 7.2.1, $z_{\alpha/2} = z_{0.025} = 1.960$. Hence

$$E = z_{\alpha/2}\sigma/\sqrt{n} = 1.96 \times 3/15 = 0.392 \quad \text{jerks}$$

so the 95% confidence interval is 7 ± 0.392 jerks, that is, we can be 95% confident that the true average response time lies between 6.608 and 7.392 jerks.

Example 7.2.2 If, in Example 7.2.1, we want the length of the 95% confidence interval not to exceed 0.5 jerks, how large a sample should we choose? (We assume the random sample of 225 values is to be enlarged.)

Solution We assume, as before, that $\bar{X}$ is normally distributed and that $s = \sigma$. Then we want

$$2E = 2z_{\alpha/2}\sigma/\sqrt{n} = 2 \times 1.96 \times 3/\sqrt{n} = 0.5$$

or

$$n = [(2 \times 1.96 \times 3)/0.5]^2 = 553.1904.$$

Thus we need a sample of size 554, or we must randomly choose 329 more values of response time.

THEOREM 7.2.2 Suppose $x_1, x_2, \ldots, x_n$ are the values of a random sample from a population determined by a normally distributed random variable X with unknown mean and variance. Then the $100(1 - \alpha)\%$ confidence interval for the mean of X is given by

$$\bar{x} \pm E \tag{7.2.5}$$

where

$$E = t_{\alpha/2}\, s/\sqrt{n}, \tag{7.2.6}$$

and $t_{\alpha/2}$ is defined by $P[T > t_{\alpha/2}] = \alpha/2$, where T has a Student-t distribution with $n - 1$ degrees of freedom. (Values of $t_{\alpha/2}$ can be read from Table 5 of Appendix A. To find $t_{\alpha/2}$ by Table 5 look under the column with *value* $\alpha/2$ and in the row labeled with the *n value* $n - 1$. This is illustrated in Example 7.2.3.)

Proof The theorem follows immediately from the fact that $(\bar{X} - E[X])/S/\sqrt{n}$ has a Student-t distribution with $n - 1$ degrees of freedom. This fact was stated in Theorem 7.1.1.

Example 7.2.3 The number of message buffers in use, X, in the central processor at Baby Blue Inc. is believed to be normally distributed. A random sample of 9 values taken during the day yields $\bar{x} = 120$ and $s = 10$. Find a 99% confidence interval for the mean number of message buffers in use.

Solution For a 99% confidence interval $\alpha = 0.01$. By Table 5 of Appendix A, $t_{0.005} = 3.355$ (for 8 degrees of freedom). Thus $E = 3.355 \times 10/3 = 11.18$ buffers, and the 99% confidence interval for the average number of buffers in use is 120 ± 11.18, or between 108.82 and 131.18 buffers.

Theorem 7.2.2 is much more satisfying than Theorem 7.2.1, because we need deal only with entities that we can measure and calculate; for Theorem 7.2.1 we had to assume a knowledge of σ which we may not have. Usually, when applying Theorem 7.2.2, we have large confidence intervals corresponding to small samples or a great deal of uncertainty; this is to be expected, since we have little information.

Figure 7.2.1 illustrates the relationship between 10 typical large sample 95% confidence intervals for a mean μ and the true value of the mean. Note that one of the confidence intervals fails to contain the mean.

The next theorem tells us how to construct a confidence interval for the variance of a normally distributed random variable.

THEOREM 7.2.3 Let $X_1, X_2, \ldots, X_n$ be a random sample from a population determined by a normally distributed random variable X with mean μ and variance σ^2. Then the random variable $Y = (n - 1)S^2/\sigma^2$ has a chi-square distribution with $n - 1$ degrees of freedom. Consequently, the $100(1 - \alpha)\%$ confidence interval for σ^2 is given by

$$(n - 1)s^2/\chi^2_{\alpha/2} \leq \sigma^2 \leq (n - 1)s^2/\chi^2_{1-\alpha/2}.$$

$\chi^2_{1-\alpha/2}$ and $\chi^2_{\alpha/2}$ can be determined from Table 4 in Appendix A for a chi-square distribution with $n - 1$ degrees of freedom.

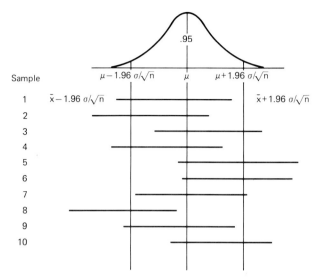

Fig. 7.2.1 Ten typical large-sample 95% confidence intervals for the mean μ.

We omit the proof which can be found in Kreyszig [1].

Example 7.2.4 The average number of lines of code per programmer-day for each of 30 programs produced at Huzunga Enterprises has been collected. The average number of lines of code per programmer-day, X, is normally distributed. If $\bar{x} = 75$ and $s^2 = 90$, find 95% confidence intervals for μ the mean of X, and for σ^2 the variance of X.

Solution By Table 4, for 29 degrees of freedom, $\chi^2_{0.975} = 16.047$ and $\chi^2_{0.025} = 45.722$. Hence, the 95% confidence interval for σ^2 is the interval from $29 \times 90/45.722 = 57.084$ to $29 \times 90/16.047 = 162.647$.

By Theorem 7.2.2, the 95% confidence interval for μ is given by $\bar{x} \pm E$, where

$$E = t_{0.025} s/\sqrt{n} = 2.045 \times \sqrt{90}/\sqrt{30} = 2.045\sqrt{3} = 3.542.$$

Hence, we are 95% confident that

$$71.458 \leq \mu \leq 78.542.$$

The large confidence interval for σ^2 is typical of the results given by Theorem 7.2.3. Nevertheless, most statistics books state that, for sample sizes of 30 or more, s^2 is a good estimate of σ^2.

We now have answered cosmic questions 1 and 4 fairly completely. We will attack cosmic questions 2 and 3 in Chapter 8.

In the next section we consider some applications of estimating to queueing systems.

7.3　ESTIMATING QUEUEING SYSTEM PARAMETERS

Not a great deal has been written about the topic of this section. Lilliefors [4], Section 7.1.1 of Gross and Harris [5], and Section 10.6 of Hillier and Lieberman [6] are the principal references.

We consider, first an M/M/1 queueing system in equilibrium, for which both λ and μ need to be estimated. Let t be a long time interval during which the system is observed. Suppose that during this time interval there are n_a arrivals, n_c service completions, and the service facility is busy for t_b time units. Then Gross and Harris [5] show that the following estimators are maximum likelihood estimators

$$\hat{\lambda} = n_a/t, \qquad \hat{\mu} = n_c/t_b, \qquad \text{and} \qquad \hat{\rho} = \hat{\lambda}/\hat{\mu}$$

(we assume that the system is stable and thus that $\hat{\rho} < 1$).

Since t is the sum of n_a independent exponentially distributed interarrival times (each with mean $1/\lambda$), by Theorem 3.2.4d, t has a gamma distribution with parameters n_a and λ. Hence, by Theorem 3.2.6, $2\lambda t$ has a chi-square distribution with $2n_a$ degrees of freedom. Similarly, $2\mu t_b$ has a chi-square distribution with $2n_c$ degrees of freedom. Thus, by Theorem 3.2.8,

$$\frac{2\mu t_b/2n_c}{2\lambda t/2n_a} = \frac{\mu n_a/t}{\lambda n_c/t_b} = \frac{\hat{\rho}}{\rho}$$

has an F-distribution with $(2n_c, 2n_a)$ degrees of freedom. Thus a $100(1 - \alpha)\%$ confidence interval can be calculated because

$$P\left[f_{1-\alpha/2;2n_c,2n_a} \le \hat{\rho}/\rho \le f_{\alpha/2;2n_c,2n_a}\right] = 1 - \alpha.$$

Thus the upper and lower ends of the $100(1 - \alpha)\%$ confidence interval for ρ are given by

$$\rho_u = \hat{\rho}/f_{1-\alpha/2;2n_c,2n_a} = \hat{\rho}f_{\alpha/2;2n_a,2n_c}$$

and

$$\rho_L = \hat{\rho}/f_{\alpha/2;2n_c,2n_a} = \hat{\rho}f_{1-\alpha/2;2n_a,2n_c},$$

since, by Theorem 3.2.8, $f_{1-\beta;n,m} = 1/f_{\beta;m,n}$.

As stated by Lilliefors [4], if $H(\rho)$ is any monotonically increasing function of ρ, then the $100(1 - \alpha)\%$ confidence interval for $H(\rho)$ is

$$H(\rho_L) \le H(\rho) \le H(\rho_u).$$

Some monotonic functions of ρ that can be considered are L, L_q, and p_0. Thus since $p_0 = 1 - \rho$ is monotonic in ρ the $100(1 - \alpha)\%$ confidence interval for p_0 is between $1 - \rho_u$ and $1 - \rho_L$.

The same reasoning can be applied to the M/M/c queueing system with $\hat{\lambda} = n_a/t$ and $\hat{\mu} = n_c/t_b$, where n_a is the total number of arrivals in a time interval of length t, n_c is the number of service completions by one server, and t_b is the total service time for the same server. Then $\hat{\rho} = \hat{\lambda}/c\hat{\mu}$. Some results worked out by Lilliefors [4] are shown in Table 7.3.1. For all the models

$$\hat{\rho} = \frac{\hat{\lambda}}{c\hat{\mu}} = \frac{n_a/t}{cn_c/t_b}.$$

For all cases of exponential service time

$$\rho_u = \hat{\rho}f_{\alpha/2;\,2n_a,\,2n_c} \quad \text{and} \quad \rho_L = \hat{\rho}/f_{\alpha/2;\,2n_c,\,2n_a}.$$

For the $M/E_k/1$ queueing system

$$\rho_u = \hat{\rho}f_{\alpha/2;\,2n_a,\,2n_ck} \quad \text{and} \quad \rho_L = \hat{\rho}/f_{\alpha/2;\,2n_ck,\,2n_a}.$$

TABLE 7.3.1

Confidence Interval Estimators for Parameters of Queueing Systems

Queueing model	Parameter	$100(1-\alpha)\%$ confidence interval	
M/M/1	$L = \dfrac{\rho}{1-\rho}$	$\dfrac{\rho_L}{1-\rho_L},$	$\dfrac{\rho_u}{1-\rho_u}$
	$L_q = \dfrac{\rho^2}{1-\rho}$	$\dfrac{\rho_L^2}{1-\rho_L},$	$\dfrac{\rho_u^2}{1-\rho_u}$
	$p_0 = 1-\rho$	$1-\rho_u,$	$1-\rho_L$
M/M/2	$L = \dfrac{2\rho}{1-\rho^2}$	$\dfrac{2\rho_L}{1-\rho_L^2},$	$\dfrac{2\rho_u}{1-\rho_u^2}$
	$L_q = \dfrac{2\rho^3}{1-\rho^2}$	$\dfrac{2\rho_L^3}{1-\rho_L^2},$	$\dfrac{2\rho_u^3}{1-\rho_u^2}$
	$p_0 = \dfrac{1-\rho}{1+\rho}$	$\dfrac{1-\rho_u}{1+\rho_u},$	$\dfrac{1-\rho_L}{1+\rho_L}$
$M/E_k/1$	$L = \dfrac{k+1}{2k}\dfrac{\rho^2}{1-\rho}+\rho$	$\dfrac{k+1}{2k}\dfrac{\rho_L^2}{1-\rho_L}+\rho_L,$	$\dfrac{k+1}{2k}\dfrac{\rho_u^2}{1-\rho_u}+\rho_u$
	$L_q = \dfrac{k+1}{2k}\dfrac{\rho^2}{1-\rho}$	$\dfrac{k+1}{2k}\dfrac{\rho_L^2}{1-\rho_L},$	$\dfrac{k+1}{2k}\dfrac{\rho_u^2}{1-\rho_u}$
	$p_0 = 1-\rho$	$1-\rho_u,$	$1-\rho_L$

Example 7.3.1 An M/M/1 queueing system is observed for 1000 time units during which 60 customers arrive, are served, and depart. The service facility is busy for 500 time units. Hence

$$\hat{\rho} = \frac{\hat{\lambda}}{\hat{\mu}} = \frac{n_a/t}{n_c/t_b} = \frac{60/1000}{60/500} = 0.5,$$

$$\rho_u = \hat{\rho} f_{0.025;120,120} = 0.5 \times 1.43 = 0.715,$$

and

$$\rho_L = \hat{\rho}/f_{0.025;120,120} = 0.5/1.43 = 0.350.$$

Hence, the maximum likelihood estimate of L is $0.5/(1-0.5) = 1$ with a 95% confidence interval of

$$\frac{0.35}{1-0.35} = 0.538, \qquad \frac{0.715}{1-0.715} = 2.509.$$

The point estimate of p_0 is $1 - 0.5 = 0.5$ with a 95% confidence interval of $1 - 0.715 = 0.285$ to $1 - 0.35 = 0.65$. The point estimate of L_q is $0.5^2/(1-0.5) = 0.5$ with a 95% confidence interval of

$$\frac{0.35^2}{1-0.35} = 0.1885 \quad \text{to} \quad \frac{0.715^2}{1-0.715} = 1.794.$$

7.4 SUMMARY

In this chapter we have considered some of the problems of statistical inference; that is, what we can determine about a random variable from a random sample of the values it assumes. We have been primarily concerned with making estimates of the parameters of a random variable and of making probability judgements about the accuracy of these estimates.

Student Sayings

A Student wouldn't have a Gosset of a chance in this class.
William Gosset will attend the student tea.

Exercises

1. [HM15] Suppose the random variable X has the density function

$$f(x) = \begin{cases} (1+\lambda)x^\lambda, & 0 < x < 1 \\ 0 & \text{otherwise.} \end{cases}$$

Show that the maximum likelihood estimate of λ based on a given random sample of size n is

$$\hat{\lambda} = -\left(1 + \frac{n}{\sum_{i=1}^{n} \ln x_i}\right).$$

2. [00] Consider a uniform random variable defined on the interval $0 \le x \le \beta$, where β is unknown. Use the method-of-moments to find an estimate of β based on a given random sample of size n.

3. [C15] If the $n = 20$ values of processing time mentioned in Example 7.1.2 are 16.39, 25.09, 16.31, 20.94, 17.58, 19.06, 17.21, 18.48, 16.88, 15.51, 25.87, 17.63, 29.13, 21.34, 11.14, 26.03, 23.28, 21.13, 18.46, 14.25, find the method-of-moments estimates of α and λ.

4. [12] Suppose you are the chief estimator for Swamp Gas Corporation. You have taken a random sample of size 121 from a normal population with unknown mean μ and variance σ^2 and calculate $\bar{x} = 12.9$ with $s = 3.2$. What is the 99% confidence interval for μ and for σ^2? If $\bar{x}$ and s were the same for a sample of size 9, what would be the 95% confidence interval for μ?

5. [C15] A sample believed to be from a normal population consists of the 20 numbers 14.56, 20.55, 14.1, 21.2, 24.57, 24.13, 9.68, 19.09, 14.51, 21.77, 14.72, 16.53, 24.92, 21.4, 14.95, 31.43, 17.86, 12.72, 18.54, and 23.92. Find a 95% confidence interval for the mean μ and the variance σ^2 of the population.

6. [18] A queueing system at the Huge Fleabite Corporation has been identified as an $M/E_2/1$ queueing system. Their system is observed for 500 time units. During this time 30 customers arrive and depart; the service facility is busy for 250 time units. Find a 90% confidence interval for ρ, L, L_q, and p_0.

References

1. E. Kreyszig, *Introductory Mathematical Statistics*. Wiley, New York, 1970.
2. H. Cramér, *Mathematical Methods of Statistics*. Princeton University Press, Princeton, New Jersey, 1946.
3. R. V. Hogg and A. T. Craig, *Introduction to Mathematical Statistics*, 3rd ed. Macmillan, New York, 1970.
4. H. W. Lilliefors, Some confidence intervals for queues, *Operations Res.* **14** (4), (1966), 723–727.
5. D. Gross and C. M. Harris, *Fundamentals of Queueing Theory*. Wiley, New York, 1974.
6. F. S. Hillier and G. J. Lieberman, *Introduction to Operations Research*, 2nd ed. Holden-Day, San Francisco, 1974.

A thousand probabilities do not make one truth.
Italian Proverb

Chapter Eight

HYPOTHESIS TESTING

8.1   INTRODUCTION

In this chapter we continue our study of statistical inference. Recall that statistical inference is the process of drawing conclusions about a population on the basis of a random sample. We, of course, assume that the population is determined by a random variable; that is, that the population consists of the possible values of a random variable. In Chapter 7 we examined the problem of estimating the values of the parameters of the random variable; we were concerned not only with making the estimates, but also with making probability judgements about the quality of our estimates.

Hypothesis testing is a procedure for determining, from information contained in a random sample from a population, whether to accept or reject a certain statement (hypothesis) about the random variable determining the population. A statistical hypothesis is usually stated as a proposition concerning the distribution of this random variable. It may be a statement about the values of one or more of the parameters of a given distribution; it also may concern the form of the distribution. Examples of statistical hypotheses follow.

1. The average response time at the specified terminal during the busiest period of the day (peak period) does not exceed 10 seconds.

2. The number of terminals in use during the first hour of the business day has a normal distribution with a mean of 2000 and a standard deviation of 300.

3. The average traffic rate into the central computer of an on-line system is 5000 messages per hour; the distribution is random.

4. The arrival pattern of inquiries to the central computer complex is Poisson.

A statistical hypothesis test is a formal, step-by-step procedure described below. It is based on the intuitively appealing idea that "Events which have low probabilities do not occur very often." That is, if a hypothesis implies a certain event to have a low probability, its occurrence is evidence *against* the hypothesis.

Hypothesis Test Procedure

1. Decide upon a null hypothesis H_0 and an alternative hypothesis H_1.

2. Select a test statistic; that is, a formula for calculating a number based upon the random sample, say $t(X_1, X_2, \ldots, X_n)$. (Common test statistics are the sample mean $\bar{x}$ or the sample variance s^2.)

3. Choose a *rejection region* (sometimes called a *critical region*) for values of the test statistic; that is, choose a set of possible test statistic values such that if H_0 is true, then the probability that the value of the statistic will fall in the rejection region is α. Here α is a preselected value, called the *level of significance of the test*. It is usually chosen to be either 0.05 or 0.01 (5% or 1%), but any value between 0 and 1 may be selected.

4. Calculate the test statistic of a random sample from the population. If this value falls in the rejection region, reject H_0 and accept H_1; otherwise accept H_0.

We illustrate the procedure in the following example.

Example 8.1.1 A manufacturer of magnetic tape for computers has discovered that the breaking strength X of the tape is normally distributed with a mean of 300 kilograms and a standard deviation of 20 kilograms. A new technique is developed for manufacturing the tape. A sample of 100 values of breaking strength of tape produced by the new process gives $\bar{x} = 290$ kilograms. At the 5% level of significance, does it appear that tape produced by the new process has a lower breaking strength than that produced by the old process?

Solution   As is customary for problems of this sort, we make the null hypothesis that there is no change. That is, we assume the average breaking strength μ is still 300 kilograms:

$$H_0 : \mu = \mu_0 = 300 \quad \text{kilograms.†}$$

Since we are concerned that the average breaking strength may have decreased, we choose

$$H_1 : \mu < 300 \quad \text{kilograms;} \qquad X \quad \text{normally distributed with} \quad \sigma = 20.$$

It should be evident that low values of $\bar{x}$ are evidence of decreased breaking strength. If H_0 is true, by the sampling theorem, Theorem 7.1.1, $\bar{X}$ is normally distributed with mean 300 and standard deviation $\sigma/\sqrt{n} = 20/\sqrt{100} = 2.0$. Thus it seems reasonable to choose as the rejection region all values less than a *critical value* $\bar{x}_c$, where $\bar{x}_c$ is determined so that $P[\bar{X} < \bar{x}_c] = 0.05$. This is illustrated in Fig. 8.1.1. If H_0 is true, we see that

$$P[\bar{X} < \bar{x}_c] = \Phi\left(\frac{\bar{x}_c - 300}{2}\right) = 0.05,$$

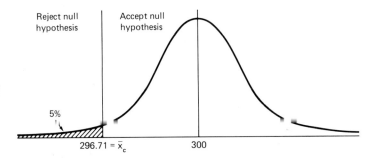

Fig. 8.1.1   Graph of density function for $\bar{X}$ in Example 8.1.1.

where Φ is the distribution function for the standard normal random variable. By Table 3 of Appendix A, this means that

$$(\bar{x}_c - 300)/2 = -1.645 \qquad \text{or} \qquad \bar{x}_c = 300 - 3.29 = 296.71.$$

Since $\bar{x} = 290 < \bar{x}_c = 296.71$, we reject H_0 and accept H_1. That is, we assume that the new process produces weaker tape than the old.

---

† The reader should note that in this chapter as in Chapter 7 we often use μ to represent the mean of a random variable; in the queueing theory sections of the book μ always represents the average service rate of a server in a service facility.

The reader has probably noted that statistical hypothesis testing is not foolproof. Thus, in Example 8.1.1, it is possible that, by pure chance, our sample of size 100 yielded a sample strength $\bar{x}$ below the *critical value* 296.71, while the mean breaking strength had not changed at all or had actually increased. This type of error is called a *Type I error*. The probability of making a Type I error is, of course, α because of the way we set up the statistical test. Another type of error we could make is accepting H_0 when it is false. This type of error is known (imaginatively) as a *Type II error*. We denote its size for a particular test by β. The concepts of Type I and Type II errors are illustrated in Table 8.1.1. In applying a given statistical test we can make only the two types of errors. However, statisticians have identified a couple more common errors made by other statisticians. They say we make a Type III error if we assume the wrong distribution for X and a Type IV error when we solve the wrong problem. These latter two types of errors are part of the statistical folklore but not of the discipline.

TABLE 8.1.1

Type I and Type II errors in testing
H_0 **against** H_1"

Decision	Unknown truth	
	H_0 true	H_1 true
Accept H_0	True decision $P = 1 - \alpha$	Type II error $P = \beta$
Accept H_1	Type I error $P = \alpha$	True decision $P = 1 - \beta$

^a P is the probability of the indicated result.

When comparing two tests of a given hypothesis H_0 with a given significance level, we would clearly choose the test which is most likely to avoid a Type II error. This is measured by the *power* of the test, which is $1 - \beta$. It is not always easy to calculate the power of a test. This is true for Example 8.1.1 because H_1 is a *composite hypothesis* rather than a *simple hypothesis*; for the latter, not only the type of distribution is assumed, but also the values of all the parameters. Suppose, in Example 8.1.1, H_1 was "The new process yields a tape with normally distributed breaking strength having a mean of 295 kilograms." (We also assume, of course, that $\sigma = 20$.) Then H_1 is simple and the test can be visualized as in Fig. 8.1.2. Here β, the probability of accepting H_0 when H_1 is true, is the area under the density function for $\bar{X}$ (when H_1 is true) to the right of 296.71.

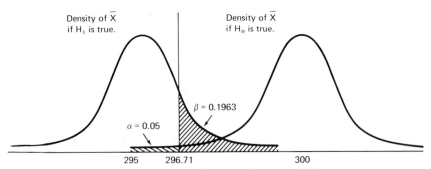

Fig. 8.1.2   Graph to illustrate Example 8.1.1.

Since, for this calculation, we assume H_1 is true, β is the area of the tail of a standard normal density to the right of the value

$$z = (296.71 - 295)/2 = 0.855.$$

By Table 3 of Appendix A, this value is 0.1963. Thus β is 0.1963 and the power of this test is 0.8037.

We can also see the relationship between α and β from this example. For, if we changed the level of significance α to 0.01, then our critical value $\bar{x}_c$ becomes $300 - 2 \times 2.326 = 295.35$. This means that β is the area under the standard normal density function to the right of $z = (295.35 - 295)/2 = 0.174$. Thus, by Table 3, $\beta = 0.43093$. Thus decreasing α (moving the critical value $\bar{x}_c$ to the left) increases β while increasing α (moving $\bar{x}_c$ to the right) decreases the value of β. Thus the level of significance, α, chosen for a test is always a compromise.

As we mentioned earlier, the "goodness" of a particular statistical test, for a fixed α and H_0, is measured by the power of the test which is

$$1 - \beta = P[\text{rejecting } H_0 | H_1 \text{ is true}].$$

In the sequel, whenever possible we shall compute the power of the tests discussed. In some cases it can be proven, mathematically, that a particular test has the maximum power of any test of a simple hypothesis H_0 at the α level of significance against an alternative composite hypothesis; such a test is called a *uniformly most powerful test*.

The reader should note that statisticians do not regard accepting the null hypothesis H_0 as being equivalent to believing it to be true. When we accept H_0 as the result of a test we merely believe that there is not sufficient evidence to reject it or that H_0 may be only approximately true. Suppose, for example, that the null hypothesis is "X is normally distributed with a mean $\mu = 30$ and a standard deviation $\sigma = 10$," while H_1 is "X is normally distributed with standard deviation $\sigma = 10$ but $\mu > 30$." Then, if the test fails to reject H_0 at a particular level of significance, say 5%, we still may believe that μ is not *exactly* 30; it may very well be 30.001 or 29.998. We are, however, reasonably sure that μ is not 35.6 or 40.2. That is, μ does not significantly exceed the value 30. In fact, the terminology used by statisticians is to say the result of a test is *significant* if the value of the test statistic falls in the critical (rejection) region.

8.2 TESTS CONCERNING MEANS

We will consider two types of hypothesis tests concerning means. The first type is a test of the null hypothesis that the mean of a population (that is, of the random variable X which determines the population) has some specified value μ_0 against an appropriate alternative hypothesis. The second type of test will be concerned with the differences between the means of two different populations.

Example 8.1.1 was an example of the first type of test. We will always have the null hypothesis $H_0 : \mu = \mu_0$. The alternative hypotheses we will consider are (a) $\mu > \mu_0$, (b) $\mu < \mu_0$, and (c) $\mu \neq \mu_0$. The resulting test is called a *one-tailed test* for alternatives (a) and (b); it is a *two-tailed test* for alternative (c). The reasons for these names will soon become apparent. We will state the hypothesis test procedure, which is a variant of the description given in Section 8.1, in the form of an algorithm.

Algorithm 8.2.1 (*Test of Value of a Mean*) Given the value of a random sample $x_1, x_2, \ldots, x_n$, from a population determined by the random variable X, this algorithm will determine at the α level of significance whether to accept or reject the null hypothesis that $\mu = E[X] = \mu_0$ against one of the alternative hypotheses (a) $\mu > \mu_0$, (b) $\mu < \mu_0$, or (c) $\mu \neq \mu_0$.

Step 1 [*Calculate the test statistic*] For the large sample case ($n \geq 30$) calculate the test statistic $z = (\bar{x} - \mu_0)/(\sigma/\sqrt{n})$ if σ is known; otherwise use the test statistic $z = (\bar{x} - \mu_0)/(s/\sqrt{n})$. If $n < 30$ and X is normally distributed (or approximately so), calculate the test statistic $t = (\bar{x} - \mu_0)/(s/\sqrt{n})$.

Step 2 [*Find the critical region*] For the large sample case and alternative hypotheses (a) or (b) find z_α such that $P[Z > z_\alpha] = \alpha$ for the standard normal random variable Z. For alternative hypothesis (c) find $z_{\alpha/2}$. The critical region for alternative hypothesis (a) is $z > z_\alpha$; for alternative hypothesis (b) it is $z < -z_\alpha$. For alternative hypothesis (c) the critical region consists of the two tails $z < -z_{\alpha/2}$ and $z > z_{\alpha/2}$. (These critical regions are shown graphically in Fig. 8.2.1.) If $n < 30$, for alternative hypotheses (a) and (b) find t_α such that $P[T > t_\alpha] = \alpha$ for a Student-t random variable T having $n - 1$ degrees of freedom. For alternative hypothesis (c) find $t_{\alpha/2}$ such that

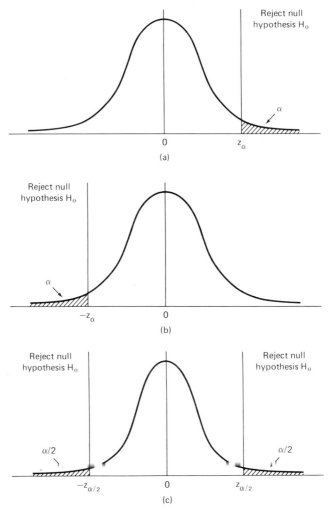

Fig. 8.2.1   Test criteria for Algorithm 8.2.1. Alternative hypotheses:
(a) $\mu > \mu_0$; **(b)** $\mu < \mu_0$; **(c)** $\mu \neq \mu_0$.

$P[T > t_{\alpha/2}] = \alpha/2$ for the random variable T described in the last sentence. The critical region for alternative hypothesis (a) is $t > t_\alpha$; for alternative hypothesis (b) it is $t < -t_\alpha$. The critical region for alternative hypothesis (c) is the two tails $t > t_{\alpha/2}$ and $t < -t_{\alpha/2}$.

Step 3   [Accept or reject H_0]   If the test statistic calculated in Step 1 falls in the critical region determined in Step 2, reject H_0; otherwise accept H_0.

We will not give a formal proof of Algorithm 8.2.1, but merely an indication of why it is reasonable to expect it to be true. A formal proof can be found in Kreyszig [1].

For the large sample case ($n \geq 30$), by the sampling theorem, Theorem 7.1.1, $(\bar{x} - \mu_0)/(\sigma/\sqrt{n})$ has approximately a standard normal distribution, and, if σ is not known, s is a good approximation to σ. Hence, each of the critical regions defined in Step 2 of the algorithm does have probability α. Similarly, by the same theorem, for small samples from a normal population, $(\bar{x} - \mu_0)/(s/\sqrt{n})$ has a Student-t distribution with $n - 1$ degrees of freedom; again, the critical regions determined in Step 2 each have probability α. (When William Gosset, using the pseudonym "A. Student," first published the small sample test (see [12] of Chapter 3) it was rumored that the nature of the sampled material, a product of the Guinness brewery, made a small sample necessary in order to obtain reliable data.) It can be shown (see Kreyszig [1]) that if X is normally distributed with σ known, then for both the one-tailed alternatives this algorithm provides the uniformly most powerful tests.

We now give some examples of the use of the algorithm.

Example 8.2.1 A random sample, of size 400, of the response time, during the peak period, of the on-line inquiry system at Hemo Globin Finance yields a sample mean $\bar{x}$ of 21 time units with a sample standard deviation s of 12 time units. A primary performance requirement for the system is that the mean response time for the peak period should not exceed 20 time units; if the system usage increases to the extent that this criterion is violated, changes are to be made to the system to bring the performance to this standard. Test, at the 1% level of significance, whether or not the performance criterion has been maintained.

Solution The null hypothesis is that $\mu = 20$; the alternative hypothesis is that $\mu > 20$. In Step 1 of Algorithm 8.2.1 we calculate $z = (21 - 20)/(12/\sqrt{400}) = 20/12 = 1.67$. In Step 2 we find the critical region is the set of all z's greater than $z_{0.01} = 2.326$. Thus we cannot reject H_0 at the 1% level of significance. Based on this test we would make no changes to the existing on-line system. It should be noted that at the 5% level of significance changes would be required, since $z_{0.05} = 1.645$.

Example 8.2.2 The average time it takes a clerk to service a customer at Holicow Savings is 3 minutes with a standard deviation of 1 minute; the service time is normally distributed. To test the feasibility of installing an on-line computer system to provide better customer service with fewer clerks, a representative clerk is trained to use such a system. This clerk, using a prototype system, processes 16 customers yielding a sample mean of 2 minutes with a sample standard deviation of 40 seconds. Holicow Savings officials are convinced that the proposed on-line system will be cost effective if with it the mean clerk service time does not exceed 110 seconds. At the 5% level of significance does the test indicate that the on-line system should be implemented?

Solution We assume the service time of the clerk remains normally distributed. H_0 is $\mu = 110$ seconds; H_1 is $\mu > 110$ seconds. In Step 1 of Algorithm 8.2.1 we calculate $t = (120 - 110)/(40/4) = 1$. In Step 2, by Table 5 of Appendix A, the critical region is the set of all $t > t_{0.05,15} = 1.753$. Hence we cannot reject the null hypothesis at the 5% level of significance; the on-line system should be installed. Of course management would be wise to repeat the experiment with other clerks and more customers—it is possible that the result is due to the Hawthorn effect. (The Hawthorn effect is the well-known effect, first documented in a famous productivity experiment at the Hawthorn Works of the Western Electric Company, in which people have a tendency to perform better while taking part in a publicized experiment than they might under more normal circumstances. See the Bell System advertisement [2].)

So far we have discussed the size α of the Type I error and the size β of the Type II error but have not emphasized the fact that the sample size n is important in both types of error; that is, both α and β decrease as n is increased. In some cases we can choose the required sample size to give ourselves the desired power in a hypothesis test. For example, suppose X is approximately normal with σ known. Suppose, further, that we want the power $1 - \beta$ when the true mean $\mu = \mu_0 + \delta$. Then we have

$$1 - \beta = P\left[\frac{\bar{x} - \mu_0}{\sigma/\sqrt{n}} > z_\alpha \Big| \mu = \mu_0 + \delta\right]$$

$$= P\left[\frac{\bar{x} - (x_0 + \delta)}{\sigma/\sqrt{n}} > z_\alpha - \frac{\delta}{\sigma/\sqrt{n}} \Big| \mu = \mu_0 + \delta\right].$$

But, if $\mu = \mu_0 + \delta$, then the test statistic

$$\frac{\bar{X} - (\mu_0 + \delta)}{\sigma/\sqrt{n}}$$

has approximately a standard normal distribution. This implies that

$$1 - \beta = P[z > z_\alpha - (\delta\sqrt{n}/\sigma)]$$

or

$$-z_\beta = z_\alpha - (\delta\sqrt{n}/\sigma) \qquad \text{(see Fig. 8.2.2).}$$

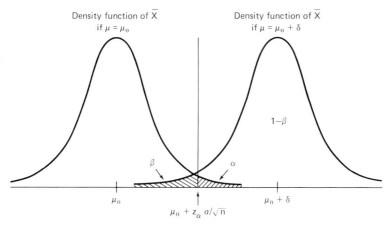

Fig. 8.2.2 Graph to illustrate the calculation of sample size needed to achieve desired power.

Hence we must have

$$n = (z_\alpha + z_\beta)^2 \sigma^2 / \delta^2.$$

The same size n is required for the alternative hypothesis $\mu < \mu_0$ (see Exercise 2). For the two-tailed test the required sample size

$$n = (z_{\alpha/2} + z_{\beta/2})^2 \sigma^2 / \delta^2 \qquad \text{(see Exercise 3)}.$$

We now consider some tests of the difference between two population means. We give these tests in the form of algorithms. These tests as well as the tests in Algorithm 8.2.1 are summarized in Table 8.5.1 at the end of this chapter.

Algorithm 8.2.2 (*Test of Difference of Two Means, Large Samples*) Given a random sample $x_1, x_2, \ldots, x_n$ from a population determined by X and a random sample $y_1, y_2, \ldots, y_m$ from a population determined by Y, where X and Y are independent, this algorithm will determine, at the α level of significance, whether to accept or reject the null hypothesis $\mu_X - \mu_Y = d_0$ against one of the alternative hypotheses (a) $\mu_X - \mu_Y > d_0$, (b) $\mu_X - \mu_Y < d_0$, or (c) $\mu_X - \mu_Y \neq d_0$. It is assumed that $n \geq 30$, $m \geq 30$, and that either (a) both σ_X and σ_Y are known or (b) s_X and s_Y are good estimates of the respective standard deviations.

Step 1 [*Calculate the test statistic*] Compute the test statistic

$$z = \frac{(\bar{x} - \bar{y}) - d_0}{\sqrt{(\sigma_X^2 / n) + (\sigma_Y^2 / m)}}.$$

If σ_X and σ_Y are not known, substitute s_X and s_Y for them.

Step 2 [*Determine the critical region*]   For alternative hypothesis (a) the critical region is the set of all $z > z_\alpha$; for alternative hypothesis (b) it is the set of all $z < -z_\alpha$; and for alternative hypothesis (c) it consists of the two tails $z < -z_{\alpha/2}$ and $z > z_{\alpha/2}$.

Step 3 [*Accept or reject H_0*]   If the test statistic of Step 1 falls in the critical region determined in Step 2, reject H_0; otherwise accept H_0.

The proof of Algorithm 8.2.2 can be found in Walpole and Myers [3].

Example 8.2.3   Hactoff Insurance has two separate on-line computer systems used to calculate insurance rates. The company wants to keep the systems balanced by assigning users to the systems in such a way that the mean response time is the same for each system. A sample of 100 response times from the first system yields $\bar{x} = 20.24$ time units with $s_X = 5.6$ time units; a sample of 120 response times from the second system yields $\bar{y} = 18.72$ time units with $s_Y = 4.2$ time units. Test the hypothesis, at the 0.025 level of significance, that $\mu_X > \mu_Y$.

Solution   We set $d_0 = 0$ in Algorithm 8.2.2. We calculate the statistic

$$z = \frac{20.24 - 18.72}{(5.6^2/100 + 4.2^2/120)^{1/2}} = 2.24.$$

Since $z_{0.025} = 1.96$ we reject the null hypothesis that $\mu_X = \mu_Y$ and shift some of the users from the first system to the second. Note that the hypothesis we wanted to test was *not* made the null hypothesis.

Sometimes we must compare means of different populations when only small samples are available. The next algorithm shows how to do this if the two populations are normally distributed.

Algorithm 8.2.3   (*Test of Difference of Two Means, Small Samples, Normal Populations*)   Given a random sample $x_1, x_2, \ldots, x_n$ from an approximately normally distributed population determined by X and a random sample $y_1, y_2, \ldots, y_m$ from an approximately normally distributed population determined by Y, where X and Y are independent with $\sigma_X = \sigma_Y$ but unknown, this algorithm will determine, at the α level of significance, whether to accept or neglect the null hypothesis $\mu_X - \mu_Y = d_0$ against one of the alternative hypotheses (a) $\mu_X - \mu_Y > d_0$, (b) $\mu_X - \mu_Y < d_0$, or (c) $\mu_X - \mu_Y \neq d_0$. It is assumed that $n < 30$ and $m < 30$, for otherwise we could use Algorithm 8.2.2.

Step 1 [*Calculate the test statistic*]   Compute

$$t = \frac{(\bar{x} - \bar{y}) - d_0}{s_p[(1/n) + (1/m)]^{1/2}} \qquad \text{where} \quad s_p = \left[\frac{(n-1)s_X^2 + (m-1)s_Y^2}{n+m-2}\right]^{1/2}.$$

Step 2 [*Determine the critical region*]   For alternative hypothesis (a) the critical region is the set of all $t > t_{\alpha, n+m-2}$; for alternative hypothesis (b) it is the set of all $t < -t_{\alpha, n+m-2}$; and for alternative hypothesis (c) it is the two tails $t < -t_{\alpha/2, n+m-2}$ and $t > t_{\alpha/2, n+m-2}$.

Step 3 [*Accept or reject H_0*]   If the test statistic calculated in Step 1 falls in the critical region found in Step 2, reject the null hypothesis; otherwise accept H_0.

The proof of Algorithm 8.2.3 can be found in Walpole and Myers [3]. The crux of the proof is the demonstration that the test statistic has a Student-t distribution with $n + m - 2$ degrees of freedom.

Example 8.2.4   The Hotstuff Chili Company has two teams of programmers independently develop the software for a new on-line application. It was found that the 9 programmers in the first team averaged 50 lines of code per programmer day with a sample standard deviation of 6.6 while the second team of 12 programmers averaged 45 lines of code per programmer day with a sample standard deviation of 5.4. Assuming that lines of code per programmer day are approximately normally distributed for both groups with about the same standard deviation, test the hypothesis that the first group is more productive than the second using $\alpha = 0.05$.

Solution   Algorithm 8.2.3 applies. (Why else would I put this example here?) Let X be the lines of code per programmer day for the first group and Y the number for the second group.

The null hypothesis is $\mu_X - \mu_Y = 0$; the alternative hypothesis is $\mu_X - \mu_Y > 0$. The test statistic t is given by

$$t = \frac{50 - 45}{s_p(1/9 + 1/12)^{1/2}} = \frac{11.3389}{5.9349} = 1.911.$$

Since $t_{0.05, 19} = 1.729$, we reject H_0 and conclude that based on this test, the first group is more productive than the second. Note that, again, the hypothesis we wanted to test is not stated as the null hypothesis.

Sometimes the situation is even more complicated than that described in Algorithm 8.2.3; the variances of the two populations may be very different or there may be no reason to suppose they are the same.

Algorithm 8.2.4   (*Test of Difference of Two Means, Small Samples, Different Variances*)   Given a random sample $x_1, x_2, \ldots, x_n$ from an approximately normally distributed population determined by X and a random sample $y_1, y_2, \ldots, y_m$ from an approximately normally distributed population determined by Y, where X and Y are independent with $\sigma_X \neq \sigma_Y$ and both unknown, this algorithm will determine, at the α level of significance, whether to accept or reject the null hypothesis $\mu_X - \mu_Y = d_0$ against one of the alternative hypotheses (a) $\mu_X - \mu_Y > d_0$, (b) $\mu_X - \mu_Y < d_0$, or (c) $\mu_X - \mu_Y \neq d_0$. It is assumed that $n < 30$ and $m < 30$, for otherwise we could use Algorithm 8.2.2.

Step 1 [*Calculate the test statistic*] Compute

$$t = \frac{(\bar{x} - \bar{y}) - d_0}{[(s_X^2/n) + (s_Y^2/m)]^{1/2}}.$$

Step 2 [*Determine the critical region*] Compute

$$v = \frac{(s_X^2/n + s_Y^2/m)^2}{(s_X^2/n)^2/(n-1) + (s_Y^2/m)^2/(m-1)}.$$

The test statistic is Student-t with v degrees of freedom so the critical region for alternative hypothesis (a) is the set of all $t > t_{\alpha,v}$; for alternative hypothesis (b) it is the set of all $t < -t_{\alpha,v}$; and for alternative hypothesis (c) it is the two tails $t < -t_{\alpha/2,v}$ and $t > t_{\alpha/2,v}$.

Step 3 [*Accept or reject H_0*] If the test statistic calculated in Step 1 falls in the critical region found in Step 2 reject H_0; otherwise accept it.

This algorithm also is proven in Walpole and Myers [3].

Example 8.2.5 ZYX Incorporated has used two different self-study courses, Course A and Course B, to teach its beginning programmers good programming techniques. The success of each course is evaluated by the scores the new programmers achieve on the ZYX Programmers Test. Nine students using Course A achieved an average test score of 89.6 with a sample standard deviation of 3.6. The seven new programmers who used Course B got an average score of 81.9 with a sample standard deviation of 12.7. Assuming all test scores are normally distributed, test the null hypothesis $\mu_X = \mu_Y$ against the alternative that $\mu_X > \mu_Y$, at the 10% level of significance, using Algorithm 8.2.4.

Solution We calculate the test statistic

$$t = \frac{89.6 - 81.9}{[(3.6^2/9) + (12.7^2/7)]^{1/2}} = 1.556.$$

$$v = \frac{(3.6^2/9 + 12.7^2/7)^2}{(3.6^2/9)^2/8 + (12.7^2/7)^2/6} = 6.75,$$

so we use $v = 7$.

Since $t_{0.1,7} = 1.415$, we reject H_0 at the 10% level of significance and decide Course A is the most effective of the two courses.

8.3 TESTS CONCERNING VARIANCES

Unfortunately the tests available concerning variances are not as extensive as those concerning means and the conditions under which the tests are valid are more restrictive. We must rejoice in what is available. The first test, stated in the form of an algorithm, as is our custom in this chapter, is an immediate consequence of Theorem 7.2.3, so no proof will be presented.

Algorithm 8.3.1 (*Test of Value of a Variance*) Given a random sample $x_1, x_2, \ldots, x_n$ from an approximately normally distributed population determined by X this algorithm will determine, at the α level of significance, whether to accept or reject the null hypothesis $\sigma^2 = \sigma_0^2$ against one of the alternative hypotheses (a) $\sigma^2 > \sigma_0^2$, (b) $\sigma^2 < \sigma_0^2$, or (c) $\sigma^2 \neq \sigma_0^2$.

Step 1 [*Calculate the test statistic*] Compute

$$\chi^2 = (n-1)s^2/\sigma_0^2.$$

(χ^2 has approximately a chi-square distribution with $n-1$ degrees of freedom.)

Step 2 [*Determine the critical region*] For alternative hypothesis (a) the critical region is all $\chi^2 > \chi^2_{\alpha,n-1}$; for alternative hypothesis (b) it is all $\chi^2 < \chi^2_{1-\alpha,n-1}$; and for alternative hypothesis (c) it consists of the two tails $\chi^2 < \chi^2_{1-\alpha/2,n-1}$ and $\chi^2 > \chi^2_{\alpha/2,n-1}$.

Step 3 [*Accept or reject H_0*] If the test statistic calculated in Step 1 falls in the critical region found in Step 2, reject H_0; otherwise accept H_0.

It is relatively easy to calculate the power of the hypothesis test of Algorithm 8.3.1. Suppose, for example, that the null and alternative hypotheses are $\sigma^2 = \sigma_0^2$ and $\sigma^2 > \sigma_0^2$, respectively. For $\sigma_1^2 > \sigma_0^2$ the power of the test is given by

$$1 - \beta = P[(n-1)s^2/\sigma_0^2 > \chi^2_{\alpha,n-1} \,|\, \sigma^2 = \sigma_1^2 > \sigma_0^2]$$

$$= P[(n-1)s^2/\sigma_1^2 > (\sigma_0^2/\sigma_1^2)\chi^2_{\alpha,n-1} \,|\, \sigma^2 = \sigma_1^2].$$

But, if $\sigma^2 = \sigma_1^2$, then $(n-1)s^2/\sigma_1^2$ has a chi-square distribution with $n-1$ degrees of freedom; we can indicate it by χ^2_{n-1}. Thus we have the power given by

$$1 - \beta = P[\chi^2_{n-1} > (\sigma_0^2/\sigma_1^2)\chi^2_{\alpha,n-1}]. \tag{8.3.1}$$

Example 8.3.1 Scores achieved by persons taking the Programmer Aptitude Test at Brownstain International have been normally distributed with mean 82.6 and variance 19.78 until two years ago. However, the test series for the 150 tests served in the last two years yielded a sample mean of

83.2 and a sample variance of 27.3. At the 5% level of significance, does Algorithm 8.3.1 indicate that the variance has increased? What is the power of the test if the actual variance is 26.0?

Solution   The null hypothesis is $\sigma^2 = 19.78$ and the alternative hypothesis is $\sigma^2 > 19.78$. The test statistic

$$\chi^2 = (149 \times 27.3)/19.78 = 205.65.$$

We can calculate $\chi^2_{0.05,149}$ by formula (26.4.17) of Abramowitz and Stegun [4] which, in our notation, is

$$\chi^2_{\alpha,k} = k[1 - (2/9k) + z_\alpha(2/9k)^{1/2}]^3, \tag{8.3.2}$$

valid for $k > 100$. Thus we calculate

$$\chi^2_{0.05,149} = 149\{1 - [2/(9 \times 149)] + 1.645[2/(9 \times 149)]^{1/2}\}^3 = 178.49.$$

Therefore, we reject H_0 and conclude that $\sigma^2 > 19.78$. By (8.3.1) the power of the test when $\sigma^2 = 26.0$ is

$$P[\chi^2_{149} > 19.78/26.0 \times 178.49] = P[\chi^2_{149} > 135.79].$$

By Theorem 3.2.6, χ^2_{149} is approximately normally distributed with mean 149 and variance 298 (standard deviation 17.263). Hence

$$P[\chi^2_{149} > 135.79] = P[z > (135.79 - 149)/17.263]$$

$$= P[z > -0.7652] = P[z < 0.7652] = 0.77792,$$

by Table 3 of Appendix A.

Our final algorithm of this section tells us how to test the difference of the variances from two independent normal populations.

Algorithm 8.3.2   (*Test of Difference of Two Variances, Normal Populations*)   Given a random sample $x_1, x_2, \ldots, x_n$ from an approximately normal population determined by X and a random sample $y_1, y_2, \ldots, y_m$ from an approximately normal population determined by Y, where X and Y are independent, this algorithm will determine, at the α level of significance, whether to accept or reject the null hypothesis $\sigma_X^2 = \sigma_Y^2$ against one of the alternative hypotheses (a) $\sigma_X^2 > \sigma_Y^2$, (b) $\sigma_X^2 < \sigma_Y^2$, or (c) $\sigma_X^2 \neq \sigma_Y^2$.

Step 1   [*Calculate the test statistic*]   Compute

$$f = s_X^2/s_Y^2.$$

Step 2   [*Determine the critical region*]   The test statistic has a Snedcor-F distribution with $v_1 = n - 1$ and $v_2 = m - 1$ degrees of freedom. Hence the critical region for alternative hypothesis (a) is the set of all $f > f_\alpha(v_1, v_2)$;

for alternative hypothesis (b) it is the set of all $f < f_{1-\alpha}(v_1, v_2)$; and for alternative hypothesis (c) it is the two tails $f < f_{1-\alpha/2}(v_1, v_2)$ and $f > f_{\alpha/2}(v_1, v_2)$.

Step 3 [*Accept or reject H_0*]   If the value of the test statistic calculated in Step 1 falls in the critical region found in Step 2, reject H_0; otherwise, accept H_0.

The proof of Algorithm 8.3.2 appears in Walpole and Myers [3].

Example 8.3.2   Use Algorithm 8.3.2 to determine, at the 5% level of significance, whether it is reasonable to suppose that the variance of the number of lines of code per programmer day is the same for the two groups of programmers in Example 8.2.4. Use the alternative hypothesis that the variance is larger for the first group.

Solution   The test statistic is

$$s_X^2/s_Y^2 = (6.6/5.4)^2 = 1.494.$$

We know that the test statistic has a Snedcor-F distribution with 8 and 11 degrees of freedom. Since

$$f_{0.05}(8, 11) = 2.95,$$

we cannot reject H_0 and conclude that it is reasonable to suppose the two populations have the same variance. Recall that, in the solution of Example 8.2.4 we found that the means did not appear to be equal.

8.4   GOODNESS-OF-FIT TESTS

In Sections 8.1–8.3 we have at least partially answered Cosmic Question 3 of Chapter 7. In this section we will consider Cosmic Question 2; that is, we will consider the problem of testing whether a given population is determined by a particular type of random variable such as exponential, normal, Poisson, Erlang-k, etc. We will, of course, have some clues from the parameters such as mean and standard deviation as to what type of random variable we are dealing with. For example, if $\bar{x} = 15$ and $s = 30$, we know X is not likely to be exponential, since the mean and standard deviation are equal for exponential distributions, but it could be hyperexponential. In many cases we have history, experience, or physical reasons for expecting a random variable to be of a certain type.

Once we have established what type of random variable we believe determines a population, we can apply a goodness-of-fit test to make a probability judgement about our choice. A goodness-of-fit test is a special hypothesis test in which the null hypothesis is that the population is determined by a particular type of random variable (normal, exponential, etc.) and the alternative hypothesis is that it is not. Unfortunately, when a goodness-of-fit test leads to the rejection of the null hypothesis, no conclusion is obtained as to what type of random variable might fit better. The two goodness-of-fit tests we will consider are the chi-square test and the Kolmogorov–Smirnov test.

Algorithm 8.4.1   (*Chi-Square Goodness-of-Fit Test*)   Each element of a given random sample $x_1, x_2, \ldots, x_n$ falls into exactly one of k categories $C_1, C_2, \ldots, C_k$. This test will determine at the α level of significance, whether or not it is reasonable to suppose the observed distribution of the n sample values into categories is consistent with the null hypothesis that X has the given distribution.

Step 1   Count the number O_i of observed elements in category C_i, for $i = 1, 2, \ldots, k$.

Step 2   On the basis of the null hypothesis, calculate E_i, the expected number of elements in category C_i, for $i = 1, 2, \ldots, k$.

Step 3   Calculate the chi-square statistic

$$\chi^2 = \sum_{i=1}^{k} \frac{(O_i - E_i)^2}{E_i}.$$

Step 4   [*Calculate the number of degrees of freedom m of the underlying chi-square distribution*]   Set $m = k - 1$. Then subtract one from m for each independent parameter that is estimated from the data to generate the E_i values in Step 2.

Step 5   Find the critical value χ_α^2 such that the probability a chi-square random variable with m degrees of freedom will exceed χ_α^2 is α. (Table 4 of Appendix A gives these values.)

Step 6   If $\chi^2 \geq \chi_\alpha^2$ then reject H_0; otherwise accept H_0.

The mathematical justification for the chi-square test is that the distribution of the χ^2 statistic approaches a chi-square distribution with m degrees of freedom as $n \to \infty$. This is proven in Cramér [7]. For small values of n the χ^2 distribution may not be closely approximated by a chi-square distribution.

It has been found, empirically, that the chi-square test works best when all the E_i are at least 5 for, if E_i is small, the division by E_i in the term $(O_i - E_i)^2/E_i$ can cause a large error in the value of χ^2. To make each $E_i \geq 5$ we sometimes must pool categories.

Example 8.4.1 Data is collected on the number of messages arriving at a message switching center during the peak period. The results are tabulated in Table 8.4.1. The table shows that, for the 500 different minutes observed, there were no message arrivals during 4 of the minutes, exactly one message arrival during 17 different observed minutes, etc. At the 5% level of significance, does the arrival pattern appear to be Poisson?

TABLE 8.4.1

Data on Number of Message Arrivals in One-Minute Period

i:	0	1	2	3	4	5	6	7	8	9	10	11	12	13	14	15	≥ 16	
O_i:	4	17	42	66	86	101	69	52	24	19	11	4	1	2	0	2	0	$\sum = 500$

Solution We use the chi-square test. Let X be the random variable which counts the number of arrivals during a one-minute period. Step 1 has already been done. To determine the E_i or expected numbers we must find the probability of i arrivals in one minute, assuming a Poisson distribution. Since the parameter λ of the assumed Poisson process was not part of the null hypothesis H_0, we must estimate it from the data. λ should be the average number of arrivals per minute so we use the estimate

$$\hat{\lambda} = \sum_{i=0}^{15} \frac{iO_i}{500},$$

that is, the total number of arrivals in 500 minutes, divided by 500. (By Examples 7.1.3 and 7.1.6 this is both the method-of-moments estimate and the maximum likelihood estimate of λ.) This calculation yields 5.022, so we use the estimate of $\hat{\lambda} = 5$ arrivals per minute. Using this λ and the Poisson formula

$$P[X = i] = e^{-\lambda} \frac{\lambda^i}{i!}, \qquad i = 0, 1, 2, \ldots,$$

we generate Table 8.4.2. We calculate the E_i by the formula

$$E_i = 500P[X = i].$$

Table 8.4.2 must be modified since there are too few expected in a number of the categories and the total probabilities do not add up to 1. We make a new category of "11 or more arrivals" and assign it the probability $1 - P[X \leq 10]$ to generate Table 8.4.3.

From Table 8.4.3 we calculate

$$\chi^2 = \sum_{i=0}^{11} \frac{(O_i - E_i)^2}{E_i} = 6.057.$$

TABLE 8.4.2

Table for Chi-Square Test

i	O_i	$P[X = i]$	E_i
0	4	0.00674	3.37
1	17	0.03369	16.845
2	42	0.08422	42.11
3	66	0.14037	70.185
4	86	0.17547	87.735
5	101	0.17547	87.735
6	69	0.14622	73.11
7	52	0.10444	52.22
8	24	0.06528	32.64
9	19	0.03627	18.135
10	11	0.01813	9.065
11	4	0.00824	4.12
12	1	0.00343	1.715
13	2	0.00132	0.66
14	0	0.00047	0.235
15	2	0.00016	0.08

We estimated the parameter λ to generate the E_i values, so the number of degrees of freedom m in our underlying chi-square distribution is 10. By Table 4 of Appendix A, the critical value $\chi^2_{0.05}$ is 18.31. Since $\chi^2 = 6.057$ is less than 18.31, we accept the hypothesis that the arrival pattern is Poisson.

Some statisticians would object to our using a category with an E_i of only 3.37. To avoid this we could pool the categories $i = 0$ and $i = 1$. Then we would have 11 categories, a χ^2 value of 5.968, and $\chi^2_{0.05} = 16.92$; the result would again be to accept the null hypothesis.

TABLE 8.4.3

Modified Table for Chi-Square Test

i	O_i	$P[X = i]$	E_i
0	4	0.00674	3.37
1	17	0.03369	16.845
2	42	0.08422	42.11
3	66	0.14037	70.185
4	86	0.17547	87.735
5	101	0.17547	87.735
6	69	0.14622	73.11
7	52	0.10444	52.22
8	24	0.06528	32.64
9	19	0.03627	18.135
10	11	0.01813	9.065
≥ 11	9	0.0137	6.85

We now consider the Kolmogorov–Smirnov goodness-of-fit test. (Contrary to what some may think, it is neither a sobriety test nor a test performed in a bar.)

Given the values of a random sample of size n, $x_1, x_2, \ldots, x_n$, we define the *sample cumulative distribution function* $S_n(\cdot)$ by $S_n(x) = i/n$, where i is the number of sample values $\leq x$. Thus $S_n(\cdot)$ is a step function which is zero for x less than the smallest x_i, has a jump of $1/n$ at each x_i, and is 1 for x greater than or equal to the largest x_i.

The Kolmogorov–Smirnov (K–S) test compares the sample cumulative distribution function $S_n(\cdot)$ with the hypothesized distribution function $F(\cdot)$. As a measure of comparison the test uses

$$D = \max_x \left| F(x) - S_n(x) \right|.$$

Since S_n is a step function, and F is monotonically increasing, it suffices to test the absolute deviation at the sample points x_i, $i = 1, \ldots, n$, and then take the maximum of these n values.

Figure 8.4.1 shows that for $F(\cdot)$ continuous we must compute two differences, $\left| F(x_i) - S_n(x_i) \right| = \left| F(x_i) - i/n \right|$ and $\left| F(x_i) - S_n(x_{i-1}) \right| = \left| F(x_i) - (i-1)/n \right|$, to find the maximum absolute deviation between $F(x)$ and $S_i(x)$ when $x = x_i$. In the figure we see that the maximum absolute deviation at x_i is $S_n(x_i) - F(x_i) = (i/n) - F(x_i)$, while at x_{i+1} it is $F(x_{i+1}) - S_n(x_i)$. If F is the distribution function of a discrete random variable we need only compute $\left| F(x_i) - S_n(x_i) \right|$ at each x_i.

The K–S test was developed for continuous distributions. When used for

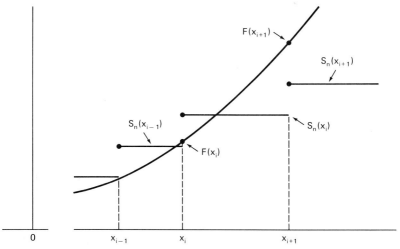

Fig. 8.4.1 Finding the maximum absolute deviation between $F(x)$ and $S_n(x)$ at a sample value x_i.

discrete distributions, it gives conservative results; this means that the test is actually operating at a lower level of significance than it was designed for so that, if the test indicates a lack of fit, the evidence is very strong. The basic form of the test assumes that the distribution $F(\cdot)$ is completely known (for example, not only that $F(\cdot)$ is normal but also that it has a given mean μ and standard deviation σ). For the basic form of the test we use Table 7 of Appendix A to find critical values of D. If $F(\cdot)$ is assumed to be exponential but the mean is estimated from the data, then Table 8 due to Liliefors [5] can be used. If $F(\cdot)$ is assumed to be normal but the mean is estimated by $\bar{x}$ and the standard deviation by s, then Table 9 due to Lilliefors [6], can be used. For other distributions with estimated means Table 7 will give very conservative results in the sense that the actual level of significance achieved will be much lower than that indicated in the table; for example, Lilliefors [6] has noted that the critical values in Table 9 for $\alpha = 0.05$ are about the same for each value of n as those in Table 7 for $\alpha = 0.20$. If the parameters are estimated but the assumed distribution is somewhat like a normal distribution (such as Erlang-10 or chi-square with 30 degrees of freedom) Table 9 should be used.

Our basic statement of the Kolmogorov–Smirnov test is given as Algorithm 8.4.2. For it we assume that either the parameters of F are known or that F is exponential. To test for the normal distribution with mean and standard deviation unknown, we have a special form of the K–S test given as Algorithm 8.4.3.

Algorithm 8.4.2 (*Kolmogorov–Smirnov Test, Parameters Known or F Exponential*)   Given a random sample $x_1, x_2, \ldots, x_n$ of size n, this test will determine, at the α level of significance, whether it is reasonable to suppose that the population distribution function is $F(\cdot)$, where the parameters of F are assumed known, or where $F(\cdot)$ is assumed to be exponentially distributed.

Step 1   [Order the sample]   Rearrange the sample values so they are in ascending order; that is, $x_1 \leq x_2 \leq \cdots \leq x_n$.

Step 2   [Calculate the test statistic D]   For each $j = 1, 2, \ldots, n$ set $D_j = \max\{|F(x_j) - j/n|, |F(x_j) - (j-1)/n|\}$, if $F(\cdot)$ is continuous. If $F(\cdot)$ is discrete, set $D_j = |F(x_j) - j/n|$. Then set $D = \max_j D_j$.

Step 3   [Find critical D]   If $F(\cdot)$ is assumed completely known, find the critical value of D in Table 7. If $F(\cdot)$ is exponential with mean estimated as $\bar{x}$, find the critical value of D in Table 8.

If the parameters of $F(\cdot)$ are estimated from the sample but $F(\cdot)$ is neither exponential nor normal, Table 7 can be used but will give extremely conservative results. To achieve better results use Table 7 but adjust the level of significance used. Thus to test at the 5% level of significance use the column in Table 7 labeled for the 20% level; to test at the 1% level use the 10% column.

Step 4 [*Apply the test statistic*] If the value of D calculated in Step 2 is greater than or equal to the critical value of D selected in Step 3, reject the null hypothesis that the population distribution function is F; otherwise accept the null hypothesis.

Algorithm 8.4.3 (*Kolmogorov–Smirnov Test for Normality with the Mean and Standard Deviation Unknown*) Given a random sample x_1, $x_2, \ldots, x_n$ of size n, this test will determine, at the α level of significance, whether or not it is reasonable to suppose that X is normally distributed. We assume that the mean μ and the standard deviation σ, of X are not known, for otherwise we could use Algorithm 8.4.2.

Step 1 [*Order the sample*] Rearrange the sample values so they are in ascending order; that is, $x_1 \leq x_2 \leq \cdots \leq x_n$.

Step 2 [*Calculate the z values*] For $j = 1, 2, \ldots, n$ set

$$z_j = (x_j - \bar{x})/s.$$

Step 3 [*Calculate the test statistic D*] For each $j = 1, 2, \ldots, n$ set

$$D_j = \max\{|\Phi(z_j) - j/n|, |\Phi(z_j) - (j - 1)/n|\},$$

where $\Phi(\cdot)$ is the standard normal distribution. Then set $D = \max_j D_j$.

Step 4 [*Find critical D*] Find the critical value of D in Table 9.

Step 5 [*Apply the test statistic*] If the value of D calculated in Step 3 is greater than or equal to the critical value of D found in Step 4, reject H_0, the hypothesis that the population is normally distributed; otherwise accept H_0.

Example 8.4.2 Marshmallow Software, Inc., claims that their information retrieval system, when installed on a brand XXX computer, has an inquiry retrieval time which is normally distributed with a mean of 40 time units and a standard deviation of 10 time units. Prospective customer YYYOY takes a sample of 30 values of retrieval time yielding 52.42, 42.17, 32.44, 46.23, 42.76, 41.10, 45.82, 27.13, 57.40, 39.05, 35.80, 37.95, 38.95, 53.40, 50.55, 44.44, 44.84, 40.13, 34.48, 54.45, 41.49, 52.21, 34.81, 26.93, 48.57, 39.29, 31.97, 30.61, 46.67, 32.37. Use the Kolmogorov–Smirnov test at the 5% level of significance to evaluate Marshmallow's claim.

Solution Algorithm 8.4.2 applies since we assume the mean and standard deviation are known. In applying the algorithm we generate Table 8.4.4. We see from the table that $D = 0.1249$. By Table 7, with $\alpha = 0.05$ and

$n = 30$, the critical value of D is 0.2417. Hence we accept the null hypothesis that the population has a normal distribution with mean 40 and standard deviation 10.

TABLE 8.4.4
Kolmogorov–Smirnov Test Data for Example 8.4.2

x_j	$j/30$	$F(x_j) = \Phi\left(\dfrac{x_j - 40}{10}\right)$	$\lvert F(x_j) - j/30 \rvert$	$\lvert F(x_j) - (j-1)/30 \rvert$
26.93	0.0333	0.0956	0.0623	0.0956
27.13	0.0667	0.0990	0.0323	0.0657
30.61	0.1000	0.1739	0.0739	0.1072
31.97	0.1333	0.2110	0.0777	0.1110
32.37	0.1667	0.2227	0.0560	0.0894
32.44	0.2000	0.2248	0.0248	0.0581
34.48	0.2333	0.2905	0.0572	0.0905
34.81	0.2667	0.3019	0.0352	0.0686
35.80	0.3000	0.3372	0.0372	0.0705
37.95	0.3333	0.4188	0.0855	0.1188
38.95	0.3667	0.4582	0.0915	0.1249
39.05	0.4000	0.4622	0.0622	0.0955
39.29	0.4333	0.4717	0.0384	0.0717
40.13	0.4667	0.5052	0.0385	0.0719
41.10	0.5000	0.5438	0.0438	0.0771
41.49	0.5333	0.5592	0.0259	0.0592
42.17	0.5667	0.5859	0.0192	0.0526
42.76	0.6000	0.6087	0.0087	0.0420
44.44	0.6333	0.6715	0.0382	0.0715
44.84	0.6667	0.6858	0.0191	0.0525
45.82	0.7000	0.7197	0.0197	0.0530
46.23	0.7333	0.7334	0.0001	0.0334
46.67	0.7667	0.7476	0.0191	0.0143
48.57	0.8000	0.8043	0.0043	0.0376
50.55	0.8333	0.8543	0.0210	0.0543
52.21	0.8667	0.8890	0.0223	0.0557
52.42	0.9000	0.8929	0.0071	0.0262
53.40	0.9333	0.9099	0.0234	0.0099
54.45	0.9667	0.9258	0.0409	0.0075
57.40	1.0000	0.9591	0.0409	0.0076

Example 8.4.3 Suppose that in Example 8.4.2 we were only given the claim that retrieval time was normally distributed. Then to test this claim, at the 5% level of significance, we could use Algorithm 8.4.3 as follows. (To ease the computation we use only the first 10 values of the sample given in the example.) For this sample of 10 values we have $\bar{x} = 42.652$ time units and

$s = 8.798$ time units. To apply Algorithm 8.4.3 we generate Table 8.4.5. The column headed $\Phi(z_j)$ was calculated directly from formula (26.2.17) of Abramowitz and Stegun [4] as was the normal probability calculation of Table 8.4.4. Readers using Table 3 of Appendix A may obtain very slightly different numbers. We see from Table 8.4.5 that $D = 0.1420$. The critical value of D from Table 9 is 0.258. Hence we accept the null hypothesis that the population is normally distributed.

TABLE 8.4.5
Kolmogorov–Smirnov Test Data for Example 8.4.3

x_j	$z_j = \dfrac{x_j - 42.652}{8.798}$	$\Phi(z_j)$	$\lvert\Phi(z_j) - (j/10)\rvert$	$\lvert\Phi(z_j) - (j-1)/10\rvert$
27.13	−1.764	0.0389	0.0611	0.0389
32.44	−1.161	0.1228	0.0772	0.0228
39.05	−0.409	0.3413	0.0413	0.1413
41.10	−0.176	0.4301	0.0301	0.1301
42.17	−0.055	0.4781	0.0219	0.0781
42.76	0.012	0.5048	0.0952	0.0048
45.82	0.360	0.6406	0.0594	0.0406
46.23	0.407	0.6580	0.1420	0.0420
52.42	1.110	0.8665	0.0335	0.0665
57.40	1.676	0.9531	0.0469	0.0531

Example 8.4.4 Yallcome Tstay Inns has an on-line computer reservation system. The queueing model they use for performance prediction assumes the arrival pattern of requests to the central computer is exponentially distributed. A random sample of 15 interarrival times yielded the values 25.98, 33.73, 32.56, 5.75, 3.67, 0.93, 7.70, 9.40, 29.98, 6.20, 14.38, 5.14, 51.93, 24.60, 14.88. (The time units are uglings, a proprietary unit of time.) At the 10% level of significance, does the interarrival time appear to be exponential?

Solution Algorithm 8.4.2 can be applied, using Table 8 for critical values of D. (It would be inappropriate to apply the chi-square test, since we have a rather small sample.) We apply the algorithm to generate Table 8.4.6. Since we do not know the population mean we use $\bar{x} = 17.7887$ as the mean in calculations of $F(x) = 1 - e^{-x/\mu}$. We see from the table that $D = 0.1492$. Since the mean is estimated from the sample, we use Table 8 to find the critical value of D is 0.244. Thus we accept the hypothesis that the interarrival time is exponential.

Another use of the Kolmogorov–Smirnov test is to develop confidence

TABLE 8.4.6

Kolmogorov–Smirnov data for Example 8.4.4

x_j	$j/15$	$F(x_j) =$ $1 - e^{-x_j/17.7887}$	$\lvert F(x_j) - j/15 \rvert$	$\lvert F(x_j) - (j-1)/15 \rvert$
0.93	0.0667	0.0509	0.0157	0.0509
3.67	0.1333	0.1864	0.0531	0.1198
5.14	0.2000	0.2509	0.0509	0.1176
5.75	0.2667	0.2762	0.0095	0.0762
6.20	0.3333	0.2943	0.0390	0.0276
7.70	0.4000	0.3513	0.0487	0.0180
9.40	0.4667	0.4105	0.0562	0.0105
14.38	0.5333	0.5544	0.0211	0.0877
14.88	0.6000	0.5668	0.0332	0.0335
24.60	0.6667	0.7492	0.0825	0.1492
25.98	0.7333	0.7679	0.0346	0.1012
29.98	0.8000	0.8146	0.0146	0.0813
32.56	0.8667	0.8396	0.0271	0.0396
33.73	0.9333	0.8499	0.0834	0.0168
51.93	1.0000	0.9460	0.0540	0.0127

limits for the true distribution function $F(\cdot)$ of the population. For example, if we take a random sample of size 100 from a population and use it to construct the sample cumulative distribution function $S_{100}(\cdot)$, then by Table 7, we can be 95% confident that the true distribution function $F(\cdot)$ does not deviate from $S_{100}(\cdot)$ by more than $1.36/\sqrt{100} = 0.136$. That is, the 95% confidence limits for $F(\cdot)$ are given by

$$S_{100}(x) - 0.136 < F(x) < S_{100}(x) + 0.136.$$

Many statisticians believe the Kolmogorov–Smirnov test is superior to the chi-square test. One advantage of the K–S test is that the exact distribution of D is known, even for small samples. By contrast, the statistic

$$\chi^2 = \sum_{i=1}^{n} \frac{(O_i - E_i)^2}{E_i}$$

is only approximately chi-square, so that, for small n, the critical value χ_α^2 from a chi-square table is not very accurate. Moreover, the categories must be set to make each $E_i \geq 5$. On the other hand, the chi-square test applies to both continuous and discrete populations. The K–S test strictly applies to continuous distributions; when used to test discrete distributions it gives very conservative results. Moreover, it is possible to adopt the chi-square test to the fact the population parameters must be estimated from the data; we do this by subtracting degrees of freedom. For the K–S test, critical values of D for the case where parameters are estimated have been published (by Lilliefors) only for the case of normal and exponential tests. It is expected that critical values of D will be developed for other useful distributions. Lilliefors [8] has worked out some tables for the gamma distribution. However, the most compelling reason for using the Kolmogorov–Smirnov test, wherever possible, is the powerful effect of the name, itself. Who can fail to be impressed by hearing that "The Kolmogorov–Smirnov test has confirmed the hypothesis"? There can be no doubt that the most sophisticated test possible has been used!

8.5 SUMMARY

In this chapter we have considered various types of statistical tests. In Section 8.2 we were concerned with tests concerning population means and in Section 8.3 with tests concerning population variances. These tests are summarized in Table 8.5.1. In Section 8.4 we considered goodness-of-fit tests, that is, tests of how well a given distribution function $F(\cdot)$ seems to fit or predict the values from a random sample. Overall, in this chapter, we have been involved in making inferences about an entire population on the basis of a random sample. That is what estimating and hypothesis testing is all about.

Student Sayings

Kolmogorov and Smirnov would make the hypothesis "Amicable relations between two consenting adults is surely fit." Chi-square would accept this statement at the 5% level of significance.

TABLE 8.5.1

Tests Concerning Means and Variances

H_0	Test statistic and conditions	H_1	Critical region
$\mu = \mu_0$	$z = \dfrac{\bar{x} - \mu_0}{\sigma/\sqrt{n}}$; σ known, $n \geq 30$, if σ not known, use s for σ. Test statistic is approximately normal for all distributions of X.	$\mu > \mu_0$ $\mu < \mu_0$ $\mu \neq \mu_0$	$z > z_\alpha$ $z < -z_\alpha$ $z < -z_{\alpha/2}$ $z > z_{\alpha/2}$
$\mu = \mu_0$	$t = \dfrac{\bar{x} - \mu_0}{s/\sqrt{n}}$; X approximately normally distributed, $n < 30$, σ unknown; statistic is Student-t with $n - 1$ degrees of freedom.	$\mu > \mu_0$ $\mu < \mu_0$ $\mu \neq \mu_0$	$t > t_{\alpha, n-1}$ $t < -t_{\alpha, n-1}$ $t < -t_{\alpha/2, n-1}$ $t > t_{\alpha/2, n-1}$
$\mu_X - \mu_Y = d_0$	$z = \dfrac{(\bar{x} - \bar{y}) - d_0}{[(\sigma_X^2/n) + (\sigma_Y^2/m)]^{\frac{1}{2}}}$; $n \geq 30$ and $m \geq 30$, σ_X and σ_Y known, if standard deviations not known substitute s_X and s_Y for the respective standard deviations.	$\mu_X - \mu_Y > d_0$ $\mu_X - \mu_Y < d_0$ $\mu_X - \mu_Y \neq d_0$	$z > z_\alpha$ $z < -z_\alpha$ $z < -z_{\alpha/2}$ $z > z_{\alpha/2}$

$\mu_X - \mu_Y = d_0$

$$t = \frac{(\bar{x} - \bar{y}) - d_0}{s_p[(1/n) + (1/m)]^{\frac{1}{2}}} \quad \text{where} \quad s_p = \left[\frac{(n-1)s_X^2 + (m-1)s_Y^2}{n+m-2}\right]^{\frac{1}{2}};$$

$\sigma_X = \sigma_Y$ but is unknown, X and Y approximately normally distributed, $n < 30$, $m < 30$. The test statistic is approximately Student-t with $n + m - 2$ degrees of freedom.

$\mu_X - \mu_Y > d_0$	$t > t_{\alpha, n+m-2}$
$\mu_X - \mu_Y < d_0$	$t < -t_{\alpha, n+m-2}$
$\mu_X - \mu_Y \neq d_0$	$t < -t_{\alpha/2, n+m-2}$
	$t > t_{\alpha/2, n+m-2}$

$\mu_X - \mu_Y = d_0$

$$t = \frac{(\bar{x} - \bar{y}) - d_0}{[(s_X^2/n) + (s_Y^2/m)]^{\frac{1}{2}}};$$

X, Y approximately normally distributed, $n < 30$, $m < 30$, $\sigma_X \neq \sigma_Y$ and are unknown. The test statistic is approximately Student-t with v degrees of freedom where

$$v = \frac{((s_X^2/n) + (s_Y^2/m))^2}{(s_X^2/n)^2/(n-1) + (s_Y^2/m)^2/(m-1)}$$

$\mu_X - \mu_Y > d_0$	$t > t_{\alpha, v}$
$\mu_X - \mu_Y < d_0$	$t < -t_{\alpha, v}$
$\mu_X - \mu_Y \neq d_0$	$t < -t_{\alpha/2, v}$
	$t > t_{\alpha/2, v}$

$\sigma^2 = \sigma_0^2$

$$\chi^2 = (n-1)s^2/\sigma_0^2;$$

X approximately normal. χ^2 has a chi-square distribution with $n - 1$ degrees of freedom, for large n.

$\sigma^2 > \sigma_0^2$	$\chi^2 > \chi^2_{\alpha, n-1}$
$\sigma^2 < \sigma_0^2$	$\chi^2 < \chi^2_{1-\alpha, n-1}$
$\sigma^2 \neq \sigma_0^2$	$\chi^2 < \chi^2_{1-\alpha/2, n-1}$
	$\chi^2 > \chi^2_{\alpha/2, n-1}$

$\sigma_X^2 = \sigma_Y^2$

$$f = s_X^2/s_Y^2;$$

X, Y are approximately normally distributed. The test statistic has a Snedcor-F distribution with $n - 1$ and $m - 1$ degrees of freedom.

$\sigma_X^2 > \sigma_Y^2$	$f > f_\alpha(n - 1, m - 1)$
$\sigma_X^2 < \delta_Y^2$	$f < f_{1-\alpha}(n - 1, m - 1)$
$\sigma_X^2 \neq \sigma_Y^2$	$f < f_{1-\alpha/2}(n - 1, m - 1)$
	$f > f_{\alpha/2}(n - 1, m - 1)$

Exercises

1. [C18] The number of terminals in use during the lunch hour at Fairlady Aircraft is normally distributed. A random sample of size 30 is taken yielding the values 21.738, 17.363, 22.604, 12.167, 23.724, 17.763, 16.46, 18.197, 15.056, 19.617, 23.223, 25.745, 19.92, 15.13, 20.07, 23.04, 19.627, 13.037, 19.141, 16.952, 22.089, 15.051, 19.121, 22.226, 12.67, 17.174, 14.27, 15.511, 18.213, 18.811. Assume that $\sigma = 5$; test the hypothesis, at the 5% level of significance, that the mean $\mu = 20$ against the alternative that $\mu \neq 20$.

2. [15] Suppose Algorithm 8.2.1 is being used by Housecold Finance to test the hypothesis $\mu = \mu_0$ against the alternative $\mu < \mu_0$ at the α level of significance, where X is approximately normal and σ is known. Show that the size of the sample needed to achieve a power of $1 - \beta$ when $\mu = \mu_0 - \delta$ is given by

$$n = (z_\alpha + z_\beta)^2 \sigma^2 / \delta^2.$$

Apply this formula to the specific case $\alpha = 0.05$, $1 - \beta = 0.95$, $\mu_0 = 100$, $\delta = 2$, $\sigma = 10$.

3. [20] Suppose the situation is exactly as in Exercise 2, except that the alternative hypothesis is $\mu \neq \mu_0$. Show that the required sample size to achieve a power of $1 - \beta$ when μ differs from μ_0 by δ ($\mu = \mu_0 \pm \delta$) is given by

$$n = (z_{\alpha/2} + z_{\beta/2})^2 \sigma^2 / \delta^2.$$

If $\alpha = 0.01$, $\mu_0 = 100$, and $\sigma = 10$, find the sample size needed to guarantee that the power is at least 0.95 if μ differs from 100 by 2 or more.

4. [C15] Fowler Heir Mining has selected a random sample of the response time at the terminal used by the VP for security and got the values 52.74, 60.18, 57.11, 58.15, 50.58, 44.02, 67.61, 48.61, 54.74, 58.83 time units. Assuming the response time is normally distributed, test the null hypothesis that the mean $\mu = 50$ time units against the alternative that $\mu \neq 50$. Use the 5% level of significance.

5. [15] The records of Cutthrote Industries show that the 200 programmers of Division A have averaged 76.21 lines of code per programmer day with a sample standard deviation of 10.37 while the 150 programmers of Division B have averaged 72.72 lines of code per programmer day with a sample standard deviation of 10.07. Test the hypothesis that the programmers of Division A are more productive than those of Division B using $\alpha = 0.01$.

6. [15] Suppose the situation is the same as in Exercise 5 except that statistics have been kept for only 6 programmers of Division A and 5 from Division B. Assume the distribution of lines of code per programmer day is normally distributed for both divisions with the same (unknown) standard deviation. Again test the hypothesis, at the 1% level of significance, that the programmers of Division A are more productive than those from Division B.

7. [C15] Consider Exercise 4. Use Algorithm 8.3.1 to determine whether or not the variance of response time at the VP's terminal is equal to 30 against the alternative that it is greater than 30. Use the 5% level of significance.

8. [15] Assuming that the distributions of Exercise 6 are normally distributed, test at the 5% level of significance, whether or not the variances of lines of code per programmer day at the two divisions are the same. Use the alternative hypothesis that they are different.

9. [C15] Consider the data of Example 8.4.2. Apply the chi-square test to this data as follows. First "normalize" the data by converting x values to z values by the formula $z_i = (x_i - \bar{x})/s$. Then partition the real line into five intervals such that one-fifth of the z_i's would be expected in each interval. Then apply the chi-square test with these intervals as categories. Use $\alpha = 0.05$ and H_0: normal distribution.

10. [C15] Use the Kolmogorov–Smirnov test to test the hypothesis, at the 10% level of significance, that the following random sample came from a normal population with mean 30 and standard deviation 5: 36.21, 26.22, 31.38, 32.91, 38.70, 27.90, 29.48, 35.27, 32.42, 27.24, 30.75, 27.40, 34.28, 25.99, 33.34, 24.34, 29.32, 31.26, 29.62, 30.95.

11. [C15] Given the sample of response times in Exercise 4, use the Kolmogorov–Smirnov test at the 20% level of significance to test the hypothesis that response time is normally distributed. Assume the mean and standard deviation is unknown.

12. [C15] Whoopdedoo Fashions has taken a random sample of 15 communication line times of their on-line order entry system to yield the values 2.31, 17.29, 26.23, 79.83, 30.35, 3.59, 1.29, 0.58, 4.81, 15.87, 28.73, 3.87, 18.99, 2.81, 62.46 time units. Use the Kolmogorov–Smirnov test at the 10% level of significance to test the hypothesis that the line time is exponential.

13. [C15] Consider the data of Exercise 1. Use the Kolmogorov–Smirnov test at the 5% level of significance to test the hypothesis that the population is normally distributed.

References

1. E. Kreyszig, *Introductory Mathematical Statistics*. Wiley, New York, 1970.
2. Bell System Advertisement, *Sci. Amer.* (August 1976), 66A, 66B.
3. R. E. Walpole and R. H. Myers, *Probability and Statistics for Engineers and Scientists*. Macmillan, New York, 1972.
4. M. Abramowitz and I. A. Stegun, *Handbook of Mathematical Functions*. National Bureau of Standards, Washington, D.C., 1964; also published by Dover, New York.
5. H. W. Lilliefors, On the Kolmogorov–Smirnov test for the exponential distribution with mean unknown, *J. Amer. Statist. Assoc.* **64** (1969), 387–389.
6. H. W. Lilliefors, On the Kolmogorov–Smirnov test for normality with mean and variance unknown, *J. Amer. Statist. Assoc.* **62** (1967), 399–402.
7. H. Cramér, *Mathematical Methods of Statistics*. Princeton University Press, Princeton, New Jersey, 1946.
8. H. W. Lilliefors, The Kolmogorov–Smirnov and other distance tests for the gamma distribution and for the extreme-value distribution when parameters must be estimated, unpublished report, The George Washington University.

Appendix A

STATISTICAL TABLES

TABLE 1
Properties of Some Common Discrete Random Variables[a]

Random variable	Parameters	Probability mass function $p(\cdot)$	z-transform $g(z)$	$E[X]$	$\mathrm{Var}[X]$
Bernoulli	$0 < p < 1$	$p(k) = \begin{cases} p & \text{if } k = 1 \\ q & \text{if } k = 0 \\ 0 & \text{otherwise} \end{cases}$	$q + pz$	p	pq
Binomial	n a positive integer $0 < p < 1$	$p(k) = b(k; n, p) = \binom{n}{k} p^k q^{n-k}$, $\quad k = 0, 1, \ldots, n$	$(q + pz)^n$	np	npq
Geometric	$0 < p < 1$	$p(k) = q^k p$, $\quad k = 0, 1, 2, \ldots$	$\dfrac{p}{1 - qz}$	$\dfrac{q}{p}$	$\dfrac{q}{p^2}$
Poisson	$\lambda > 0$	$p(k) = e^{-\lambda} \dfrac{\lambda^k}{k!}$, $\quad k = 0, 1, 2, \ldots$	$e^{\lambda(z-1)}$	λ	λ

[a] $q = 1 - p$.

TABLE 2

Properties of Some Common Continuous Random Variables

Random variable	Parameters	Density function $f(\cdot)$	Moment generating function $\psi(\cdot)$	$E[X]$	$\mathrm{Var}[X]$
Uniform over interval a to b	$a < b$	$f(x) = \begin{cases} \dfrac{1}{b-a}, & a < x < b \\ 0, & \text{otherwise} \end{cases}$	$\dfrac{e^{\theta b} - e^{\theta a}}{\theta(b-a)}$	$\dfrac{a+b}{2}$	$\dfrac{(b-a)^2}{12}$
Normal	μ real $\sigma > 0$	$f(x) = \dfrac{1}{\sigma\sqrt{2\pi}} e^{-\frac{1}{2}((x-\mu)/\sigma)^2}$	$e^{\theta\mu + \frac{1}{2}\theta^2\sigma^2}$	μ	σ^2
Exponential	$\lambda > 0$	$f(x) = \begin{cases} \lambda e^{-\lambda x}, & x > 0 \\ 0, & x \le 0 \end{cases}$	$\dfrac{\lambda}{\lambda - \theta}$	$\dfrac{1}{\lambda}$	$\dfrac{1}{\lambda^2}$
Gamma	$\alpha > 0$ $\lambda > 0$	$f(x) = \begin{cases} \dfrac{\lambda(\lambda x)^{\alpha-1}}{\Gamma(\alpha)} e^{-\lambda x}, & x > 0 \\ 0, & x \le 0 \end{cases}$	$\left(\dfrac{\lambda}{\lambda - \theta}\right)^{\alpha}$	$\dfrac{\alpha}{\lambda}$	$\dfrac{\alpha}{\lambda^2}$
Erlang-k	k a positive integer $\lambda > 0$	$f(x) = \begin{cases} \dfrac{\lambda k(\lambda k x)^{k-1} e^{-\lambda k x}}{(k-1)!}, & x > 0 \\ 0, & x \le 0 \end{cases}$	$\left(\dfrac{k\lambda}{k\lambda - \theta}\right)^{k}$	$\dfrac{1}{\lambda}$	$\dfrac{1}{k\lambda^2}$
Chi-square	n a positive integer	$f(x) = \begin{cases} \dfrac{x^{(n/2)-1} e^{-x/2}}{2^{n/2}\Gamma(n/2)}, & x > 0 \\ 0, & x \le 0 \end{cases}$	$\left(\dfrac{1}{1 - 2\theta}\right)^{n/2}$	n	$2n$
Student-t	n a positive integer	$f(x) = \dfrac{\Gamma[(n+1)/2]}{\sqrt{n\pi}\,\Gamma(n/2)}\left(1 + \dfrac{x^2}{n}\right)^{-(n+1)/2},\ x$ real	does not exist	0 for $n > 1$	$\dfrac{n}{n-2}$ for $n > 2$
Snedcor-F	n, m positive integers	$f(x) = \begin{cases} \dfrac{(n/m)^{n/2}\Gamma((n+m)/2)x^{(n/2)-1}}{\Gamma(n/2)\Gamma(m/2)(1+(n/m)x)^{(n+m)/2}}, & x > 0 \\ 0, & x \le 0 \end{cases}$	does not exist	$\dfrac{m}{m-2}$ if $m > 2$	$\dfrac{m^2(2m+2n-4)}{n(m-2)^2(m-4)}$ if $m > 4$

TABLE 3

The Normal Distribution Function $\Phi(z) = \int_{-\infty}^{z} \dfrac{e^{-t^2/2}}{\sqrt{2\pi}}\, dt$

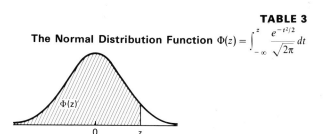

z	0.00	0.01	0.02	0.03	0.04	0.05	0.06	0.07	0.08	0.09
0.0	.50000	.50399	.50798	.51197	.51595	.51994	.52392	.52790	.53188	.53586
0.1	.53983	.54380	.54776	.55172	.55567	.55962	.56356	.56749	.57142	.57535
0.2	.57926	.58317	.58706	.59095	.59483	.59871	.60257	.60642	.61026	.61409
0.3	.61791	.62172	.62552	.62930	.63307	.63683	.64058	.64431	.64803	.65173
0.4	.65542	.65910	.66276	.66640	.67003	.67364	.67724	.68082	.68439	.68793
0.5	.69146	.69497	.69847	.70194	.70540	.70884	.71226	.71566	.71904	.72240
0.6	.72575	.72907	.73237	.73565	.73891	.74215	.74537	.74857	.75175	.75490
0.7	.75804	.76115	.76424	.76730	.77035	.77337	.77637	.77935	.78230	.78524
0.8	.78814	.79103	.79389	.79673	.79955	.80234	.80511	.80785	.81057	.81327
0.9	.81594	.81859	.82121	.82381	.82639	.82894	.83147	.83398	.83646	.83891
1.0	.84134	.84375	.84614	.84849	.85083	.85314	.85543	.85769	.85993	.86214
1.1	.86433	.86650	.86864	.87076	.87286	.87493	.87698	.87900	.88100	.88298
1.2	.88493	.88686	.88877	.89065	.89251	.89435	.89617	.89796	.89973	.90147
1.3	.90320	.90490	.90658	.90824	.90988	.91149	.91308	.91466	.91621	.91774
1.4	.91924	.92073	.92220	.92364	.92507	.92647	.92785	.92922	.93056	.93189
1.5	.93319	.93448	.93574	.93699	.93822	.93943	.94062	.94179	.94295	.94408
1.6	.94520	.94630	.94738	.94845	.94950	.95053	.95154	.95254	.95352	.95449
1.7	.95543	.95637	.95728	.95818	.95907	.95994	.96080	.96164	.96246	.96327
1.8	.96407	.96485	.96562	.96638	.96712	.96784	.96856	.96926	.96995	.97062
1.9	.97128	.97193	.97257	.97320	.97381	.97441	.97500	.97558	.97615	.97670
2.0	.97725	.97778	.97831	.97882	.97932	.97982	.98030	.98077	.98124	.98169
2.1	.98214	.98257	.98300	.98341	.98382	.98422	.98461	.98500	.98537	.98574
2.2	.98610	.98645	.98679	.98713	.98745	.98778	.98809	.98840	.98870	.98899
2.3	.98928	.98956	.98983	.99010	.99036	.99061	.99086	.99111	.99134	.99158
2.4	.99180	.99202	.99224	.99245	.99266	.99286	.99305	.99324	.99343	.99361
2.5	.99379	.99396	.99413	.99430	.99446	.99461	.99477	.99492	.99506	.99520
2.6	.99534	.99547	.99560	.99573	.99585	.99598	.99609	.99621	.99632	.99643
2.7	.99653	.99664	.99674	.99683	.99693	.99702	.99711	.99720	.99728	.99736
2.8	.99744	.99752	.99760	.99767	.99774	.99781	.99788	.99795	.99801	.99807
2.9	.99813	.99819	.99825	.99831	.99836	.99841	.99846	.99851	.99856	.99861
3.0	.99865	.99869	.99874	.99878	.99882	.99886	.99889	.99893	.99896	.99900
3.1	.99903	.99906	.99910	.99913	.99916	.99918	.99921	.99924	.99926	.99929
3.2	.99931	.99934	.99936	.99938	.99940	.99942	.99944	.99946	.99948	.99950
3.3	.99952	.99953	.99955	.99957	.99958	.99960	.99961	.99962	.99964	.99965
3.4	.99966	.99968	.99969	.99970	.99971	.99972	.99973	.99974	.99975	.99976
3.5	.99977	.99978	.99978	.99979	.99980	.99981	.99981	.99982	.99983	.99983
3.6	.99984	.99985	.99985	.99986	.99986	.99987	.99987	.99988	.99988	.99989
3.7	.99989	.99990	.99990	.99990	.99991	.99991	.99992	.99992	.99992	.99992
3.8	.99993	.99993	.99993	.99994	.99994	.99994	.99994	.99995	.99995	.99995

TABLE 4

Critical Values of the Chi-Square Distribution[a]

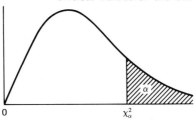

α n^b	0.995	0.990	0.975	0.950	0.05	0.025	0.010	0.005
1	0.04393	0.03157	0.03982	0.02393	3.8415	5.0239	6.6349	7.8794
2	0.0100	0.0201	0.0506	0.1026	5.9915	7.3778	9.2103	10.597
3	0.0717	0.1148	0.2158	0.3518	7.8147	9.3484	11.345	12.838
4	0.2070	0.2971	0.4844	0.7107	9.4877	11.143	13.277	14.860
5	0.4117	0.5543	0.8312	1.1455	11.071	12.833	15.086	16.750
6	0.6757	0.8721	1.2373	1.6354	12.592	14.449	16.812	18.548
7	0.9893	1.2390	1.6899	2.1674	14.067	16.013	18.475	20.278
8	1.3444	1.6465	2.1797	2.7326	15.507	17.535	20.090	21.955
9	1.7350	2.0879	2.7004	3.3251	16.920	19.023	21.666	23.589
10	2.1559	2.5582	3.2470	3.9403	18.307	20.483	23.209	25.188
11	2.6032	3.0535	3.8158	4.5748	19.675	21.920	24.725	26.757
12	3.0738	3.5706	4.4038	5.2260	21.026	23.337	26.217	28.300
13	3.5650	4.1069	5.0087	5.8919	22.362	24.736	27.688	29.819
14	4.0747	4.6604	5.6287	6.5706	23.685	26.119	29.141	31.319
15	4.6009	5.2294	6.2621	7.2609	24.996	27.488	30.578	32.801
16	5.1422	5.8122	6.9077	7.9616	26.296	28.845	32.000	34.267
17	5.6972	6.4078	7.5642	8.6718	27.587	30.191	33.409	35.719
18	6.2648	7.0149	8.2308	9.3905	28.869	31.526	34.805	37.156
19	6.8440	7.6327	8.9066	10.117	30.144	32.852	36.191	38.582
20	7.4339	8.2604	9.5908	10.851	31.410	34.170	37.566	39.997
21	8.0337	8.8972	10.283	11.591	32.671	35.479	38.932	41.401
22	8.6427	9.5425	10.982	12.338	33.924	36.781	40.289	42.796
23	9.2604	10.196	11.689	13.091	35.173	38.076	41.638	44.181
24	9.8862	10.856	12.401	13.848	36.415	39.364	42.980	45.559
25	10.520	11.524	13.120	14.611	37.653	40.647	44.314	46.928
26	11.160	12.198	13.844	15.379	38.885	41.923	45.642	48.290
27	11.808	12.879	14.573	16.151	40.113	43.194	46.963	49.645
28	12.461	13.565	15.308	16.928	41.337	44.461	48.278	50.993
29	13.121	14.257	16.047	17.708	42.557	45.722	49.588	52.336
30	13.787	14.954	16.791	18.493	43.773	46.980	50.892	53.672
40	20.707	22.164	24.433	26.509	55.759	59.342	63.691	66.766
50	27.991	29.707	32.357	34.764	67.505	71.420	76.154	79.490
60	35.535	37.485	40.482	43.188	79.082	83.298	88.380	91.952
70	43.275	45.442	48.758	51.739	90.531	95.023	100.425	104.215
80	51.172	53.540	57.153	60.392	101.879	106.629	112.329	116.321
90	59.196	61.754	65.647	69.126	113.145	118.136	124.116	128.299
100	67.328	70.065	74.222	77.930	124.342	129.561	135.807	140.169
z_α	-2.5758	-2.3263	-1.9600	-1.6449	$+1.6449$	$+1.9600$	$+2.3263$	$+2.5758$

[a] Adapted from *Biometrika Tables for Statisticians*, (E. S. Pearson and H. O. Hartley, eds.), Vol. 1, 4th ed. Cambridge University Press, Cambridge, 1966, by permission of Biometrika Trustees.

[b] For $n > 100$ use

$$\chi_\alpha^2 = n\left\{ 1 - \frac{2}{9n} + z_\alpha\sqrt{\frac{2}{9n}}\right\}^3$$

where z_α is given on the bottom line of the table.

TABLE 5
Critical Values of the Student-t Distribution[a]

n \ α	0.10	0.05	0.025	0.01	0.005
1	3.078	6.314	12.706	31.821	63.657
2	1.886	2.920	4.303	6.965	9.925
3	1.638	2.353	3.182	4.541	5.841
4	1.533	2.132	2.776	3.747	4.604
5	1.476	2.015	2.571	3.365	4.032
6	1.440	1.943	2.447	3.143	3.707
7	1.415	1.895	2.365	2.998	3.499
8	1.397	1.860	2.306	2.896	3.355
9	1.383	1.833	2.262	2.821	3.250
10	1.372	1.812	2.228	2.764	3.169
11	1.363	1.796	2.201	2.718	3.106
12	1.356	1.782	2.179	2.681	3.055
13	1.350	1.771	2.160	2.650	3.012
14	1.345	1.761	2.145	2.624	2.977
15	1.341	1.753	2.131	2.602	2.947
16	1.337	1.746	2.120	2.583	2.921
17	1.333	1.740	2.110	2.567	2.898
18	1.330	1.734	2.101	2.552	2.878
19	1.328	1.729	2.093	2.539	2.861
20	1.325	1.725	2.086	2.528	2.845
21	1.323	1.721	2.080	2.518	2.831
22	1.321	1.717	2.074	2.508	2.819
23	1.319	1.714	2.069	2.500	2.807
24	1.318	1.711	2.064	2.492	2.797
25	1.316	1.708	2.060	2.485	2.787
26	1.315	1.706	2.056	2.479	2.779
27	1.314	1.703	2.052	2.473	2.771
28	1.313	1.701	2.048	2.467	2.763
29	1.311	1.699	2.045	2.462	2.756
30	1.310	1.697	2.042	2.457	2.750
40	1.303	1.684	2.021	2.423	2.704
60	1.296	1.671	2.000	2.390	2.660
120	1.289	1.658	1.980	2.358	2.617
∞	1.282	1.645	1.960	2.326	2.576

TABLE 6

Critical Values of the F Distribution[a]

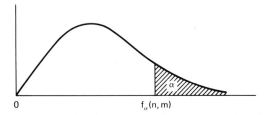

$f_\alpha(n, m)$

Denominator	Numerator	$f_{0.05}(n, m)$								
m	n	1	2	3	4	5	6	7	8	9
1		161.40	199.50	215.70	224.60	230.20	234.00	236.80	238.90	240.50
2		18.51	19.00	19.16	19.25	19.30	19.33	19.35	19.35	19.38
3		10.13	9.55	9.28	9.12	9.01	8.94	8.89	8.85	8.81
4		7.71	6.94	6.59	6.39	6.26	6.16	6.09	6.04	6.00
5		6.61	5.79	5.41	5.19	5.05	4.95	4.88	4.82	4.77
6		5.99	5.14	4.76	4.53	4.39	4.28	4.21	4.15	4.10
7		5.59	4.74	4.35	4.12	3.97	3.87	3.79	3.73	3.68
8		5.32	4.46	4.07	3.84	3.69	3.58	3.50	3.44	3.39
9		5.12	4.26	3.86	3.63	3.48	3.37	3.29	3.23	3.18
10		4.96	4.10	3.71	3.48	3.33	3.22	3.14	3.07	3.02
11		4.84	3.98	3.59	3.36	3.20	3.09	3.01	2.95	2.90
12		4.75	3.89	3.49	3.26	3.11	3.00	2.91	2.85	2.80
13		4.67	3.81	3.41	3.18	3.03	2.92	2.83	2.77	2.71
14		4.60	3.74	3.34	3.11	2.96	2.85	2.76	2.70	2.65
15		4.54	3.68	3.29	3.06	2.90	2.79	2.71	2.64	2.59
16		4.49	3.63	3.24	3.01	2.85	2.74	2.66	2.59	2.54
17		4.45	3.59	3.20	2.96	2.81	2.70	2.61	2.55	2.49
18		4.41	3.55	3.16	2.93	2.77	2.66	2.58	2.51	2.46
19		4.38	3.52	3.13	2.90	2.74	2.63	2.54	2.48	2.42
20		4.35	3.49	3.10	2.87	2.71	2.60	2.51	2.45	2.39
21		4.32	3.47	3.07	2.84	2.68	2.57	2.49	2.42	2.37
22		4.30	3.44	3.05	2.82	2.66	2.55	2.46	2.40	2.34
23		4.28	3.42	3.03	2.80	2.64	2.53	2.44	2.37	2.32
24		4.26	3.40	3.01	2.78	2.62	2.51	2.42	2.36	2.30
25		4.24	3.39	2.99	2.76	2.60	2.49	2.40	2.34	2.28
26		4.23	3.37	2.98	2.74	2.59	2.47	2.39	2.32	2.27
27		4.21	3.35	2.96	2.73	2.57	2.46	2.37	2.31	2.25
28		4.20	3.34	2.95	2.71	2.56	2.45	2.36	2.29	2.24
29		4.18	3.33	2.93	2.70	2.55	2.43	2.35	2.28	2.22
30		4.17	3.32	2.92	2.69	2.53	2.42	2.33	2.27	2.21
40		4.08	3.23	2.84	2.61	2.45	2.34	2.25	2.18	2.12
60		4.00	3.15	2.76	2.53	2.37	2.25	2.17	2.10	2.04
120		3.92	3.07	2.68	2.45	2.29	2.17	2.09	2.02	1.96
∞		3.84	3.00	2.60	2.37	2.21	2.10	2.01	1.94	1.88

[a] Adapted from *Biometrika Tables for Statisticians* (E. S. Pearson and H. O. Hartley, eds.), Vol. 1, 4th ed. Cambridge University Press, Cambridge, 1966, by permission of Biometrika Trustees.

TABLE 6
Critical Values of the F Distribution *(Continued)*

Denominator m	Numerator n	$f_{0.05}(n, m)$									
		10	12	15	20	24	30	40	60	120	∞
1		241.9	243.9	245.9	248.0	249.1	250.1	251.1	252.2	253.3	254.3
2		19.40	19.41	19.43	19.45	19.45	19.46	19.47	19.48	19.49	19.50
3		8.79	8.74	8.70	8.66	8.64	8.62	8.59	8.57	8.55	8.53
4		5.96	5.91	5.86	5.80	5.77	5.75	5.72	5.69	5.66	5.63
5		4.74	4.68	4.62	4.56	4.53	4.50	4.46	4.43	4.40	4.36
6		4.06	4.00	3.94	3.87	3.84	3.81	3.77	3.74	3.70	3.67
7		3.64	3.57	3.51	3.44	3.41	3.38	3.34	3.30	3.27	3.23
8		3.35	3.28	3.22	3.15	3.12	3.08	3.04	3.01	2.97	2.93
9		3.14	3.07	3.01	2.94	2.90	2.86	2.83	2.79	2.75	2.71
10		2.98	2.91	2.85	2.77	2.74	2.70	2.66	2.62	2.58	2.54
11		2.85	2.79	2.72	2.65	2.61	2.57	2.53	2.49	2.45	2.40
12		2.75	2.69	2.62	2.54	2.51	2.47	2.43	2.38	2.34	2.30
13		2.67	2.60	2.53	2.46	2.42	2.38	2.34	2.30	2.25	2.21
14		2.60	2.53	2.46	2.39	2.35	2.31	2.27	2.22	2.18	2.13
15		2.54	2.48	2.40	2.33	2.29	2.25	2.20	2.16	2.11	2.07
16		2.49	2.42	2.35	2.28	2.24	2.19	2.15	2.11	2.06	2.01
17		2.45	2.38	2.31	2.23	2.19	2.15	2.10	2.06	2.01	1.96
18		2.41	2.34	2.27	2.19	2.15	2.11	2.06	2.02	1.97	1.92
19		2.38	2.31	2.23	2.16	2.11	2.07	2.03	1.98	1.93	1.88
20		2.35	2.28	2.20	2.12	2.08	2.04	1.99	1.95	1.90	1.84
21		2.32	2.25	2.18	2.10	2.05	2.01	1.96	1.92	1.87	1.81
22		2.30	2.23	2.15	2.07	2.03	1.98	1.94	1.89	1.84	1.78
23		2.27	2.20	2.13	2.05	2.01	1.96	1.91	1.86	1.81	1.76
24		2.25	2.18	2.11	2.03	1.98	1.94	1.89	1.84	1.79	1.73
25		2.24	2.16	2.09	2.01	1.96	1.92	1.87	1.82	1.77	1.71
26		2.22	2.15	2.07	1.99	1.95	1.90	1.85	1.80	1.75	1.69
27		2.20	2.13	2.06	1.97	1.93	1.88	1.84	1.79	1.73	1.67
28		2.19	2.12	2.04	1.96	1.91	1.87	1.82	1.77	1.71	1.65
29		2.18	2.10	2.03	1.94	1.90	1.85	1.81	1.75	1.70	1.64
30		2.16	2.09	2.01	1.93	1.89	1.84	1.79	1.74	1.68	1.62
40		2.08	2.00	1.92	1.84	1.79	1.74	1.69	1.64	1.58	1.51
60		1.99	1.92	1.84	1.75	1.70	1.65	1.59	1.53	1.47	1.39
120		1.91	1.83	1.75	1.66	1.61	1.55	1.50	1.43	1.35	1.25
∞		1.83	1.75	1.67	1.57	1.52	1.46	1.39	1.32	1.22	1.00

TABLE 6
Critical Values of the F Distribution *(Continued)*

Denom-inator	Numer-ator	$f_{0.01}(n, m)$								
m	n	1	2	3	4	5	6	7	8	9
1		4052	4999.5	5403	5625	5764	5859	5928	5982	6022
2		98.50	99.00	99.17	99.25	99.30	99.33	99.36	99.37	99.39
3		34.12	30.82	29.46	28.71	28.24	27.91	27.67	27.49	27.35
4		21.20	18.00	16.69	15.98	15.52	15.21	14.98	14.80	14.66
5		16.26	13.27	12.06	11.39	10.97	10.67	10.46	10.29	10.16
6		13.75	10.92	9.78	9.15	8.75	8.47	8.26	8.10	7.98
7		12.25	9.55	8.45	7.85	7.46	7.19	6.99	6.84	6.72
8		11.26	8.65	7.59	7.01	6.63	6.37	6.18	6.03	5.91
9		10.56	8.02	6.99	6.42	6.06	5.80	5.61	5.47	5.35
10		10.04	7.56	6.55	5.99	5.64	5.39	5.20	5.06	4.94
11		9.65	7.21	6.22	5.67	5.32	5.07	4.89	4.74	4.63
12		9.33	6.93	5.95	5.41	5.06	4.82	4.64	4.50	4.39
13		9.07	6.70	5.74	5.21	4.86	4.62	4.44	4.30	4.19
14		8.86	6.51	5.56	5.04	4.69	4.46	4.28	4.14	4.03
15		8.68	6.36	5.42	4.89	4.56	4.32	4.14	4.00	3.89
16		8.53	6.23	5.29	4.77	4.44	4.20	4.03	3.89	3.78
17		8.40	6.11	5.18	4.67	4.34	4.10	3.93	3.79	3.68
18		8.29	6.01	5.09	4.58	4.25	4.01	3.84	3.71	3.60
19		8.18	5.93	5.01	4.50	4.17	3.94	3.77	3.63	3.52
20		8.10	5.85	4.94	4.43	4.10	3.87	3.70	3.56	3.46
21		8.02	5.78	4.87	4.37	4.04	3.81	3.64	3.51	3.40
22		7.95	5.72	4.82	4.31	3.99	3.76	3.59	3.45	3.35
23		7.88	5.66	4.76	4.26	3.94	3.71	3.54	3.41	3.30
24		7.82	5.61	4.72	4.22	3.90	3.67	3.50	3.36	3.26
25		7.77	5.57	4.68	4.18	3.85	3.63	3.46	3.32	3.22
26		7.72	5.53	4.64	4.14	3.82	3.59	3.42	3.29	3.18
27		7.68	5.49	4.60	4.11	3.78	3.56	3.39	3.26	3.15
28		7.64	5.45	4.57	4.07	3.75	3.53	3.36	3.23	3.12
29		7.60	5.42	4.54	4.04	3.73	3.50	3.33	3.20	3.09
30		7.56	5.39	4.51	4.02	3.70	3.47	3.30	3.17	3.07
40		7.31	5.18	4.31	3.83	3.51	3.29	3.12	2.99	2.89
60		7.08	4.98	4.13	3.65	3.34	3.12	2.95	2.82	2.72
120		6.85	4.79	3.95	3.48	3.17	2.96	2.79	2.66	2.56
∞		6.63	4.61	3.78	3.32	3.02	2.80	2.64	2.51	2.41

TABLE 6
Critical Values of the F Distribution *(Continued)*

Denominator m	Numerator n	$f_{0.01}(n, m)$									
		10	12	15	20	24	30	40	60	120	∞
1		6056	6106	6157	6209	6235	6261	6287	6313	6339	6366
2		99.40	99.42	99.43	99.45	99.46	99.47	99.47	99.48	99.49	99.50
3		27.23	27.05	26.87	26.69	26.60	26.50	26.41	26.32	26.22	26.13
4		14.55	14.37	14.20	14.02	13.93	13.84	13.75	13.65	13.56	13.46
5		10.05	9.89	9.72	9.55	9.47	9.38	9.29	9.20	9.11	9.02
6		7.87	7.72	7.56	7.40	7.31	7.23	7.14	7.06	6.97	6.88
7		6.62	6.47	6.31	6.16	6.07	5.99	5.91	5.82	5.74	5.65
8		5.81	5.67	5.52	5.36	5.28	5.20	5.12	5.03	4.95	4.86
9		5.26	5.11	4.96	4.81	4.73	4.65	4.57	4.48	4.40	4.31
10		4.85	4.71	4.56	4.41	4.33	4.25	4.17	4.08	4.00	3.91
11		4.54	4.40	4.25	4.10	4.02	3.94	3.86	3.78	3.69	3.60
12		4.30	4.16	4.01	3.86	3.78	3.70	3.62	3.54	3.45	3.36
13		4.10	3.96	3.82	3.66	3.59	3.51	3.43	3.34	3.25	3.17
14		3.94	3.80	3.66	3.51	3.43	3.35	3.27	3.18	3.09	3.00
15		3.80	3.67	3.52	3.37	3.29	3.21	3.13	3.05	2.96	2.87
16		3.69	3.55	3.41	3.26	3.18	3.10	3.02	2.93	2.84	2.75
17		3.59	3.46	3.31	3.16	3.08	3.00	2.92	2.83	2.75	2.65
18		3.51	3.37	3.23	3.08	3.00	2.92	2.84	2.75	2.66	2.57
19		3.43	3.30	3.15	3.00	2.92	2.84	2.76	2.67	2.58	2.49
20		3.37	3.23	3.09	2.94	2.86	2.78	2.69	2.61	2.52	2.42
21		3.31	3.17	3.03	2.88	2.80	2.72	2.64	2.55	2.46	2.36
22		3.26	3.12	2.98	2.83	2.75	2.67	2.58	2.50	2.40	2.31
23		3.21	3.07	2.93	2.78	2.70	2.62	2.54	2.45	2.35	2.26
24		3.17	3.03	2.89	2.74	2.66	2.58	2.49	2.40	2.31	2.21
25		3.13	2.99	2.85	2.70	2.62	2.54	2.45	2.36	2.27	2.17
26		3.09	2.96	2.81	2.66	2.58	2.50	2.42	2.33	2.23	2.13
27		3.06	2.93	2.78	2.63	2.55	2.47	2.38	2.29	2.20	2.10
28		3.03	2.90	2.75	2.60	2.52	2.44	2.35	2.26	2.17	2.06
29		3.00	2.87	2.73	2.57	2.49	2.41	2.33	2.23	2.14	2.03
30		2.98	2.84	2.70	2.55	2.47	2.39	2.30	2.21	2.11	2.01
40		2.80	2.66	2.52	2.37	2.29	2.20	2.11	2.02	1.92	1.80
60		2.63	2.50	2.35	2.20	2.12	2.03	1.94	1.84	1.73	1.60
120		2.47	2.34	2.19	2.03	1.95	1.86	1.76	1.66	1.53	1.38
∞		2.32	2.18	2.04	1.88	1.79	1.70	1.59	1.47	1.32	1.00

TABLE 6
Critical Values of the F Distribution *(Continued)*

Denom- inator m	Numer- ator n	$f_{0.025}(n, m)$								
		1	2	3	4	5	6	7	8	9
1		647.8	799.5	864.2	899.6	921.8	937.1	948.2	956.7	963.3
2		38.51	39.00	39.17	39.25	39.30	39.33	39.36	39.37	39.39
3		17.44	16.04	15.44	15.10	14.88	14.73	14.62	14.54	14.47
4		12.22	10.65	9.98	9.60	9.36	9.20	9.07	8.98	8.90
5		10.01	8.43	7.76	7.39	7.15	6.98	6.85	6.76	6.68
6		8.81	7.26	6.60	6.23	5.99	5.82	5.70	5.60	5.52
7		8.07	6.54	5.89	5.52	5.29	5.12	4.99	4.90	4.82
8		7.57	6.06	5.42	5.05	4.82	4.65	4.53	4.43	4.36
9		7.21	5.71	5.08	4.72	4.48	4.32	4.20	4.10	4.03
10		6.94	5.46	4.83	4.47	4.24	4.07	3.95	3.85	3.78
11		6.72	5.26	4.63	4.28	4.04	3.88	3.76	3.66	3.59
12		6.55	5.10	4.47	4.12	3.89	3.73	3.61	3.51	3.44
13		6.41	4.97	4.35	4.00	3.77	3.60	3.48	3.39	3.31
14		6.30	4.86	4.24	3.89	3.66	3.50	3.38	3.29	3.21
15		6.20	4.77	4.15	3.80	3.58	3.41	3.29	3.20	3.12
16		6.12	4.69	4.08	3.73	3.50	3.34	3.22	3.12	3.05
17		6.04	4.62	4.01	3.66	3.44	3.28	3.16	3.06	2.98
18		5.98	4.56	3.95	3.61	3.38	3.22	3.10	3.01	2.93
19		5.92	4.51	3.90	3.56	3.33	3.17	3.05	2.96	2.88
20		5.87	4.46	3.86	3.51	3.29	3.13	3.01	2.91	2.84
21		5.83	4.42	3.82	3.48	3.25	3.09	2.97	2.87	2.80
22		5.79	4.38	3.78	3.44	3.22	3.05	2.93	2.84	2.76
23		5.75	4.35	3.75	3.41	3.18	3.02	2.90	2.81	2.73
24		5.72	4.32	3.72	3.38	3.15	2.99	2.87	2.78	2.70
25		5.69	4.29	3.69	3.35	3.13	2.97	2.85	2.75	2.68
26		5.66	4.27	3.67	3.33	3.10	2.94	2.82	2.73	2.65
27		5.63	4.24	3.65	3.31	3.08	2.92	2.80	2.71	2.63
28		5.61	4.22	3.63	3.29	3.06	2.90	2.78	2.69	2.61
29		5.59	4.20	3.61	3.27	3.04	2.88	2.76	2.67	2.59
30		5.57	4.18	3.59	3.25	3.03	2.87	2.75	2.65	2.57
40		5.42	4.05	3.46	3.13	2.90	2.74	2.62	2.53	2.45
60		5.29	3.93	3.34	3.01	2.79	2.63	2.51	2.41	2.33
120		5.15	3.80	3.23	2.89	2.67	2.52	2.39	2.30	2.22
∞		5.02	3.69	3.12	2.79	2.57	2.41	2.29	2.19	2.11

TABLE 6

Critical Values of the F Distribution *(Continued)*

Denominator m	Numerator n	$f_{0.025}(n, m)$									
		10	12	15	20	24	30	40	60	120	∞
1		968.6	976.7	984.9	993.1	997.2	1001	1006	1010	1014	1018
2		39.40	39.41	39.43	39.45	39.46	39.46	39.47	39.48	39.49	39.50
3		14.42	14.34	14.25	14.17	14.12	14.08	14.04	13.99	13.95	13.90
4		8.84	8.75	8.66	8.56	8.51	8.46	8.41	8.36	8.31	8.26
5		6.62	6.52	6.43	6.33	6.28	6.23	6.18	6.12	6.07	6.02
6		5.46	5.37	5.27	5.17	5.12	5.07	5.01	4.96	4.90	4.85
7		4.76	4.67	4.57	4.47	4.42	4.36	4.31	4.25	4.20	4.14
8		4.30	4.20	4.10	4.00	3.95	3.89	3.84	3.78	3.73	3.67
9		3.96	3.87	3.77	3.67	3.61	3.56	3.51	3.45	3.39	3.33
10		3.72	3.62	3.52	3.42	3.37	3.31	3.26	3.20	3.14	3.08
11		3.53	3.43	3.33	3.23	3.17	3.12	3.06	3.00	2.94	2.88
12		3.37	3.28	3.18	3.07	3.02	2.96	2.91	2.85	2.79	2.72
13		3.25	3.15	3.05	2.95	2.89	2.84	2.78	2.72	2.66	2.60
14		3.15	3.05	2.95	2.84	2.79	2.73	2.67	2.61	2.55	2.49
15		3.06	2.96	2.86	2.76	2.70	2.64	2.59	2.52	2.46	2.40
16		2.99	2.89	2.79	2.68	2.63	2.57	2.51	2.45	2.38	2.32
17		2.92	2.82	2.72	2.62	2.56	2.50	2.44	2.38	2.32	2.25
18		2.87	2.77	2.67	2.56	2.50	2.44	2.38	2.32	2.26	2.19
19		2.82	2.72	2.62	2.51	2.45	2.39	2.33	2.27	2.20	2.13
20		2.77	2.68	2.57	2.46	2.41	2.35	2.29	2.22	2.16	2.09
21		2.73	2.64	2.53	2.42	2.37	2.31	2.25	2.18	2.11	2.04
22		2.70	2.60	2.50	2.39	2.33	2.27	2.21	2.14	2.08	2.00
23		2.67	2.57	2.47	2.36	2.30	2.24	2.18	2.11	2.04	1.97
24		2.64	2.54	2.44	2.33	2.27	2.21	2.15	2.08	2.01	1.94
25		2.61	2.51	2.41	2.30	2.24	2.18	2.12	2.05	1.98	1.91
26		2.59	2.49	2.39	2.28	2.22	2.16	2.09	2.03	1.95	1.88
27		2.57	2.47	2.36	2.25	2.19	2.13	2.07	2.00	1.93	1.85
28		2.55	2.45	2.34	2.23	2.17	2.11	2.05	1.98	1.91	1.83
29		2.53	2.43	2.32	2.21	2.15	2.09	2.03	1.96	1.89	1.81
30		2.51	2.41	2.31	2.20	2.14	2.07	2.01	1.94	1.87	1.79
40		2.39	2.29	2.18	2.07	2.01	1.94	1.88	1.80	1.72	1.64
60		2.27	2.17	2.06	1.94	1.88	1.82	1.74	1.67	1.58	1.48
120		2.16	2.05	1.94	1.82	1.76	1.69	1.61	1.53	1.43	1.31
∞		2.05	1.94	1.83	1.71	1.64	1.57	1.48	1.39	1.27	1.00

TABLE 7
Critical Values of D in the Kolmogorov–Smirnov Test
with Parameters of F Known[a]

Sample size n	Level of a significance for $D = \max \lvert F(x) - S_n(x) \rvert			
	$\alpha = 0.20$	$\alpha = 0.10$	$\alpha = 0.05$	$\alpha = 0.01$
1	0.9000	0.9500	0.9750	0.9950
2	0.6838	0.7764	0.8419	0.9293
3	0.5648	0.6360	0.7076	0.8290
4	0.4927	0.5652	0.6239	0.7342
5	0.4470	0.5095	0.5633	0.6685
6	0.4104	0.4680	0.5193	0.6166
7	0.3815	0.4361	0.4834	0.5758
8	0.3583	0.4096	0.4543	0.5418
9	0.3391	0.3875	0.4300	0.5133
10	0.3226	0.3687	0.4093	0.4889
11	0.3083	0.3524	0.3912	0.4677
12	0.2958	0.3382	0.3754	0.4491
13	0.2847	0.3255	0.3614	0.4325
14	0.2748	0.3142	0.3489	0.4176
15	0.2659	0.3040	0.3376	0.4042
16	0.2578	0.2947	0.3273	0.3920
17	0.2504	0.2863	0.3180	0.3809
18	0.2436	0.2785	0.3094	0.3706
19	0.2374	0.2714	0.3014	0.3612
20	0.2316	0.2647	0.2941	0.3524
25	0.2079	0.2377	0.2640	0.3166
30	0.1903	0.2176	0.2417	0.2899
35	0.1766	0.2019	0.2243	0.2690
Over 35	$1.07/\sqrt{n}$	$1.22/\sqrt{n}$	$1.36/\sqrt{n}$	$1.63/\sqrt{n}$

[a] Adapted from Table 1 of L. H. Miller, Table of percentage points of Kolmogorov statistics, *J. Amer. Statist. Assoc.* **51** (1956), 113, and Table 1 of F. J. Massey, Jr., The Kolmogorov–Smirnov test for goodness-of-fit, *J. Amer. Statist. Assoc.* **46** (1951), 70, with permission of the authors and publisher.

TABLE 8

Critical Values of D in the Kolmogorov–Smirnov Test for the Exponential Distribution with Mean Unknown[a]

Sample size n	Level of significance for $D = \max\lvert F(x) - S_n(x)\rvert$				
	$\alpha = 0.20$	$\alpha = 0.15$	$\alpha = 0.10$	$\alpha = 0.05$	$\alpha = 0.01$
3	0.451	0.479	0.511	0.551	0.600
4	0.396	0.422	0.449	0.487	0.548
5	0.359	0.382	0.406	0.442	0.504
6	0.331	0.351	0.375	0.408	0.470
7	0.309	0.327	0.350	0.382	0.442
8	0.291	0.308	0.329	0.360	0.419
9	0.277	0.291	0.311	0.341	0.399
10	0.263	0.277	0.295	0.325	0.380
11	0.251	0.264	0.283	0.311	0.365
12	0.241	0.254	0.271	0.298	0.351
13	0.232	0.245	0.261	0.287	0.338
14	0.224	0.237	0.252	0.277	0.326
15	0.217	0.229	0.244	0.269	0.315
16	0.211	0.222	0.236	0.261	0.306
17	0.204	0.215	0.229	0.253	0.297
18	0.199	0.210	0.223	0.246	0.289
19	0.193	0.204	0.218	0.239	0.283
20	0.188	0.199	0.212	0.234	0.278
25	0.170	0.180	0.191	0.210	0.247
30	0.155	0.164	0.174	0.192	0.226
Over 30	$0.86/\sqrt{n}$	$0.91/\sqrt{n}$	$0.96/\sqrt{n}$	$1.06/\sqrt{n}$	$1.25/\sqrt{n}$

[a] Adapted from Table 1 of H. W. Lilliefors, On the Kolmogorov–Smirnov test for the exponential with mean unknown, *J. Amer. Statist. Assoc.* **64** (1969), 388, with permission of the author and publisher.

TABLE 9
Critical Values of D in the Kolmogorov–Smirnov Test for Normality with Mean and Variance Unknown[a]

Sample size n	Level of significance for $D = \max \lvert \Phi(z) - z \rvert$				
	$\alpha = 0.20$	$\alpha = 0.15$	$\alpha = 0.10$	$\alpha = 0.05$	$\alpha = 0.01$
4	0.300	0.319	0.352	0.381	0.417
5	0.285	0.299	0.315	0.337	0.405
6	0.265	0.277	0.294	0.319	0.364
7	0.247	0.258	0.276	0.300	0.348
8	0.233	0.244	0.261	0.285	0.331
9	0.223	0.233	0.249	0.271	0.311
10	0.215	0.224	0.239	0.258	0.294
11	0.206	0.217	0.230	0.249	0.284
12	0.199	0.212	0.223	0.242	0.275
13	0.190	0.202	0.214	0.234	0.268
14	0.183	0.194	0.207	0.227	0.261
15	0.177	0.187	0.201	0.220	0.257
16	0.173	0.182	0.195	0.213	0.250
17	0.169	0.177	0.189	0.206	0.245
18	0.166	0.173	0.184	0.200	0.239
19	0.163	0.169	0.179	0.195	0.235
20	0.160	0.166	0.174	0.190	0.231
25	0.142	0.147	0.158	0.173	0.200
30	0.131	0.136	0.144	0.161	0.187
Over 30	$0.736/\sqrt{n}$	$0.768/\sqrt{n}$	$0.805/\sqrt{n}$	$0.886/\sqrt{n}$	$1.031/\sqrt{n}$

[a] Adapted from Table 1 of H. W. Lilliefors, On the Kolmogorov–Smirnov test for normality with mean and variance unknown, *J. Amer. Statist. Assoc.* **62** (1967), 400, with permission of the author and publisher. Correction for $n = 25$ provided by Professor Lilliefors.

Appendix B

APL PROGRAMS

All the APL programs mentioned in this book are listed here. The listing was done on the printer of an IBM 5100 computer so the characters may look slightly different than those printed out at a terminal such as an IBM 2741, 3270, or 3767. The programs are all "naive" programs in the sense that no attempt has been made to optimize the efficiency of execution or to make it difficult for the user to read and understand the program. The queueing theory programs are usually a direct translation of the equations of Appendix C. No attempt has been made to organize the calculations to minimize round-off error or to avoid numerical instabilities. Therefore, no guarantee can be made that a given program will always produce correct results.

All the programs have been executed both on an IBM 5100 computer and on an IBM System/370 using VS APL for CMS.

Space prevents us from writing an introduction to APL here. For those having an IBM 5100, [1] is an excellent introduction to this computer and the use of APL on the machine. Reference [2] gives the details of APL in all versions supported by IBM. References [3] and [4] are two excellent introductory books on APL.

```
   1 ∇ L←K ALGΔM X
[ 1] ⍝ THE FUNCTION ALGΔM IMPLEMENTS  ALGORITHM M TO FIND
[ 2] ⍝ THE MAXIMUM NUMBER OF REQUESTS PER SECOND THAT K DISK
[ 3] ⍝ DRIVES ON ONE CHANNEL CAN SUPPORT IF THE AVERAGE CHANNEL
[ 4] ⍝ TIME PER REQUEST IS ES AND THE AVERAGE TIME TO USE A
[ 5] ⍝ DISK DRIVE, EXCLUSIVE OF CHANNEL TIME, IS EO ,THE CALL IS
[ 6] ⍝ '' K ALGΔM EO ,ES ''.  TIME UNITS SHOULD BE SECONDS.
[ 7] EO←X[1]
[ 8] ES←X[2]
[ 9] Z←EO÷ES
[10] RHO←1-K BCU Z
[11] L←RHO÷ES
   ∇


   2 ∇ Z←C BCU U
[ 1] ⍝ BCU CALCULATES THE VALUE OF ERLANG'S LOSS
[ 2] ⍝ FORMULA (SOMETIMES CALLED ERLANG'S B FORMULA).
[ 3] ⍝ THE CALLING SEQUENCE IS ''C BCU U''.
[ 4] K1←¯1+⍳C+1
[ 5] K2←(U*K1)÷(!K1)
[ 6] Z←K2[C+1]÷+/K2
   ∇


   3 ∇ Z←NP BINOMIAL K
[ 1] ⍝ '' NP BINOMIAL K '' COMPUTES THE PROBABILITY
[ 2] ⍝ THAT A BINOMIALLY DISTRIBUTED RANDOM VARIABLE
[ 3] ⍝ WITH PARAMETERS N AND P ASSUMES THE VALUE K.
[ 4] ⍝ NP IS THE VECTOR OBTAINED BY CATENATING N AND P.
[ 5] N←NP[1]
[ 6] P←NP[2]
[ 7] Z←(K!N)×(P*K)×(1-P)*(N-K)
   ∇


   4 ∇ Z←NP BINΔSUM IJ;Y;X
[ 1] ⍝ '' NP BINΔSUM IJ '' COMPUTES THE PROBABILITY
[ 2] ⍝ THAT A BINOMIAL RANDOM VARIABLE WITH PARAMETERS
[ 3] ⍝ N AND P WILL ASSUME A VALUE BETWEEN I AND J,
[ 4] ⍝ INCLUSIVE.  NP AND IJ ARE THE VECTORS FORMED
[ 5] ⍝ BY CATENATING THE RESPECTIVE PARAMETERS.
[ 6] I←IJ[1]
[ 7] J←IJ[2]
[ 8] Y←((I≤X)∧(X≤J))/X←¯1+⍳J+1
[ 9] Z←+/NP BINOMIAL Y
   ∇


   5 ∇ P CONDEXPECT EXΔYEX2ΔY
[ 1] ⍝CONDEXPECT CALCULATES THE MEAN ,EX, THE SECOND MOMENT, EX2,
[ 2] ⍝THE VARIANCE , VARX, AND THE STANDARD DEVIATION ,SIGX,
[ 3] ⍝WHERE THE FIRST AND SECOND MOMENTS OF X GIVEN Y ARE KNOWN.
[ 4] ⍝TO CALL TYPE ' P CONDEXPECT EXΔYEX2ΔY ' WHERE P IS THE SET
[ 5] ⍝OF VALUES OF THE PROBABILITY MASS FUNCTION OF Y AND
[ 6] ⍝EXΔYES2ΔY IS THE VECTOR OF CONDITIONAL EXPECTATIONS OF
[ 7] ⍝X GIVEN Y CATENATED WITH THE VECTOR OF THE SECOND CON -
[ 8] ⍝DITIONAL MOMENTS OF X GIVEN Y.
[ 9] N←ρP
[10] EX←+/P×N↑EXΔYEX2ΔY
[11] EX2←+/P×N↓EXΔYEX2ΔY
[12] VARX←EX2-EX*2
[13] SIGX←VARX*0.5
[14] 'THE MEAN OF X IS   ',⍕EX
[15] 'THE SECOND MOMENT OF X  IS  ',⍕EX2
[16] 'THE VARIANCE OF X IS    ',⍕VARX
[17] 'THE STANDARD DEVIATION IS   ';⍕SIGX
   ∇
```

```
   6 ∇ LAMBDA DMΔMΔ1ΔK ESΔKΔT
[ 1] ⍝ THIS FUNCTION COMPUTES THE DISTRIBUTION FUNCTION
[ 2] ⍝ FOR BOTH QUEUEING TIME AND SYSTEM TIME FOR
[ 3] ⍝ THE M/M/1/K QUEUEING SYSTEM.  THE CALLING FORMAT
[ 4] ⍝ IS '' LAMBDA DMΔMΔ1ΔK ES,K,T ''.  THE FUNCTION
[ 5] ⍝ PARTIALΔSUM IS USED BY THIS FUNCTION.
[ 6] ES←ESΔKΔT[1]
[ 7] K←ESΔKΔT[2]
[ 8] T←ESΔKΔT[3]
[ 9] U←LAMBDAxES
[10] P0←(1-U)÷1-U*K+1
[11] P←P0xU*(⁻1+⍳K+1)
[12] Q←P[⍳K]÷(1-P[K+1])
[13] MUΔT←T÷ES
[14] WT←1-+/QX(MUΔT PARTIALΔSUM K-1)
[15] WQT←1-+/(1↓Q)x(MUΔT PARTIALΔSUM K-2)
[16] 'THE PROBABILITY THE WAITING TIME IN THE SYSTEM,'
[17] 'W, IS NOT GREATER THAN T, WT, IS  ',⍕WT
[18] 'THE PROBABILITY THE QUEUEING TIME , Q, '
[19] 'DOES NOT EXCEED T, WQT, IS  ',⍕WQT
    ∇
```

```
   7 ∇ C DWΔMΔMΔC UΔESΔT;CCU;C1;C2
[ 1] ⍝DWΔMΔMΔC CALCULATES THE VALUE OF
[ 2] ⍝THE DESTRIBUTION FUNCTION OF W,
[ 3] ⍝THE WAITING TIME IN THE SYSTEM,
[ 4] ⍝AND OF THE DISTRIBUTION FUNCTION
[ 5] ⍝OF Q, THE WAITING TIME IN THE
[ 6] ⍝QUEUE, AT TIME T FOR THE M/M/C
[ 7] ⍝QUEUEING SYSTEM.  THE CALLING
[ 8] ⍝FORMAT IS ''C DWΔMΔMΔC U,ES,T ''
[ 9] U←UΔESΔT[1]
[10] ES←UΔESΔT[2]
[11] T←UΔESΔT[3]
[12] CCU←C ERLANGΔC U
[13] WQT←1-CCUx*-(C-U)xT÷ES
[14] →EXCEPTx⍳U=C-1
[15] C1←((U-C)+1-CCU)÷C-1+U
[16] C2←CCU÷C-1+U
[17] WT←1+(C1x*-T÷ES)+C2x*-(C-U)xT÷ES
[18] →PRINT
[19] EXCEPT:
[20] WT←1-(1+CCUxT÷ES)x*-T÷ES
[21] PRINT:
[22] 'THE PROBABILITY THE WAITING TIME'
[23] 'IN QUEUE, Q , IS LESS THAN T IS'
[24] WQT
[25] 'THE PROBABILITY THE WAITING TIME'
[26] 'IN THE SYSTEM, W, IS LESS THAN'
[27] 'T IS  ',⍕WT
    ∇
```

```
   8 ∇ Z←C ERLANGΔC U;X;Y
[ 1] X←(U*C)÷!C
[ 2] Y←⁻1+⍳C
[ 3] Z←X÷(1-U÷C)x((+/(U*Y)÷!Y)+X÷(1-U÷C))
    ∇
```

```
      9 ∇ PIO GI∆M∆1 LAMBDA∆ES
[ 1]  ⍝THIS FUNCTION COMPUTES THE PERFORMANCE STATISTICS
[ 2]  ⍝FOR THE GI/M/1 QUEUEING SYSTEM. THE CALL IS
[ 3]  ⍝PIO GI∆M∆1 LAMBDA ES WHERE PIO IS THE PROBABILITY
[ 4]  ⍝AN ARRIVING CUSTOMER WILL FIND THE SYSTEM EMPTY.
[ 5]  LAMBDA←LAMBDA∆ES[1]
[ 6]  ES←LAMBDA∆ES[2]
[ 7]  RHO←LAMBDA×ES
[ 8]  'SERVER UTILIZATION, ρ, IS  ',⍕RHO
[ 9]  W←ES÷PIO
[10]  WQ←W-ES
[11]  VARQ←(1-PIO*2)×(ES÷PIO)*2
[12]  SIGQ←VARQ*0.5
[13]  C2Q←VARQ÷WQ*2
[14]  'THE AVERAGE QUEUEING TIME, WQ , IS  ',⍕WQ
[15]  'WITH STANDARD DEVIATION, SIGQ, ',⍕SIGQ
[16]  'C-SQUARED FOR Q, C2Q, IS  ',⍕C2Q
[17]  '90TH PERCENTILE QUEUEING TIME, PIQ90, IS'
[18]  ⎕←PIQ90←(W×⍟10×(1-PIO))⌈0
[19]  '95TH PERCENTILE QUEUEING TIME, PI95, IS'
[20]  ⎕←PIQ95←(W×⍟20×(1-PIO))⌈0
[21]  'THE DISTRIBUTION OF QUEUEING TIME FOR THOSE WHO'
[22]  'MUST IS THE SAME AS W.  THAT IS, EXPONENTIAL.'
[23]  'THE AVERAGE WAITING TIME IN THE SYSTEM, W, IS'
[24]  W
[25]  'WITH STANDARD DEVIATION  ',⍕W
[26]  'C-SQUARED FOR W IS 1, SINCE W IS EXPONENTIAL  '
[27]  PIW90←W×⍟10
[28]  '90TH PERCENTILE WAITING TIME IN THE SYSTEM, PIW90,'
[29]  ' IS ',⍕PIW90
[30]  PIW95←W×⍟20
[31]  '95TH PERCENTILE WAITING TIME IN THE SYSTEM, PIW95,'
[32]  ' IS  ',⍕PIW95
[33]  LQ←LAMBDA×WQ
[34]  L←LAMBDA×W
[35]  VARN←RHO×(2-(PIO+RHO))÷PIO*2
[36]  VARNQ←RHO×(1-PIO)×((2-PIO)-RHO×(1-PIO))÷PIO*2
[37]  SIGN←VARN*0.5
[38]  SIGNQ←VARNQ*0.5
[39]  'THE AVERAGE NUMBER IN THE QUEUE, LQ, IS'
[40]  LQ
[41]  'THE STANDARD DEVIATION OF NQ, SIGNQ, IS'
[42]  SIGNQ
[43]  'THE AVERAGE NUMBER IN THE SYSTEM, L, IS'
[44]  L
[45]  'THE STANDARD DEVIATION OF N, SIGN, IS'
[46]  SIGN
      ∇
```

```
   11 ∇ X MACHΔREP Y
[ 1] ⍝ MACHΔREP CALCULATES THE STATISTICS FOR THE MACHINE REPAIR
[ 2] ⍝ QUEUEING MODEL WITH K MACHINES AND C REPAIRMEN.  THE CALL
[ 3] ⍝ IS ''KΔC MACHΔREP EO,ES'' WHERE EO IS THE AVERAGE TIME
[ 4] ⍝ A MACHINE IS IN OPERAION AFTER BEING REPAIRED, ES IS
[ 5] ⍝ AVERAGE REPAIR TIME, AND KΔC IS A VECTOR OF K AND C.
[ 6] K←X[1]
[ 7] C←X[2]
[ 8] EO←Y[1]
[ 9] ES←Y[2]
[10] Z1←÷Z←EO÷ES
[11] PNΔP0←1,(Y!K)×Z1*(Y←⍳C)
[12] PNΔP0←PNΔP0,(!N)×(÷!C)×(÷(C*(N-C)))×(N!K)×Z1*(N←C+⍳K-C)
[13] P←PNΔP0×P0←÷+/PNΔP0
[14] ⍝P IS THE VECTOR OF PROBABILITIES THERE ARE N
[15] ⍝MACHINES DOWN FOR N = 0,1,...,K
[16] LQ←+/P[(C+1)+⍳K-C]×⍳K-C
[17] WQ←LQ×(EO+ES)÷(K-LQ)
[18] LAMBDA←K÷ETB←EO+WQ+ES
[19] ⍝D IS THE PROBABILITY OF DELAY, THAT IS,
[20] ⍝THAT ALL THE REPAIRMEN ARE BUSY.
[21] D←+/P[C+⍳K-C-1]
[22] ⍝WQΔD IS THE AVERAGE WAITING TIME FOR A
[23] ⍝MACHINE WHICH IS DELAYED.
[24] WQΔD←WQ÷D
[25] RHO←(LAMBDA×ES)÷C
[26] ⍝PΔDOWN IS THE PROBABILITY THAT ANY GIVEN
[27] ⍝MACHINE IS DOWN.
[28] PΔDOWN←(WQ+ES)÷ETB
[29] W←WQ+ES
[30] L←LAMBDA×W
[31] N2←+/P×(¯1+⍳K+1)*2
[32] SIGN←(VARN←N2-L*2)*0.5
[33] C2N←VARN÷L*2
[34] 'THE UTILIZATION OF EACH REPAIRMAN IS  ',⍕RHO
[35] 'LAMBDA, THE AVERAGE ARRIVAL RATE INTO THE SYSTEM IS  ',⍕LAMBDA
[36] 'THE AVERAGE QUEUEING TIME,WQ, IS  ',⍕WQ
[37] 'THE AVERAGE NUMBER QUEUEING, LQ, IS  ',⍕LQ
[38] 'THE AVERAGE NUMBER IN THE SYSTEM, L, IS  ',⍕L
[39] 'THE AVERAGE QUEUEING  TIME FOR THOSE WHO MUST IS '
[40] WQΔD
[41] 'THE PROBABILITY OF QUEUEING  FOR SERVICE IS  ',⍕D
[42] 'THE AVERAGE TIME IN THE QUEUEING SYSTEM IS  ',⍕W
[43] 'THE PROBABILITY THAT ANY GIVEN MACHINE IS DOWN IS'
[44] PΔDOWN
[45] 'C-SQUARED FOR THE NUMBER IN THE SYSTEM IS  ',⍕C2N
[46] 'FOR THE PROBABILITIES THAT 0,1,...,K  ARE IN THE'
[47] 'SYSTEM TYPE ''P'''
     ∇
```

```
    12 ∇ M MPOLL P
[  1]  ⍝ '' M MPOLL P '' CALCULATES THE AVERAGE, VARIANCE,
[  2]  ⍝ STANDARD DEVIATION, AND SECOND MOMENT OF THE
[  3]  ⍝ NUMBER OF POLLS TAKEN ON A LINE CONTAINING M
[  4]  ⍝ TERMINALS IF P IS THE VECTOR OF PROBABILIES
[  5]  ⍝ THAT ZERO THROUGH M TERMINALS ARE READY TO
[  6]  ⍝ TRANSMIT.  POLLING IS ASSUMED COMPLETE WHEN
[  7]  ⍝ A READY TERMINAL IS FOUND.  THIS FUNCTION
[  8]  ⍝ USES THE FUNCTIONS POLLM AND POLL2M. ⍝
[  9]  N←0
[10]  EXY←M
[11]  EX2Y←M×M
[12]  LOOP:→PRINT×⍳M<N←N+1
[13]  EXY←EXY,M POLLM N
[14]  →LOOP,EX2Y←EX2Y,M POLL2M N
[15]  PRINT:EX←+/P×EXY
[16]  EX2←+/P×EX2Y
[17]  VARX←EX2-EX*2
[18]  SIGX←VARX*0.5
[19]  'THE AVERAGE VALUE OF X IS  ',⍕EX
[20]  'THE VARIANCE OF X IS  ',⍕VARX
[21]  'THE STANDARD DEVIATION IS  ',⍕SIGX
[22]  'THE SECOND MOMENT IS  ',⍕EX2
    ∇
```

```
    13 ∇ K M∆EK∆1 X
[  1]  LAMBDA←X[1]
[  2]  ES←X[2]
[  3]  RHO←LAMBDA×ES
[  4]  R1←1-RHO
[  5]  ES2←(1+÷K)×ES*2
[  6]  ES3←(1+2÷K)×ES2×ES
[  7]  LQ←(ES2×LAMBDA*2)÷2×R1
[  8]  WQ←LQ÷LAMBDA
[  9]  WQ∆Q←WQ÷RHO
[10]  EQ2←((÷3×R1)×LAMBDA×ES3)+(÷2)×(C1←LAMBDA×ES2÷R1)*2
[11]  VARQ←EQ2-WQ*2
[12]  SIGQ←VARQ*0.5
[13]  L←LQ+RHO
[14]  W←L÷LAMBDA
[15]  EW2←EQ2+ES2÷R1
[16]  VARW←EW2-W*2
[17]  SIGW←VARW*0.5
[18]  VARN←((÷3×R1)×ES3×LAMBDA*3)+(0.25×(LAMBDA*2)×C1*2)
[19]  VARN←VARN+((3-2×RHO)×C1×LAMBDA÷2)+RHO×R1
[20]  SIGN←VARN*0.5
[21]  'THE AVERAGE NUMBER QUEUEING,LQ, IS  ',⍕LQ
[22]  'THE AVERAGE TIME WAITING IN QUEUE, WQ, IS  ',⍕WQ
[23]  'WITH STANDARD DEVIATION, SIGQ,  ',⍕SIGQ
[24]  'THE AVERAGE QUEUEING TIME FOR THOSE WHO MUST'
[25]  'WAIT, IS  ',⍕WQ∆Q
[26]  'THE AVERAGE NUMBER IN THE SYSTEM, L, IS  ',⍕L
[27]  'WITH STANDARD DEVIATION  ',⍕SIGN
[28]  'THE AVERAGE TIME IN THE SYSTEM,W, IS  ',⍕W
[29]  'WITH STANDARD DEVIATION  ',⍕SIGW
[30]  'THE MARTIN RULE ESTIMATE OF THE 90TH PERCENTILE'
[31]  'TIME IN THE SYSTEM IS  ',⍕W90←W+1.3×SIGW
[32]  'THE C-SQUARED VALUE FOR WAITING TIME IS  ',⍕C2Q←VARQ÷WQ*2
[33]  'THE C-SQUARED VALUE FOR SYSTEM TIME IS  ',⍕C2W←VARW÷W*2
[34]  'THE C-SQUARED VALUE FOR NUMBER IN THE SYSTEM IS  ',⍕C2N←VARN÷L*2
    ∇
```

```
14 ∇ X MΔGΔ1 ES
[ 1] ⍝ MΔGΔ1 CALCULATES THE USUAL STATISTICS FOR THE
[ 2] ⍝ M/G/1 MODEL USING THE POLLACZEK-KHINTCHINE
[ 3] ⍝ EQUATIONS.  THE CALLING SEQUENCE IS ''LAMBDA SIG MΔGΔ1 ES''
[ 4] ⍝ WHERE LAMBDA IS THE AVERAGE ARRIVAL RATE, SIG IS THE STANDARD
[ 5] ⍝ DEVIATION OF SERVICE TIME, AND ES IS AVERAGE SERVICE TIME.
[ 6] LAMBDA←X[1]
[ 7] SIG←X[2]
[ 8] RHO←LAMBDA×ES
[ 9] LQ←(((LAMBDA×SIG)*2)+RHO*2)÷2×(1-RHO)
[10] L←LQ+RHO
[11] WQ←LQ÷LAMBDA
[12] W←WQ+ES
[13] 'THE SERVER UTILIZATION, RHO, IS  ',⍕RHO
[14] 'THE AVERAGE NUMBER IN THE SYSTEM IS  ',⍕L
[15] 'THE AVERAGE NUMBER WAITING FOR SERVICE IS  ',⍕LQ
[16] 'THE AVERAGE TIME IN THE SYSTEM IS  ',⍕W
[17] 'THE AVERAGE WAITING TIME IS  ',⍕WQ
[18] 'THE AVERAGE WAITING TIME FOR CUSTOMERS DELAYED IS  ',⍕WQ÷RHO
    ∇


15 ∇ LAMBDA MΔGΔ1ΔEXT X
[ 1] ⍝ MΔGΔ1ΔEXT CALCULATES THE USUAL STATISTICS FOR THE
[ 2] ⍝ M/G/1 MODEL, BUT ALSO  CALCULATES THE STANDARD DEVIATIONS
[ 3] ⍝ AND THE C-SQUARED VALUES FOR MANY OF THE VARIABLES SO
[ 4] ⍝ THAT ERLANG APPROXIMATIONS CAN BE MADE.  THE CALLING
[ 5] ⍝ SEQUENCE IS '' LAMBDA MΔGΔ1ΔEXT ES ES2 ES3 ''
[ 6] ⍝ WHERE ES ES2 ES3 ARE THE FIRST THREE MOMENTS OF
[ 7] ⍝ THE SERVICE TIME DISTRIBUTION.
[ 8] ES←X[1]
[ 9] ES2←X[2]
[10] ES3←X[3]
[11] RHO←LAMBDA×ES
[12] LQ←(LAMBDA*2)×ES2÷2×(1-RHO)
[13] L←LQ+RHO
[14] VARN←(LAMBDA*3)×ES3÷3×(1-RHO)
[15] VARN←VARN+(LAMBDA*2)×(ES2*2)÷4×(1-RHO)*2
[16] VARN←VARN+(LAMBDA*2)×(3-2×RHO)×ES2÷2×(1-RHO)
[17] VARN←VARN+RHO×(1-RHO)
[18] SIGN←VARN*0.5
[19] WQ←LQ÷LAMBDA
[20] EQ2←(LAMBDA×ES3÷3×(1-RHO))+2×WQ*2
[21] VARQ←EQ2-WQ*2
[22] SIGQ←VARQ*0.5
[23] W←WQ+ES
[24] EW2←EQ2+ES2÷1-RHO
[25] EW2←EW2
[26] VARW←EW2-W*2
[27] SIGW←VARW*0.5
[28] 'SERVER UTILIZATION,RHO, IS  ',⍕RHO
[29] 'THE AVERAGE NUMBER IN THE SYSTEM, L, IS  ',⍕L
[30] 'WITH STANDARD DEVIATION  ',⍕SIGN
[31] 'C-SQUARED FOR NUMBER IN THE SYSTEM IS  ',⍕C2N←VARN÷L*2
[32] 'THE AVERAGE NUMBER QUEUEING ,LQ, IS  ',⍕LQ
[33] 'THE AVERAGE TIME IN THE SYSTEM,W, IS  ',⍕W
[34] 'WITH STANDARD DEVIATION  ',⍕SIGW
[35] 'C-SQUARED FOR TIME IN THE SYSTEM IS  ',⍕C2W←VARW÷W*2
[36] 'THE AVERAGE QUEUEING TIME,WQ, IS  ',⍕WQ
[37] 'WITH STANDARD DEVIATION  ',⍕SIGQ
[38] 'C-SQUARED FOR QUEUEING TIME IS  ',⍕C2Q←VARQ÷WQ*2
[39] ' THE MARTIN ESTIMATES OF 90TH AND 95TH'
[40] ' PERCENTILES OF TIME IN THE SYSTEM ARE'
[41] □←PIW90←W+1.3×SIGW
[42] □←PIW95←W+2×SIGW
    ∇
```

```
   16 ∇ U MΔMΔC X
[ 1]  ⍝ MΔMΔC IS A FUNCTION FOR CALCULATING THE STATISTICS OF THE
[ 2]  ⍝ M/M/C QUEUEING SYSTEM.  IT USES THE FUNCTION ERLANGΔC THE
[ 3]  ⍝ CALLING SEQUENCE IS '' U MΔMΔC C,ES''   WHERE C IS THE
[ 4]  ⍝ NUMBER OF SERVERS, ES IS THE AVERAGE SERVICE TIME OF EACH
[ 5]  ⍝ SERVER, AND U IS THE TRAFFIC INTENSITY.
[ 6]  C←X[1]
[ 7]  ES←X[2]
[ 8]  RHO←U÷C
[ 9]  CCU←C ERLANGΔC U
[10]  LQ←RHO×CCU÷1-RHO
[11]  VARNQ←LQ×(1+RHO-RHO×CCU)÷1-RHO
[12]  SIGNQ←VARNQ*0.5
[13]  LAMBDA←U÷ES
[14]  WQ←LQ÷LAMBDA
[15]  VARQ←(2-CCU)×CCU×(ES*2)÷(((1-RHO)*2)×C*2)
[16]  SIGQ←VARQ*0.5
[17]  W←WQ+ES
[18]  L←LAMBDA×W
[19]  →LΔUΔCΔ1×⍳U=C-1
[20]  EW2←(2×CCU×ES*2)÷(U+1-C)
[21]  EW2←EW2×(1-(C-U)*2)÷(C-U)*2
[22]  →LΔSIGΔW,EW2←EW2+2×ES*2
[23]  LΔUΔCΔ1:EW2←((4×CCU)+2)×ES*2
[24]  LΔSIGΔW:VARW←EW2-W*2
[25]  SIGW←VARW*0.5
[26]  K←ES÷C-U
[27]  WQ90←0⌈K×⍟10×CCU
[28]  WQ95←0⌈K×⍟20×CCU
[29]  'THE AVERAGE ARRIVAL RATE, LAMBDA, IS  ',⍕LAMBDA
[30]  'THE PROBABILITY THAT ALL THE SERVERS ARE BUSY IS  ',⍕CCU
[31]  'THE SERVER UTILIZATION, RHO = U÷C, IS  ',⍕RHO
[32]  'AVERAGE NUMBER IN THE QUEUE IS  ',⍕LQ
[33]  'WITH STANDARD DEVIAION  ',⍕SIGNQ
[34]  'AVERAGE TIME IN THE QUEUE IS  ',⍕WQ
[35]  'WTH STANDARD DEVIATION  ',⍕SIGQ
[36]  'AVERAGE QUEUEING TIME FOR CUSTOMERS DELAYED IS   ',⍕WQΔD←WQ÷CCU
[37]  'AVERAGE NUMBER IN THE SYSTEM IS  ',⍕L
[38]  'AVERAGE TIME IN THE SYSTEM IS  ',⍕W
[39]  'WTH STANDARD DEVIATION  ',⍕SIGW
[40]  'THE 90-TH PERCENTILE QUEUEING TIME IS  ',⍕WQ90
[41]  'THE 95-TH PERCENTILE   QUEUEING TIME IS  ',⍕WQ95
[42]  'C-SQUARED FOR SYSTEM TIME IS  ',⍕C2W←VARW÷W*2
[43]  'C-SQUARED  FOR QUEUEING TIME IS  ',⍕C2WQ←VARQ÷WQ*2
[44]  'THE MARTIN ESTIMATE OF 90-TH PERCENTILE SYSTEM TIME IS:'
[45]  ⎕←W90←W+1.3×SIGW
[46]  'THE MARTIN ESTIMATE OF 95-TH PERCENTILE SYSTEM TIME IS:'
[47]  ⎕←W95←W+2×SIGW
    ∇


   17 ∇ C MΔMΔCΔLOSS U;W1;W2
[ 1]  ⍝ MΔMΔCΔLOSS CALCULATES THE STATISTICS FOR THE M/M/C LOSS SYSTEM,
[ 2]  ⍝ THAT IS, THE M/M/C SYSTEM WITH NO WAITING LINE.  THE CALLING
[ 3]  ⍝ FORMAT IS ''C MΔMΔCΔLOSS U '', WHERE C IS THE NUMBER OF
[ 4]  ⍝ SERVERS AND U IS THE TRAFFIC INTENSITY.
[ 5]  P0←÷/(U*W1)÷W2←!W1←¯1+⍳C+1
[ 6]  P←P0×(U*W1)÷W2
[ 7]  'PROBABILITY AN ARRIVING CUSTOMER IS LOST,PC, IS  ',⍕PC←P[C+1]
[ 8]  'THE AVERAGE NUMBER OF SERVERS OCCUPIED,L, IS   ',⍕L←U×(1-PC)
[ 9]  'THE TRUE SERVER UTILIZATION,ρ, IS   ',⍕RHO←L÷C
    ∇
```

```
 18 ∇ LAMBDA MΔMΔ1 ES
[ 1]  ⍝MΔMΔ1 CALCULATES THE STATISTICS FOR THE CLASSICAL M/M/1
[ 2]  ⍝MODEL.  IT IS INVOKED BY TYPING '' LAMBDA MΔMΔ1 ES '',
[ 3]  ⍝WHERE LAMBDA IS THE AVERAGE ARRIVAL RATE AND ES IS
[ 4]  ⍝THE AVERAGE SERVICE TIME.
[ 5]  RHO←LAMBDA×ES
[ 6]  'THE SERVER UTILIZATION, ρ, IS  ',⍕RHO
[ 7]  'THE MEAN NUMBER IN THE SYSTEM, L, IS  ',⍕L←RHO÷(1-RHO)
[ 8]  'WITH STANDARD DEVIATION  ',⍕SIGN←(L÷(1-RHO))*0.5
[ 9]  'THE MEAN NUMBER IN THE QUEUE ,LQ, IS  ',⍕LQ←RHO×L
[10]  'WITH STANDARD DEVIATION  ',⍕SIGNQ←((L*2)×(1+RHO-RHO*2))*0.5
[11]  'THE MEAN TIME IN THE SYSTEM, W, IS  ',⍕W←ES÷(1-RHO)
[12]  'WITH STANDARD DEVIATION  ',⍕W
[13]  'THE MEAN TIME IN THE QUEUE, WQ, IS  ',⍕WQ←RHO×W
[14]  'WITH STANDARD DEVIATION  ',⍕SIGQ←SIGN×ES×(2-RHO)*0.5
[15]  'THE MEAN QUEUEING TIME FOR THOSE WHO MUST QUEUE IS  ',⍕W
[16]  'THE MEAN NUMBER QUEUEING, WHEN THE QUEUE IS '
[17]  'NOT EMPTY IS  ',⍕÷1-RHO
[18]  '90TH PERCENTILE TIME IN THE SYSTEM IS  ',⍕W90←W×⍟10
[19]  '95TH PERCENTILE TIME IN THE SYSTEM IS  ',⍕W95←W×⍟20
[20]  '90TH PERCENTILE QUEUEING TIME IS  ',⍕Q90←(W×⍟10×RHO)⌈0
[21]  '95TH PERCENTILE QUEUEING TIME IS  ',⍕Q95←(W×⍟20×RHO)⌈0
     ∇

 19 ∇ LAMBDA MΔMΔ1ΔK X
[ 1]  ⍝ MΔMΔ1ΔK CALCULATES THE STATISTICS FOR THE MΔMΔ1
[ 2]  ⍝ MODEL RESTRICTED SO THAT NO MORE THAN K CUSTOMERS
[ 3]  ⍝ ARE ALLOWED IN THE SYSTEM.
[ 4]  ES←X[1]
[ 5]  K←X[2]
[ 6]  U←LAMBDA×ES
[ 7]  P0←(1-U)÷XN1←1-U*K+1
[ 8]  P←P0×U*(⁻1+⍳K+1)
[ 9]  PK←P0×U*K
[10]  'THE PROBABILITY, PSUBZERO, THAT THE SYSTEM IS EMPTY IS'
[11]  P0
[12]  'THE PROBABILITY THAT AN ARRIVING CUSTOMER IS TURNED'
[13]  'AWAY, PSUBK, IS  ',⍕PK
[14]  'THE TRAFFIC INTENSITY, U, IS  ',⍕U
[15]  'THE SERVER UTILIZATION, RHO, IS  ',⍕(1-PK)×U
[16]  L←(U×1+((K×U)-(K+1))×U*K)÷(XU←1-U)×XN1
[17]  EN2←+/P×(⁻1+⍳K+1)*2
[18]  VARN←EN2-L*2
[19]  C2N←VARN÷L*2
[20]  SIGN←VARN*0.5
[21]  LQ←L-(1-P0)
[22]  ENQ2←+/P[2+⍳K-1]×(⍳K-1)*2
[23]  VARNQ←ENQ2-LQ*2
[24]  C2NQ←VARNQ÷LQ*2
[25]  SIGNQ←VARNQ*0.5
[26]  LAMBDAA←LAMBDA×(1-PK)
[27]  WQ←LQ÷LAMBDAA
[28]  W←WQ+ES
[29]  WQΔQ←WQ÷(1-P0)
[30]  'THE AVERAGE NUMBER IN THE SYSTEM, L, IS  ',⍕L
[31]  'WITH STANDARD DEVIATION  ',⍕SIGN
[32]  'C-SQUARED FOR NUMBER IN THE SYSTEM, N, IS  ',⍕C2N
[33]  'THE AVERAGE NUMBER WAITING, LQ, IS  ',⍕LQ
[34]  'WITH STANDARD DEVIATION  ',⍕SIGNQ
[35]  'C-SQUARED FOR NUMBER IN THE QUEUE, NQ, IS  ',⍕C2NQ
[36]  'THE AVERAGE TIME IN THE SYSTEM, W, IS  ',⍕W
[37]  'THE AVERAGE WAITING TIME, WQ, IS  ',⍕WQ
[38]  'THE AVERAGE WAITING TIME FOR THOSE WHO MUST WAIT IS   ,⍕WQΔQ
     ∇
```

```
 20 ∇ P←NDIST T;S;R;Z
[ 1] ⍝'' NDIST T '' EVALUATES THE STANDARD NORMAL DISTRIBUTION
[ 2] ⍝ FUNCTION AT THE POINT OR VECTOR T, USING FORMULA 26.2.17
[ 3] ⍝ IN ABRAMOWITZ AND STEGUN.
[ 4] R←⍴T
[ 5] S←(T<0)/⍳⍴T←,T
[ 6] T[S]←-T[S]
[ 7] P←(÷(○2)*0.5)×*-(T*2)÷2
[ 8] T←÷1+0.2316419×T
[ 9] Z←¯0.356563782+T×1.781477937+T×¯1.821255978+T×1.330274429
[10] P←1-P×T×0.31938153+T×Z
[11] P[S]←1-P[S]
[12] P←R⍴P
     ∇
```

```
 21 ∇ X PARAM P
[ 1] ⍝PARAM CALCULATES THE MEAN, EX, AND THE STANDARD
[ 2] ⍝DEVIATION, SIGX, FOR THE DESCRETE RANDOM VAR-
[ 3] ⍝IABLE WITH THE VECTOR X OF POSSIBLE VALUES AND
[ 4] ⍝THE VECTOR P OF CORRESPONDING PROBABILITIES.
[ 5] EX←+/X×P
[ 6] VARX←+/((X-EX)*2)×P
[ 7] SIGX←VARX*0.5
[ 8] 'THE MEAN VALUE, EX, IS  ';EX
[ 9] 'THE STANDARD DEVIATION, SIGX, IS  ';SIGX
     ∇
```

```
 22 ∇ Z←MU PARTIALΔSUM K
[ 1] ⍝ THIS FUNCTION IS NEEDED BY THE FUNCTION
[ 2] ⍝ DMΔMΔ1ΔK.  IT COMPUTES A SEQUENCE OF  POISSON
[ 3] ⍝ SUMS.
[ 4] N←0
[ 5] Z←*-MU        .
[ 6] START:→0×⍳K<N←N+1
[ 7] →START,Z←Z,MU POISSONΔDIST N
     ∇
```

```
 23 ∇ Z←LAMBDA POISSON K
[ 1] ⍝'' LAMBDA POISSON K '' COMPUTES THE PROBABILITY
[ 2] ⍝THAT A POISSON DISTRIBUTED RANDOM VARIABLE WITH
[ 3] ⍝PARAMETER LAMBDA WILL ASSUME THE VALUE K.
[ 4] Z←(*-LAMBDA)×(LAMBDA*K)÷!K
     ∇
```

```
 24 ∇ Z←LAMBDA POISSONΔDIST N
[ 1] ⍝THIS FUNCTION IS THE POISSON DISTRIBUTION
[ 2] ⍝FUNCTION.  THAT IS, GIVEN THAT X IS A POISSON
[ 3] ⍝RANDOM VARIABLE WITH AVERAGE VALUE LAMBDA, THEN
[ 4] ⍝'' LAMBDA POISSONΔDIST N '' WILL PRODUCE THE
[ 5] ⍝PROBABILITY THAT X IS NOT GREATER THAN N.
[ 6] Z←+/LAMBDA POISSON ¯1+⍳N+1
     ∇
```

```
   25 ∇ M POLL N
[ 1] ⍝POLL ASSUMES A LINE WITH M TERMINALS, N OF
[ 2] ⍝WHICH ARE READY TO TRANSMIT.  POLL CALCULATES
[ 3] ⍝THE PROBABILITY THAT 1,2,...,M-N+1 POLLS ARE
[ 4] ⍝REQUIRED TO FIND THE FIRST READY TERMINAL.
[ 5] ⍝THE AVERAGE, EX, VARIANCE, VARX, AND THE
[ 6] ⍝STANDARD DEVIATION, SIGX, ALSO ARE COMPUTED.
[ 7] P←((N-1)!M-⍳(M-N)+1)÷(N!M)
[ 8] X←⍳(M-N)+1
[ 9] EX←+/X×P
[10] VARX←+/((X-EX)*2)×P
[11] SIGX←VARX*0.5
[12] 'THE AVERAGE NUMBER OF POLLS REQUIRED IS:'
[13] EX
[14] 'THE STANDARD DEVIATION, SIGX, IS  ';SIGX
[15] 'TO SEE THE PROBABILITY THAT 1, 2, ..., M+N+1'
[16] 'POLLS ARE REQUIRED, TYPE P.'
     ∇
```

```
   26 ∇ Z←M POLLM N;P;X
[ 1] ⍝ THIS FUNCTION IS USED BY MPOLL.
[ 2] P←((N-1)!M-⍳(M-N)+1)÷(N!M)
[ 3] X←⍳(M-N)+1
[ 4] Z←+/P×X
     ∇
```

```
   27 ∇ Z←M POLL2M N;P;X
[ 1] ⍝ THIS FUNCTION IS USED BY MPOLL.
[ 2] P←((N-1)!M-⍳(M-N)+1)÷(N!M)
[ 3] X←⍳(M-N)+1
[ 4] Z←+/P×X×X
     ∇
```

```
    28 ∇ PLΔPES PRΔQUEUE PES2
[ 1]  ⍝ PRΔQUEUE IS A FUNCTION FOR CALCULATING THE STATISTICS FOR
[ 2]  ⍝ A M/G/1 PREEMPTIVE-RESUME QUEUEING SYSTEM.  THE CALLING FORMAT
[ 3]  ⍝ IS '' PLΔPES PRΔQUEUE PES2 '' WHERE PLΔPES IS THE VECTOR, PL,
[ 4]  ⍝ OF THE AVERAGE ARRIVAL RATES FOR EACH PRIORITY CLASS, CATENATED
[ 5]  ⍝ WITH THE VECTOR, PES, OF THE CORRESPONDING AVERAGE SERVICE TIMES.
[ 6]  ⍝ PES2 IS THE VECTOR OF THE CORRESPONDING SECOND MOMENTS OF THE
[ 7]  ⍝ SERVICE TIMES.
[ 8]  N←ρPES2
[ 9]  PLAMBDA←N↑PLΔPES
[10]  PES←(-N)↑PLΔPES
[11]  LAMBDA←+/PLAMBDA
[12]  PROB←PLAMBDA÷LAMBDA
[13]  ES←+/PROB×PES
[14]  ES2←+/PROB×PES2
[15]  PU←0,+\PLAMBDA×PES
[16]  DIV1←N↑(1-PU)
[17]  DIV2←(-N)↑(1-PU)
[18]  PV←+\PLAMBDA×PES2
[19]  PW←(PES+PV÷2×DIV2)÷DIV1
[20]  PL←PLAMBDA×PW
[21]  PWQ←PW-PES
[22]  PLQ←PLAMBDA×PWQ
[23]  WQ←+/PROB×PWQ
[24]  LQ←LAMBDA×WQ
[25]  W←+/PROB×PW
[26]  L←LAMBDA×W
[27]  'ES =   ',⍕ES
[28]  'ES2 =  ',⍕ES2
[29]  'LAMBDA =  ',⍕LAMBDA
[30]  'RHO =  ',⍕RHO←LAMBDA×ES
[31]  'WQ =  ',⍕WQ
[32]  'W =  ',⍕W
[33]  'L =  ',⍕L
[34]  'LQ =  ',⍕LQ
[35]  'THE AVERAGE QUEUEING TIMES FOR THE
[36]  'RESPECTIVE PRIORITY CLASSES ARE:'
[37]  PWQ
[38]  'THE CORRESPONDING SYSTEM TIMES ARE:'
[39]  PW
[40]  'THE AVERAGE NUMBER IN THE SYSTEM FROM'
[41]  'THE RESPECTIVE PRIORITY CLASSES IS:'
[42]  PL
[43]  'THE CORRESPONDING AVERAGE NUMBER QUEUEING IS:'
[44]  PLQ
    ∇
```

```
 29 ∇ PLΔPES PΔQUEUE PES2
[ 1]  ⍝ PΔQUEUE IS A FUNCTION FOR CALCULATING THE STATISTICS FOR
[ 2]  ⍝ A M/G/1 NON-PREEMTIVE QUEUEING SYSTEM, THAT IS , AN HOL SYSTEM.
[ 3]  ⍝ THE CALLING FORMAT IS '' PLΔPES PΔQUEUE PES2 '',  HERE
[ 4]  ⍝ PLΔPES IS THE VECTOR, PL, OF THE AVERAGE ARRIVAL RATES FOR
[ 5]  ⍝ EACH PRIORITY CLASS, CATENATED WITH THE VECTOR, PES, OF THE
[ 6]  ⍝ CORRESPONDING AVERAGE SERVICE TIMES.  PES2 IS THE VECTOR
[ 7]  ⍝ OF THE CORRESPONDING SECOND MOMENTS OF THE SERVICE TIMES.
[ 8]  N←ρPES2
[ 9]  PLAMBDA←N↑PLΔPES
[10]  PES←(-N)↑PLΔPES
[11]  LAMBDA←+/PLAMBDA
[12]  PROB←PLAMBDA÷LAMBDA
[13]  ES←+/PROB×PES
[14]  ES2←+/PROB×PES2
[15]  PU←0,+\PLAMBDA×PES
[16]  DIV1←N↑(1-PU)
[17]  DIV2←(-N)↑(1-PU)
[18]  PWQ←0.5×LAMBDA×ES2÷DIV1×DIV2
[19]  PW←PWQ+PES
[20]  PL←PLAMBDA×PW
[21]  PLQ←PLAMBDA×PWQ
[22]  WQ←+/PROB×PWQ
[23]  LQ←LAMBDA×WQ
[24]  W←WQ+ES
[25]  L←LAMBDA×W
[26]  'ES=   ',⍕ES
[27]  'ES2=  ',⍕ES2
[28]  'LAMBDA=   ',⍕LAMBDA
[29]  'RHO=   ',⍕RHO←LAMBDA×ES
[30]  'WQ=   ',⍕WQ
[31]  'W=   ',⍕W
[32]  'LQ=   ',⍕LQ
[33]  'L=   ',⍕L
[34]  'THE AVERAGE QUEUEING TIMES FOR'
[35]  'THE RESPECTIVE PRIORITY CLASSES ARE:'
[36]  PWQ
[37]  'THE CORRESPONDING SYSTEM TIMES ARE:'
[38]  PW
[39]  'THE AVERAGE NUMBER IN THE SYSTEM FROM'
[40]  'THE RESPECTIVE PRIORITY CLASSES IS:'
[41]  PL
[42]  'THE CORRESPONDING AVERAGE NUMBER QUEUEING IS:'
[43]  PLQ
        ∇
```

References

1. IBM 5100 APL Reference Manual, SA21-9213, IBM General Systems Division, 5775D Glenridge Drive N.E., Atlanta, Georgia 30301.
2. APL Language, GC 26-3847, IBM Data Processing Division, 1133 Westchester Avenue, White Plains, New York 10604.
3. D. P. Geller and D. P. Freedman, *Structured Programming in APL*, Winthrop, Cambridge, Massachusetts, 1976.
4. L. Gilman and A. J. Rose, *APL An Interactive Approach*, 2nd ed. Wiley, New York, 1974.

QUEUEING THEORY DEFINITIONS
AND FORMULAS

In Figs. 5.1.1 and 5.1.2, reproduced from Chapter 5, we indicate the elements and random variables used in queueing theory models. Table 1 is a compendium of the queueing theory definitions and notation used in this book. The remainder of Appendix C consists of tables of queueing theory formulas for the most useful models and figures to help with the calculations. APL functions are displayed in Appendix B to implement the formulas for most of the queueing models.

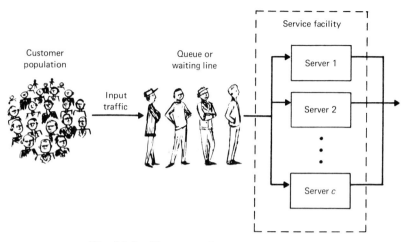

Fig. 5.1.1   Elements of a queueing system.

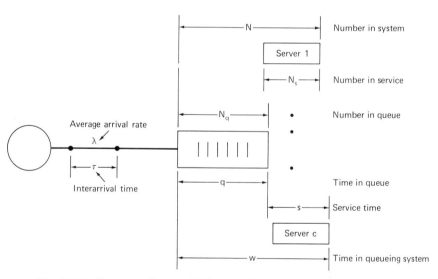

Fig. 5.1.2   Some random variables used in queueing theory models.

TABLE 1
Queueing Theory Notation and Definitions

$A(t)$	Distribution of interarrival time, $A(t) = P[\tau \le t]$.
$B(c, u)$	Erlang's B formula or the probability all c servers are busy in an $M/M/c/c$ queueing system.
$C(c, u)$	Erlang's C formula or the probability all c servers are busy in an $M/M/c$ queueing system.
c	Symbol for the number of servers in the service facility of a queueing system.
D	Symbol for constant (deterministic) interarrival or service time distribution.
$E[N]$	Expected (average or mean) number of customers in the steady state queueing system. The letter L is also used for $E[N]$.
$E[N_q]$	Expected (average or mean) number of customers in the queue (waiting line) when the system is in the steady state. The symbol L_q is also used for $E[N_q]$.
$E[N_s]$	Expected (average or mean) number of customers receiving service when the system is in the steady state.
$E[q]$	Expected (average or mean) queueing time (does not include service time) when the system is in the steady state. The symbol W_q is also used for $E[q]$.
$E[s]$	Expected (average or mean) service time for one customer. The symbol W_s is also used for $E[s]$.
$E[\tau]$	Expected (average or mean) interarrival time. $E[\tau] = 1/\lambda$, where λ is average arrival rate.
$E[w]$	Expected (average or mean) waiting time in the system (this includes both queueing time and service time) when the system is in the steady state. The letter W is also used for $E[w]$.
E_k	Symbol for Erlang-k distribution of interarrival or service time.
$E[N_q \mid N_q > 0]$	Expected (average or mean) queue length of nonempty queues when the system is in the steady state.
$E[q \mid q > 0]$	Expected (average or mean) waiting time in queue for customers delayed when the system is in the steady state. Same as $W_{q\mid q > 0}$.
FCFS	Symbol for, "first come, first served," queue discipline.
FIFO	Symbol for "first in, first out," queue discipline which is identical with FCFS.
G	Symbol for general probability distribution of service time. Independence usually assumed.
GI	Symbol for general independent interarrival time distribution.
K	Maximum number allowed in queueing system, including both those waiting for service and those receiving service. Also size of population in finite population models.
L	$E[N]$, expected (average or mean) number in the queueing system when the system is in the steady state.
$\ln (\cdot)$	The natural logarithm function or the logarithm to the base e.
L_q	$E[N_q]$, expected (average or mean) number in the queue, not including those in service, for steady state system.
LCFS	Symbol for "last come, first served," queue discipline.
LIFO	Symbol for "last in, first out," queue discipline which is identical to LCFS.
λ	Average (mean) arrival rate to queueing system. $\lambda = 1/E[\tau]$, where $E[\tau] =$ average interarrival time.

TABLE 1 *(Continued)*

λ_T	Average throughput of a computer system measured in jobs or interactions per unit time.
M	Symbol for exponential interarrival or service time distribution.
μ	Average (mean) service rate per server. Average service rate $\mu = 1/E[s]$, where $E[s]$ is the average (mean) service time.
N	Random variable describing number in queueing system when system is in the steady state.
N_q	Random variable describing number of customers in the steady state queue.
N_s	Random variable describing number of customers receiving service when the system is in the steady state.
$\bigcirc$	Operating time of a machine in the machine repair queueing model (Sections 5.2.6 and 5.2.7). $\bigcirc$ is the time a machine remains in operation after repair before repair again is necessary.
$p_n(t)$	Probability that there are n customers in the queueing system at time t.
p_n	Steady state probability that there are n customers in the queueing system.
PRI	Symbol for priority queueing discipline.
PS	Abbreviation for "processor-sharing queue discipline." See Section 6.2.1.
$\pi_q(r)$	Symbol for rth percentile queueing time; that is, the queueing time that r percent of the customers do not exceed.
$\pi_w(r)$	Symbol for rth percentile waiting time in the system; that is, the time in the system (queueing time plus service time) that r percent of the customers do not exceed.
q	Random variable describing the time a customer spends in the queue (waiting line) before receiving service.
RSS	Symbol for queue discipline with "random selection for service."
ρ	Server utilization = traffic intensity/$c = \lambda E[s]/c = (\lambda/\mu)/c$. The probability that any particular server is busy.
s	Random variable describing service time for one customer.
SIRO	Symbol for queue discipline, "service in random order" which is identical with RSS. It means that each waiting customer has the same probability of being served next.
τ	Random variable describing interarrival time.
u	Traffic intensity = $E[s]/E[\tau] = \lambda E[s] = \lambda/\mu$. Unit of measure is the erlang.
w	Random variable describing the total time a customer spends in the queueing system, including both service time and time spent queueing for service.
$W(t)$	Distribution function for w, $W(t) = P[w \le t]$.
W	$E[w]$, expected (average or mean) time in the steady state system.
$W_q(t)$	Distribution function for time in the queue, $W_q(t) = P[q \le t]$.
W_q	$E[q]$, expected (average or mean) time in the queue (waiting line), excluding service time, for steady state system.
$W_{q|q>0}$	Expected (average or mean) queueing time for those who must queue. Same as $E[q \mid q > 0]$.
$W_s(t)$	Distribution function for service time, $W_s(t) = P[s \le t]$.
W_s	$E[s]$, expected (average or mean) service time, $1/\mu$.

TABLE 2
Relationships Between Random Variables of Queueing Theory Models

$u = E[s]/E[\tau] = \lambda E[s] = \lambda/\mu$	Traffic intensity in erlangs.
$\rho = u/c = \lambda E[s]/c = \lambda/c\mu$	Server utilization. The probability any particular server is busy.
$w = q + s$	Total waiting time in the system, including waiting in queue and service time.
$W = E[w] = E[q] + E[s] = W_q + W_s$	Average total waiting time in the steady state system.
$N = N_q + N_s$	Number of customers in the steady state system.
$L = E[N] = E[N_q] + E[N_s] = \lambda E[w] = \lambda W$	Average number of customers in the steady state system. $L = \lambda W$ is known as "Little's formula."
$L_q = E[N_q] = \lambda E[q] = \lambda W_q$	Average number in the queue for service for steady state system. $L_q = \lambda W_q$ is also called "Little's formula."

TABLE 3
Steady State Formulas for M/M/1 Queueing System

$p_n = P[N = n] = (1 - \rho)\rho^n, \qquad n = 0, 1, 2, \ldots.$

$P[N \geq n] = \sum\limits_{k=n}^{\infty} p_k = \rho^n, \qquad n = 0, 1, 2, \ldots.$

$L = E[N] = \rho/(1 - \rho), \qquad \sigma_N^2 = \rho/(1 - \rho)^2.$

$L_q = E[N_q] = \rho^2/(1 - \rho), \qquad \sigma_{N_q}^2 = \rho^2(1 + \rho - \rho^2)/(1 - \rho)^2.$

$E[N_q | N_q > 0] = 1/(1 - \rho), \qquad \mathrm{Var}[N_q | N_q > 0] = \rho/(1 - \rho)^2.$

$W(t) = P[w \leq t] = 1 - e^{-t/W}, \qquad P[w > t] = e^{-t/W}.$

$W = E[w] = E[s]/(1 - \rho), \qquad \sigma_w = W.$

$\pi_w(90) = W \ln 10 \approx 2.3W, \qquad \pi_w(95) = W \ln 20 \approx 3W.$

$\pi_w(r) = W \ln [100/(100 - r)].$

$W_q(t) = P[q \leq t] = 1 - \rho e^{-t/W}, \qquad P[q > t] = \rho e^{-t/W}.$

$W_q = E[q] = \rho E[s]/(1 - \rho).$

$\sigma_q^2 = (2 - \rho)\rho E[s]^2/(1 - \rho)^2.$

$E[q | q > 0] = W, \qquad \mathrm{Var}[q | q > 0] = W^2.$

$\pi_q(90) = W \ln(10\rho), \qquad \pi_q(95) = W \ln(20\rho).$

$\pi_q(r) = W \ln \left(\dfrac{100\rho}{100 - r} \right).$

TABLE 4
Steady State Formulas for M/M/1/K Queueing System

$(K \geq 1$ and $N \leq K)$

$$p_n = P[N = n] = \begin{cases} \dfrac{(1 - u)u^n}{1 - u^{K+1}} & \text{if } \lambda \neq \mu \text{ and } n = 0, 1, ..., K \\[2mm] \dfrac{1}{K + 1} & \text{if } \lambda = \mu \text{ and } n = 0, 1, ..., K. \end{cases}$$

$p_K = P[N = K]$.   Probability an arriving customer is lost.

$\lambda_a = (1 - p_K)\lambda$. λ_a is the actual arrival rate at which customers enter the system.

$$L = E[N] = \begin{cases} \dfrac{u[1 - (K + 1)u^K + Ku^{K+1}]}{(1 - u)(1 - u^{K+1})} & \text{if } \lambda \neq \mu \\[2mm] \dfrac{K}{2} & \text{if } \lambda = \mu. \end{cases}$$

$L_q = E[N_q] = L - (1 - p_0)$

$q_n = \dfrac{p_n}{1 - p_K}, \qquad n = 0, 1, 2, ..., K - 1.$

q_n is the probability that there are n customers in the system just before a customer enters.

$W(t) = P[w \leq t] = 1 - \displaystyle\sum_{n=0}^{K-1} q_n \sum_{k=0}^{n} e^{-\mu t} \dfrac{(\mu t)^k}{k!}$.

$W = E[w] = L/\lambda_a$.

$W_q(t) = P[q \leq t] = 1 - \displaystyle\sum_{n=0}^{K-2} q_{n+1} \sum_{k=0}^{n} e^{-\mu t} \dfrac{(\mu t)^k}{k!}$.

$W_q = E[q] = L_q/\lambda_a$.

$E[q \,|\, q > 0] = W_q/(1 - p_0)$.

$\rho = (1 - p_K)u$.

ρ is the true server utilization (fraction of time the server is busy).

TABLE 5
Steady State Formulas for M/M/c Queueing System

$u = \lambda/\mu = \lambda E[s], \qquad \rho = u/c.$

$$p_0 = P[N = 0] = \left[\sum_{n=0}^{c-1} \frac{u^n}{n!} + \frac{u^c}{c! \, (1 - \rho)} \right]^{-1}.$$

$$p_n = \begin{cases} \dfrac{u^n}{n!} p_0 & \text{if } n = 0, 1, \ldots, c \\[2ex] \dfrac{u^n p_0}{c! \, c^{n-c}} & \text{if } n \geq c. \end{cases}$$

$$L_q = E[N_q] = \lambda W_q = \frac{uC(c, u)}{c(1 - \rho)}, \qquad \sigma_{N_q}^2 = \frac{\rho C(c, u)[1 + \rho - \rho C(c, u)]}{(1 - \rho)^2},$$

where $C(c, u) = P[N \geq c] =$ probability all c servers are busy is called Erlang's C formula.

$$C(c, u) = \frac{u^c}{c!} \bigg/ \left[\frac{u^c}{c!} + (1 - \rho) \sum_{n=0}^{c-1} \frac{u^n}{n!} \right].$$

$$L = E[N] = L_q + u = \lambda W.$$

$$W_q(0) = P[q = 0] = 1 - \frac{p_c}{1 - \rho} = 1 - C(c, u).$$

$$W_q(t) = P[q \leq t] = 1 - \frac{p_c}{1 - \rho} e^{-ct(1 - \rho)/E[s]} = 1 - C(c, u)e^{-ct(1 - \rho)/E[s]}.$$

$$W_q = E[q] = \frac{C(c, u)E[s]}{c(1 - \rho)}, \qquad E[q \,|\, q > 0] = \frac{E[s]}{c(1 - \rho)}.$$

$$\sigma_q^2 = \frac{[2 - C(c, u)]C(c, u)E[s]^2}{c^2(1 - \rho)^2}.$$

$$\pi_q(r) = \frac{E[s]}{c(1 - \rho)} \ln \left(\frac{100C(c, u)}{100 - r} \right).$$

$$\pi_q(90) = \frac{E[s]}{c(1 - \rho)} \ln(10C(c, u)), \qquad \pi_q(95) = \frac{E[s]}{c(1 - \rho)} \ln(20C(c, u)).$$

$$W(t) = P[w \leq t] = \begin{cases} 1 + C_1 e^{-\mu t} + C_2 e^{-c\mu t(1 - \rho)} & \text{if } u \neq c - 1 \\ 1 - [1 + C(c, u)\mu t]e^{-\mu t} & \text{if } u = c - 1, \end{cases}$$

where $C_1 = \dfrac{u - c + W_q(0)}{c - 1 - u} \quad$ and $\quad C_2 = \dfrac{C(c, u)}{c - 1 - u}.$

$$W = E[q] + E[s].$$

$$E[w^2] = \begin{cases} \dfrac{2C(c, u)E[s]^2}{u + 1 - c} \left| \dfrac{1 - c^2(1 - \rho)^2}{c^2(1 - \rho)^2} \right| + 2E[s]^2, & u \neq c - 1, \\[2ex] 4C(c, u)E[s]^2 + 2E[s]^2, & u = c - 1. \end{cases}$$

$$\sigma_w^2 = E[w^2] - E[w]^2$$

TABLE 6

Steady State Formulas for M/M/2 Queueing System

$\rho = \lambda E[s]/2 = u/2.$

$p_0 = P[N = 0] = (1 - \rho)/(1 + \rho).$

$p_n = P[N = n] = 2p_0\rho^n = \dfrac{2(1 - \rho)\rho^n}{(1 + \rho)}, \qquad n = 1, 2, 3, \ldots.$

$L_q = E[N_q] = \dfrac{2\rho^3}{1 - \rho^2}, \qquad \sigma_{N_q}^2 = \dfrac{2\rho^3[(\rho + 1)^2 - 2\rho^3]}{(1 - \rho^2)^2}.$

$C(2, u) = P[\text{both servers busy}] = 2\rho^2/(1 + \rho).$

$L = E[N] = L_q + u = 2\rho/(1 - \rho^2).$

$W_q(0) = P[q = 0] = (1 + \rho - 2\rho^2)/(1 + \rho).$

$W_q(t) = P[q \le t] = 1 - [(2\rho^2)/(1 + \rho)]e^{-2\mu t(1 - \rho)}.$

$W_q = E[q] = \rho^2 E[s]/(1 - \rho^2), \qquad E[q \mid q > 0] = E[s]/2(1 - \rho).$

$\sigma_q^2 = \rho^2(1 + \rho - \rho^2)E[s]^2/(1 - \rho^2)^2.$

$\pi_q(r) = \dfrac{E[s]}{2(1 - \rho)}\ln\left(\dfrac{200\rho^2}{(100 - r)(1 + \rho)}\right).$

$\pi_q(90) = \dfrac{E[s]}{2(1 - \rho)}\ln\left(\dfrac{20\rho^2}{1 + \rho}\right), \qquad \pi_q(95) = \dfrac{E[s]}{2(1 - \rho)}\ln\left(\dfrac{40\rho^2}{1 + \rho}\right).$

$$W(t) = P[w \le t] = \begin{cases} 1 - \dfrac{(1 - \rho)}{1 - \rho - 2\rho^2}e^{-\mu t} + \dfrac{2\rho^2}{1 - \rho - 2\rho^2}e^{-2\mu t(1 - \rho)} & \text{if } u \ne 1 \\[2ex] 1 - \left[1 + \dfrac{\mu t}{3}\right]e^{-\mu t} & \text{if } u = 1. \end{cases}$$

$W = E[s]/(1 - \rho^2).$

$$E[w^2] = \begin{cases} \dfrac{\rho^2 E[s]^2[1 - 4(1 - \rho)^2]}{(2\rho - 1)(1 - \rho)(1 - \rho^2)} + 2E[s]^2, & u \ne 1, \\[2ex] \frac{10}{3}E[s]^2, & u = 1. \end{cases}$$

$\sigma_w^2 = E[w^2] - E[w]^2.$

TABLE 7
Steady State Formulas for M/M/c/c Queueing System (M/M/c Loss)

$$P_n = P[N = n] = \frac{u^n/n!}{1 + u + u^2/2! + \cdots + u^c/c!}, \qquad n = 0, 1, 2, \ldots, c.$$

$B(c, u) = $ probability all servers busy is called "Erlang's B formula."

Thus $B(c, u) = \dfrac{u^c/c!}{1 + u + u^2/2! + \cdots + u^c/c!}.$

$\lambda_a = \lambda(1 - B(c, u)).$ $\quad \lambda_a$ is the average traffic rate experienced by the system.

$\rho = \lambda_a E[s]/c = (1 - B(c, u))u/c.$

$L = E[N] = u(1 - B(c, u)).$

$W = E[w] = E[s].$

$W(t) = P[w \le t] = 1 - e^{-t/E[s]}.$

All the formulas except the last are true for an M/G/c/c queueing system. For general service time we have

$W(t) = P[w \le t] = W_s(t),$

where $W_s(t)$ is the distribution function for service time s.

TABLE 8
Steady State Formulas for M/M/c/K Queueing System

$$p_n = \begin{cases} \dfrac{u^n}{n!} p_0 & n = 1, 2, \dots, c \\[2ex] \dfrac{u^c}{c!} \left(\dfrac{u}{c}\right)^{n-c} p_0 & n = c+1, \dots, K, \end{cases}$$

where

$$p_0 = \left[\sum_{n=0}^{c} \frac{u^n}{n!} + \frac{u^c}{c!} \sum_{n=1}^{K-c} \left(\frac{u}{c}\right)^n \right]^{-1}$$

$$\lambda_a = \lambda(1 - p_K), \qquad \rho = (1 - p_K)u/c.$$

$$L_q = \frac{u^c p_0 u/c}{c! \, (1 - u/c)^2} \left[1 - \left(\frac{u}{c}\right)^{K-c+1} - (K - c + 1)\left(\frac{u}{c}\right)^{K-c}\left(1 - \frac{u}{c}\right) \right]$$

$$L = L_q + E[N_s] = L_q + \sum_{n=0}^{c-1} np_n + c\left(1 - \sum_{n=0}^{c-1} p_n\right)$$

$$W_q = L_q/\lambda_a, \qquad E[q \,|\, q > 0] = W_q \Big/ \left(1 - \sum_{n=0}^{c-1} p_n\right)$$

$$W = L/\lambda_a$$

TABLE 9

Steady State Formulas for M/M/ ∞ Queueing System

$p_n = e^{-u}u^n/n!$, $n = 0, 1, 2, 3, \ldots$.

$L = E[N] = u$, $\sigma_N^2 = u$.

$L_q = 0$, $\sigma_{N_q} = 0$.

$W(t) = P[w \le t] = 1 - e^{-t/E[s]}$.

All the formulas except the last are also true for the M/G/∞ queueing system. For general service time $W(t) = P[w \le t] = W_s(t)$, where $W_s(t)$ is the distribution function for service time s.

TABLE 10
Steady State Formulas for M/M/1/K/K (Machine Repair) Queueing System
($K > 1$)

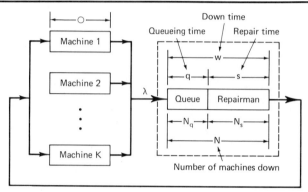

Fig. 5.2.7 Machine repair queueing system with one repairman. (M/M/1/K/K queueing system.)

$E[\bigcirc] = 1/\alpha$ = average operating time per machine.

$E[s] = 1/\mu$ = average repair time per machine (by one repairman).

$p_n = P[n$ machines out of service$] = P[N = n]$

$$= \frac{K!}{(K - n)!}\left(\frac{E[s]}{E[\bigcirc]}\right)^n p_0, \qquad n = 0, 1, 2, \ldots, K,$$

where

$$p_0 = \left[\sum_{n=0}^{K} \frac{K!}{(K - n)!}\left(\frac{E[s]}{E[\bigcirc]}\right)^n\right]^{-1}$$

$\rho = 1 - p_0$

$\lambda = \rho/E[s] = \rho\mu.$

$W_q = \dfrac{K}{\lambda} - E[\bigcirc] - E[s], \qquad E[q \mid q > 0] = W_q/\rho$

$W = W_q + E[s].$

$L_q = \lambda W_q, \qquad L = \lambda W.$

$P[$machine n is down$] = W/(W + E[\bigcirc]), \qquad n = 1, \ldots, K.$

$P[$down machine must wait for repair$] = \rho.$

TABLE 11
Steady State Formulas for M/M/c/K/K (Machine Repair With Multiple Repairmen) Queueing System ($K > c$)

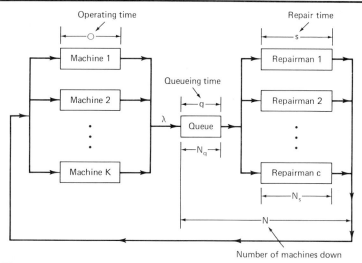

Fig. 5.2.9 Machine repair queueing system with c repairmen. (M/M/c/K/K queueing system.)

$E[\bigcirc] = 1/\alpha$ = average operating time per machine.

$E[s] = 1/\mu$ = average repair time per machine (by one repairman).

$p_n = P[N = n]$.

$$\frac{p_n}{p_0} = \begin{cases} \binom{K}{n}\left(\dfrac{E[s]}{E[\bigcirc]}\right)^n & n = 1, 2, \ldots, c \\[2ex] \dfrac{n!}{c!\,c^{n-c}}\binom{K}{n}\left(\dfrac{E[s]}{E[\bigcirc]}\right)^n & n = c + 1, \ldots, K. \end{cases}$$

$$p_0 = \left[1 + \sum_{n=1}^{K}\left(\frac{p_n}{p_0}\right)\right]^{-1}.$$

$$L_q = \sum_{n=c+1}^{K}(n - c)p_n.$$

$W_q = L_q(E[\bigcirc] + E[s])/(K - L_q)$.

$D = P[\text{down machine must wait for repair}] = \displaystyle\sum_{n=c}^{K} p_n$.

$E[q \mid q > 0] = W_q/D$.

$W = W_q + E[s]$.

$\lambda = K/(E[\bigcirc] + W_q + E[s])$, $\rho = \dfrac{\lambda E[s]}{c}$.

$L = \lambda W$ = average number of machines down.

$P[\text{machine } n \text{ is down}] = W/(W + E[\bigcirc])$, $n = 1, 2, \ldots, K$.

TABLE 12
Steady State Formulas for M/G/1 Queueing System

Let $p_n = P[N = n] = P[n \text{ customers in system}]$, $n = 0, 1, 2, \ldots$.
Then

$$P(z) = \sum_{n=0}^{\infty} p_n z^n = \frac{(1 - \rho)(1 - z)W_s^*[\lambda(1 - z)]}{W_s^*[\lambda(1 - z)] - z}, \tag{5.3.30}$$

where $W_s^*(\theta)$ is the Laplace–Stieltjes transform of the service time s.
   The Laplace–Stieltjes transforms of w and q are given by

$$W^*(\theta) = \frac{(1 - \rho)\theta W_s^*(\theta)}{\theta - \lambda + W_s^*(\theta)}, \tag{5.3.32}$$

and

$$W_q^*(\theta) = \frac{(1 - \rho)\theta}{\theta - \lambda + W_s^*(\theta)}, \tag{5.3.33}$$

respectively.
   Equations (5.3.30), (5.3.32), and (5.3.33) are each called the *Pollaczek–Khintchine transform equation* by various authors.

$$p_0 = P[\text{no customers in system}] = 1 - \rho, \quad \text{where} \quad \rho = \lambda E[s]. \tag{5.3.15}$$

$$P[\text{server busy}] = P[N \geq 1] = \rho.$$

$$W_q = E[q] = \frac{\lambda E[s^2]}{2(1 - \rho)} = \frac{\rho E[s]}{1 - \rho}\left(\frac{1 + C_s^2}{2}\right) \quad \text{(Pollaczek's formula)}. \tag{5.3.42}$$

$$E[q \mid q > 0] = W_q/\rho = \frac{E[s]}{1 - \rho}\left(\frac{1 + C_s^2}{2}\right).$$

$$E[q^2] = 2E[q]^2 + \frac{\lambda E[s^3]}{3(1 - \rho)}. \tag{5.3.43}$$

$$\sigma_q^2 = E[q^2] - E[q]^2$$

$$W = E[w] = E[q] + E[s]. \tag{5.3.44}$$

$$E[w^2] = E[q^2] + \frac{E[s^2]}{1 - \rho}. \tag{5.3.45}$$

$$\sigma_w^2 = E[w^2] - E[w]^2$$

$$\begin{aligned}
\pi_w(90) &\approx E[w] + 1.3\sigma_w \\
\pi_w(95) &\approx E[w] + 2\sigma_w
\end{aligned} \quad \text{Martin's estimate} \qquad \begin{aligned} &(5.3.47) \\ &(5.3.48) \end{aligned}$$

$$L_q = E[N_q] = \lambda W_q = \frac{\rho^2}{1 - \rho}\left(\frac{1 + C_s^2}{2}\right) \tag{5.3.46}$$

$$L = E[N] = \lambda W = L_q + \rho \tag{5.3.23}$$

$$\sigma_N^2 = \frac{\lambda^3 E[s^3]}{3(1 - \rho)} + \left(\frac{\lambda^2 E[s^2]}{2(1 - \rho)}\right)^2 + \frac{\lambda^2(3 - 2\rho)E[s^2]}{2(1 - \rho)} + \rho(1 - \rho).$$

TABLE 13
Steady State Formulas for M/Gamma/1 Queueing System
(Special case of M/G/1)

We assume the service time s has a gamma distribution with parameters α and β so that $E[s] = 1/\mu$, $E[s^2] = E[s]^2 + 1/\beta\mu$, $\sigma_s^2 = 1/\beta\mu$, $C_s^2 = \mu/\beta$, and

$$E[s^3] = \frac{\beta^2 + 3\beta\mu + 2\mu^2}{\beta^2} E[s]^3. \text{ Then}$$

$$W_q = E[q] = \frac{\rho E[s]}{1 - \rho}\left(\frac{1 + \mu/\beta}{2}\right) \qquad \text{(Pollaczek's formula)}.$$

$$E[q \mid q > 0] = W_q/\rho = \frac{E[s]}{1 - \rho}\left(\frac{1 + \mu/\beta}{2}\right).$$

$$E[q^2] = 2E[q]^2 + \frac{\rho E[s]^2}{3(1 - \rho)}\left(\frac{\beta^2 + 3\beta\mu + 2\mu^2}{\beta^2}\right).$$

$$\sigma_q^2 = E[q^2] - E[q]^2.$$

$$W = E[w] = E[q] + E[s].$$

$$E[w^2] = E[q^2] + \frac{E[s]^2 + 1/\beta\mu}{1 - \rho}.$$

$$\sigma_w^2 = E[w^2] - E[w]^2.$$

$$\left.\begin{array}{l}\pi_w(90) \approx E[w] + 1.3\sigma_w \\ \pi_w(95) \approx E[w] + 2\sigma_w\end{array}\right\} \text{ Martin's estimate.}$$

$$L_q = E[N_q] = \lambda W_q = \frac{\rho^2}{1 - \rho}\left(\frac{1 + \mu/\beta}{2}\right).$$

$$L = E[N] = \lambda W = L_q + \rho.$$

$$\sigma_N^2 = \frac{\rho^3(\beta^2 + 3\beta\mu + 2\mu^2)}{3(1 - \rho)\beta^2} + \left(\frac{\rho^2 + \lambda\rho/\beta}{2(1 - \rho)}\right)^2 + \frac{(\rho^2 + \lambda\rho/\beta)(3 - 2\rho)}{2(1 - \rho)} + \rho(1 - \rho).$$

TABLE 14
Steady State Formulas for M/E$_k$/1 Queueing System

(This is a special case of the M/Gamma/1 queueing system of Table 13, with $\alpha = k$, $\beta = k\mu$, that is, s is Erlang-k.)

$$W_q = E[q] = \frac{\rho E[s]}{1-\rho}\left(\frac{1+1/k}{2}\right) \qquad \text{(Pollaczek's formula)}.$$

$$E[q\,|\,q>0] = W_q/\rho = \frac{E[s]}{1-\rho}\left(\frac{1+1/k}{2}\right).$$

$$E[q^2] = 2E[q]^2 + \frac{\rho E[s]^2(k+1)(k+2)}{3(1-\rho)k^2}.$$

$$\sigma_q^2 = E[q^2] - E[q]^2.$$

$$W = E[w] = E[q] + E[s].$$

$$E[w^2] = E[q^2] + \frac{E[s]^2}{(1-\rho)}\left(1+\frac{1}{k}\right).$$

$$\sigma_w^2 = E[w^2] - E[w]^2.$$

$$\left.\begin{array}{l} \pi_w(90) \approx E[w] + 1.3\sigma_w \\ \pi_w(95) \approx E[w] + 2\sigma_w \end{array}\right\} \quad \text{Martin's estimate.}$$

$$L_q = E[N_q] = \lambda W_q = \frac{\rho^2}{1-\rho}\left(\frac{1+1/k}{2}\right).$$

$$L = E[N] = \lambda W = L_q + \rho.$$

$$\sigma_N^2 = \frac{\rho^3(k+1)(k+2)}{3(1-\rho)k^2} + \left[\frac{\rho^2(1+1/k)}{2(1-\rho)}\right]^2 + \frac{\rho^2(1+1/k)(3-2\rho)}{2(1-\rho)} + \rho(1-\rho).$$

TABLE 15
Steady State Formulas for M/D/1 Queueing System

(This is a special case of $M/E_k/1$ with $k = \infty$, or s is constant.)

$$W_q = E[q] = \frac{\rho E[s]}{2(1 - \rho)} \qquad \text{(Pollaczek's formula)}.$$

$$E[q \,|\, q > 0] = W_q/\rho = \frac{E[s]}{2(1 - \rho)}.$$

$$E[q^2] = 2E[q]^2 + \frac{\rho E[s]^2}{3(1 - \rho)}.$$

$$\sigma_q^2 = E[q^2] - E[q]^2.$$

$$W = E[w] = E[q] + E[s].$$

$$E[w^2] = E[q^2] + \frac{E[s]^2}{(1 - \rho)}.$$

$$\sigma_w^2 = E[w^2] - E[w]^2.$$

$$\left.\begin{array}{l} \pi_w(90) \approx E[w] + 1.3\sigma_w \\ \pi_w(95) \approx E[w] + 2\sigma_w \end{array}\right\} \text{ Martin's estimate.}$$

$$L_q = E[N_q] = \lambda W_q = \frac{\rho^2}{2(1 - \rho)}.$$

$$L = E[N] = \lambda W = L_q + \rho.$$

$$\sigma_N^2 = \frac{\rho^3}{3(1 - \rho)} + \left(\frac{\rho^2}{2(1 - \rho)}\right)^2 + \frac{\rho^2(3 - 2\rho)}{2(1 - \rho)} + \rho(1 - \rho).$$

TABLE 16

Steady State Formulas for GI/M/1 Queueing System

Let π_n, $n = 0, 1, 2, \ldots$ be the steady state number of customers an arriving customer finds in the system. Then π_0, the probability that an arriving customer finds the system empty, is the unique solution of the equation

$$1 - \pi_0 = A^*(\mu\pi_0), \tag{5.3.50}$$

(where $A^*(\theta)$ is the Laplace–Stieltjes transform of the interarrival time) such that $0 < \pi_0 < 1$, and

$$\pi_n = \pi_0(1 - \pi_0)^n, \qquad n = 0, 1, 2, 3, \ldots. \tag{5.3.49}$$

Thus, X, the random variable describing the steady state number of customers an arriving customer finds in the system, has a geometric distribution with

$$E[X] = \frac{1 - \pi_0}{\pi_0}, \qquad \mathrm{Var}[X] = \frac{1 - \pi_0}{\pi_0{}^2}. \tag{5.3.51}$$

The steady state number of customers in the system, N, has the distribution $\{p_n\}$, given by

$$p_0 = P[N = 0] = 1 - \rho, \tag{5.3.65}$$

$$p_n = P[N = n] = \rho\pi_0(1 - \pi_0)^{n-1}, \qquad n = 1, 2, 3, \ldots, \tag{5.3.66}$$

with

$$L = E[N] = \frac{\rho}{\pi_0}, \qquad \mathrm{Var}[N] = \frac{\rho(2 - \pi_0 - \rho)}{\pi_0{}^2}.$$

$$L_q = E[N_q] = \lambda W_q = \frac{(1 - \pi_0)\rho}{\pi_0}, \qquad \mathrm{Var}[N_q] = \frac{\rho(1 - \pi_0)\{2 - \pi_0 - \rho(1 - \pi_0)\}}{\pi_0{}^2}.$$

$$E[N_q \mid N_q > 0] = L_q/P[N_q > 0] = 1/\pi_0.$$

The waiting time in the system, w, has an exponential distribution with

$$W = E[w] = E[s]/\pi_0, \tag{5.3.57}$$

and

$$W(t) = P[w \le t] = 1 - e^{-t/W}, \qquad P[w > t] = e^{-t/W}. \tag{5.3.56}$$

$$\pi_w(r) = W \ln \frac{100}{100 - r}. \tag{5.3.59}$$

$$\pi_w(90) = W \ln 10 \approx 2.3W, \qquad \pi_w(95) = W \ln 20 \approx 3W. \tag{5.3.60}$$

$$W_q = E[q] = (1 - \pi_0)E[s]/\pi_0. \tag{5.3.53}$$

$$\sigma_q{}^2 = (1 - \pi_0{}^2)(E[s]/\pi_0)^2. \tag{5.3.55}$$

$$W_q(t) = P[q \le t] = 1 - (1 - \pi_0)e^{-t/W}. \tag{5.3.52}$$

$$\pi_q(r) = W \ln \frac{100(1 - \pi_0)}{100 - r} \qquad \pi_q(90) = W \ln(10(1 - \pi_0)). \qquad \pi_q(95) = W \ln(20(1 - \pi_0)). \tag{5.3.52}$$

$$P[q' \le t] = P[w \le t] = 1 - e^{-t/W}, \qquad t > 0, \tag{5.3.63}$$

where q' is queueing time for those who must.

$$E[q \mid q > 0] = W, \qquad \mathrm{Var}[q \mid q > 0] = W^2. \tag{5.3.64}$$

All formulas for percentile values of q yield negative values when $1 - \pi_0$ is small; replace all negative values with zero.

TABLE 17

π_0 as a Function of ρ for the GI/M/1 Queueing System[a,b]

ρ	E_2	E_3	U	D	H_2	H_2
0.100	0.970820	0.987344	0.947214	0.999955	0.815535	0.810575
0.200	0.906226	0.940970	0.887316	0.993023	0.662348	0.624404
0.300	0.821954	0.868115	0.817247	0.959118	0.536805	0.444949
0.400	0.724695	0.776051	0.734687	0.892645	0.432456	0.281265
0.500	0.618034	0.669467	0.639232	0.796812	0.343070	0.154303
0.600	0.504159	0.551451	0.531597	0.675757	0.263941	0.081265
0.700	0.384523	0.424137	0.412839	0.533004	0.191856	0.044949
0.800	0.260147	0.289066	0.284028	0.371370	0.124695	0.024404
0.900	0.131782	0.147390	0.146133	0.193100	0.061057	0.010575
0.950	0.066288	0.074362	0.074048	0.098305	0.030252	0.004999
0.980	0.026607	0.029899	0.029849	0.039732	0.012039	0.001941
0.999	0.001333	0.001500	0.001500	0.001999	0.000600	0.000095

[a] π_0 is the probability an arriving customer finds the system empty.

[b] The hyperexponential distribution described in the next to last column has $\alpha_1 = 0.4$, $\alpha_2 = 0.6$, $\mu_1 = 0.5\lambda$, $\mu_2 = 3\lambda$, where λ is the average arrival rate, in the notation of Section 3.2.9. The hyperexponential distribution described in the last column above was generated by Algorithm 6.2.1 for $C^2 = 20$; thus $\alpha_1 = 0.024405$, $\alpha_2 = 0.975595$, $\mu_1 = 2\alpha_1\lambda$, and $\mu_2 = 2\alpha_2\lambda$.

TABLE 18
**Steady-State Formulas for an M/G/1 Nonpreemptive Priority (HOL)
Queueing System**

$$\lambda = \lambda_1 + \lambda_2 + \cdots + \lambda_n$$

$$E[s] = \frac{\lambda_1}{\lambda} E[s_1] + \frac{\lambda_2}{\lambda} E[s_2] + \cdots + \frac{\lambda_n}{\lambda} E[s_n]$$

$$E[s^2] = \frac{\lambda_1}{\lambda} E[s_1{}^2] + \frac{\lambda_2}{\lambda} E[s_2{}^2] + \cdots + \frac{\lambda_n}{\lambda} E[s_n{}^2]$$

$$u_j = \lambda_1 E[s_1] + \cdots + \lambda_j E[s_j], \qquad j = 1, 2, \ldots, n$$

$$u_n = u = \lambda E[s]$$

$$W_{q_j} = E[q_j] = \frac{\lambda E[s^2]}{2(1 - u_{j-1})(1 - u_j)}, \qquad j = 1, 2, \ldots, n, \quad u_0 = 0$$

$$W_q = E[q] = \frac{\lambda_1}{\lambda} E[q_1] + \frac{\lambda_2}{\lambda} E[q_2] + \cdots + \frac{\lambda_n}{\lambda} E[q_n]$$

$$W_j = E[w_j] = E[q_j] + E[s_j], \qquad j = 1, 2, \ldots, n$$

$$W = E[w] = E[q] + E[s]$$

$$L_q = E[N_q] = \lambda E[q]$$

$$L = E[N] = \lambda E[w]$$

TABLE 19

Steady-State Formulas for an M/G/1 Preemptive-Resume Priority
Queueing System

$$\lambda = \lambda_1 + \lambda_2 + \cdots + \lambda_n$$

$$E[s] = \frac{\lambda_1}{\lambda} E[s_1] + \frac{\lambda_2}{\lambda} E[s_2] + \cdots + \frac{\lambda_n}{\lambda} E[s_n]$$

$$E[s^2] = \frac{\lambda_1}{\lambda} E[s_1^2] + \frac{\lambda_2}{\lambda} E[s_2^2] + \cdots + \frac{\lambda_n}{\lambda} E[s_n^2]$$

$$u_j = \lambda_1 E[s_1] + \cdots + \lambda_j E[s_j], \qquad j = 1, 2, \ldots, n$$

$$u = u_n = \lambda E[s]$$

$$W_j = E[w_j] = \frac{1}{1 - u_{j-1}} \left[E[s_j] + \frac{\sum_{i=1}^{j} \lambda_i E[s_i^2]}{2(1 - u_j)} \right], \qquad u_0 = 0, \quad j = 1, 2, \ldots, n$$

$$W_{q_j} = E[q_j] = E[w_j] - E[s_j]$$

$$W_q = E[q] = \frac{\lambda_1}{\lambda} E[q_1] + \frac{\lambda_2}{\lambda} E[q_2] + \cdots + \frac{\lambda_n}{\lambda} E[q_n]$$

$$L_{q_j} = E[N_{q_j}] = \lambda_j E[q_j], \qquad j = 1, 2, \ldots, n$$

$$L_q = E[N_q] = \lambda E[q] = \lambda W_q$$

$$W = E[w] = E[q] + E[s]$$

$$L = E[N] = \lambda E[w] = \lambda W$$

TABLE 20

Steady-State Formulas for M/G/1 Processor-Sharing Queueing System

This model is exactly like the M/G/1 queueing system except for the queueing discipline. Each arriving customer immediately starts receiving an equal share of the service facility. Thus, if after a customer arrives, there are n customers present, each is serviced at the average rate μ/n.

$$\rho = \lambda E[s] = \lambda/\mu.$$

$$p_n = P[N = n] = (1 - \rho)\rho^n, \qquad n = 0, 1, 2, \ldots.$$

$$L = E[N] = \frac{\rho}{1 - \rho}.$$

$$E[w \mid s = t] = \frac{t}{1 - \rho}.$$

$$E[w] = \frac{E[s]}{1 - \rho},$$

$$E[q \mid s = t] = \frac{\rho t}{1 - \rho}.$$

(This value represents the average customer delay for a customer needing t units of service time due to not having the full resources of the server available and thus is $E[w \mid s = t] - t$.)

$$E[q] = \frac{\rho E[s]}{1 - \rho}.$$

(This value is the average delay of a customer due to not having the full resources of the server available and thus is $E[w] - E[s]$.)

The output of this system is a Poisson stream.

TABLE 21

Jackson-Type Two-Stage Cyclic Queueing Model of Multiprogramming

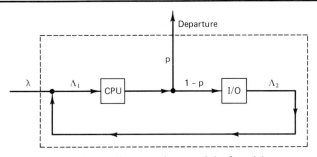

Fig. 6.2.3 Two-stage cyclic queueing model of multiprogramming.

There is a Poisson arrival process with average rate λ to the multiprogrammed computer system (inside the dashed lines). The CPU and the I/O unit provide exponential service with average rate μ_1 and μ_2, respectively. A program completing a CPU service leaves the system (with probability p) or queues for an I/O service (with probability $1 - p$). Upon completion of I/O service a program rejoins the CPU queue. This cycle is repeated until the program completes execution in the CPU and departs. All queues have infinite capacity. Λ_1 and Λ_2 are the respective average arrival rates at the CPU and I/O unit.

$$\Lambda_1 = \lambda/p.$$

$$\Lambda_2 = (1 - p)\Lambda_1 = (1 - p)\lambda/p.$$

$$\rho_1 = \frac{\Lambda_1}{\mu_1} = \frac{\lambda}{p\mu_1} \qquad \text{(CPU utilization)}.$$

$$\rho_2 = \frac{\Lambda_2}{\mu_2} = \frac{(1 - p)\lambda}{p\mu_2} \qquad (I/O \text{ utilization}).$$

$$\lambda_T = p\Lambda_1 = \lambda \qquad \text{(average throughput in jobs per unit time)}.$$

$$p(n, m) = p(n)p(m) = (1 - \rho_1)\rho_1{}^n(1 - \rho_2)\rho_2{}^m.$$

$$L = E[N] = \frac{\rho_1}{1 - \rho_1} + \frac{\rho_2}{1 - \rho_2} \qquad \text{(average number in the system)}.$$

$$W = L/\lambda \qquad \text{(average response time or time a job is in the system)}.$$

TABLE 22
Two-Stage Cyclic Queueing Model of a Multiprogrammed Computer System with a Fixed Level K of Multiprogramming and All Service Times Exponential

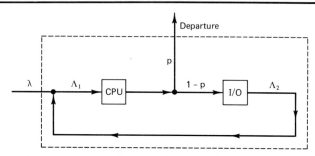

Fig. 6.2.3 Two-stage cyclic queueing model of multiprogramming.

The CPU and the I/O unit provide exponential service with average rates μ_1 and μ_2, respectively. There are exactly K customers (jobs) in the queueing network inside the dotted lines. A customer completing a service at the CPU leaves the system with probability p (to be replaced by another customer, instantly). p_n is the probability there are n jobs in the CPU queue or in CPU service. Then

$$p_n = \left(\frac{1-r}{1-r^{K+1}}\right) r^n, \qquad n = 0, 1, 2, \ldots, K, \qquad (6.2.18)$$

where

$$r = \frac{\mu_2}{(1-p)\mu_1},$$

ρ_1 is the server utilization of the CPU, and ρ_2 is the server utilization of the I/O unit.

$$\rho_1 = 1 - p_0 = \frac{r - r^{K+1}}{1 - r^{K+1}}. \qquad (6.2.19)$$

$$\rho_2 = 1 - p_K = \frac{1 - r^K}{1 - r^{K+1}}. \qquad (6.2.20)$$

The average response time (average time a job is in the computer system) is

$$\frac{K}{\mu_1 p \rho_1}. \qquad (6.2.21)$$

The average throughput (jobs per unit of time) λ_T is given by

$$\lambda_T = \mu_1 p \rho_1. \qquad (6.2.22)$$

TABLE 23

**Two-Stage Cyclic Queueing Model of Multiprogramming With Fixed
Level K of Multiprogramming, General Service Time for
One Server, Exponential Service for Second**

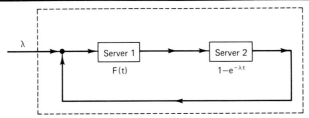

Fig. 6.2.4 Two-stage cyclic queueing model of multiprogramming with
general service time distribution for first server, exponential for the second,
fixed number of jobs (customers) in system.

Server 1 has mean service rate μ with general service time distribution. Server 2 has exponential service time distribution with average rate λ. $\rho = \lambda/\mu$ can be greater than one.

$$p(n, K - n) = P[n \text{ jobs in Queue 1 and } K - n \text{ jobs in Queue 2}] = p_K(n), \quad (6.2.23)$$

where $p_K(n)$ is given by (6.2.24). "n jobs in Queue 1" means n jobs either in waiting line or in service.

$$p_K(n) = \begin{cases} C_K \hat{p}(n) & 0 \le n < K \\ 1 - [1 - C_K(1 - \rho)]/\rho & n = K, \end{cases} \quad (6.2.24)$$

where

$$C_K = \left\{ 1 - \rho \left[1 - \sum_{n=0}^{K-1} \hat{p}(n) \right] \right\}^{-1}, \quad (6.2.25)$$

and the values of $\hat{p}(n)$ can be calculated by

$$\hat{p}(n) = \frac{1}{n!} \left. \frac{d^n P(z)}{dz^n} \right|_{z=0}. \quad (6.2.26)$$

In (6.2.26), $P(z)$ is defined by

$$P(z) = \frac{(1 - \rho)(1 - z)W_s^*[\lambda(1 - z)]}{W_s^*[\lambda(1 - z)] - z}, \quad (6.2.27)$$

where $W_s^*(\theta)$ is the Laplace–Stieltjes transform of the service time of Server 1 (see (2.9.19)).
The utilization of Server 1 is

$$\rho_1 = 1 - p_K(0), \quad (6.2.29)$$

and

$$\rho_2 = 1 - p_K(K) \quad (6.2.30)$$

is the utilization of Server 2.

TABLE 23 *(Continued)*

The average number of jobs in queue 1 and queue 2 is

$$E[N_1] = \sum_{n=1}^{K} np_K(n), \qquad E[N_2] = K - E[N_1]. \tag{6.2.31}$$

The average cycle time of a job is

$$\frac{K}{\rho_1 \mu} = \frac{K}{\rho_2 \lambda}. \tag{6.2.32}$$

If the average number of cycles required for a job to complete execution is m, then the average response time

$$W = \frac{mK}{\rho_1 \mu} = \frac{mK}{\rho_2 \lambda}, \tag{6.2.33}$$

and the average throughput λ_T is given by

$$\lambda_T = \frac{\rho_1 \mu}{m} = \frac{\rho_2 \lambda}{m}. \tag{6.2.34}$$

If the service time of the 1st server is either two-stage hyperexponential or Erlang-2, the $\hat{p}(n)$'s for (6.2.24) can be calculated as follows:

$$\hat{p}(n) = C_1 z_1^{-n} + C_2 z_2^{-n}, \qquad \text{for all } n, \tag{6.2.39}$$

where

$$C_1 = (1 - \rho z_2)(1 - z_1)/(z_2 - z_1), \tag{6.2.40}$$

$$C_2 = (1 - \rho z_1)(1 - z_2)/(z_1 - z_2). \tag{6.2.41}$$

Here z_1 and z_2 are the roots of the following polynomials:

$$\rho^2 z^2 - \rho(\rho + 4)z + 4 = 0 \tag{6.2.42}$$

for Erlang-2 service, and

$$\beta_1 \beta_2 z^2 - (\beta_1 + \beta_2 + \beta_1 \beta_2)z + 1 + \beta_1 + \beta_2 - \rho = 0 \tag{6.2.43}$$

for two-stage hyperexponential service. In (6.2.43), $\beta_1 = \lambda/\mu_1$ and $\beta_2 = \lambda/\mu_2$, where μ_1 and μ_2 are parameters of the hyperexponential distribution with density function

$$f(t) = \alpha_1 \mu_1 e^{-\mu_1 t} + \alpha_2 \mu_2 e^{-\mu_2 t}, \qquad t > 0, \tag{3.2.58}$$

such that

$$\frac{\alpha_1}{\mu_1} + \frac{\alpha_2}{\mu_2} = \frac{1}{\mu}.$$

TABLE 24

**Steady State Formulas for the Finite Population
Queueing Model of Interactive Computing With
Processor-Sharing**

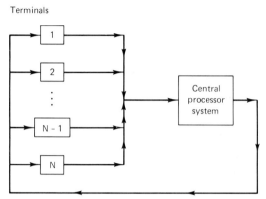

Terminals

Fig. 6.3.1   Finite population queueing model of interactive computer system. Special case in which the central processor system consists of a single CPU with processor-sharing queue discipline.

The CPU operates with the processor-sharing queue discipline. CPU service time is general with the restriction that the Laplace–Stieltjes transform must be rational. The same restriction holds on think time. $E[t] = 1/\alpha$ is the average think time with $E[s] = 1/\mu$ the average CPU service time. Then

$$p_0 = \left[\sum_{n=0}^{N} \frac{N!}{(N-n)!} \left(\frac{E[s]}{E[t]} \right)^n \right]^{-1} = \left[\sum_{n=0}^{N} \frac{N!}{(N-n)!} \left(\frac{\alpha}{\mu} \right)^n \right]^{-1}.$$

The CPU utilization

$$\rho = 1 - p_0,$$

and the average throughput

$$\lambda_{\mathrm{T}} = \frac{\rho}{E[s]} = \frac{1 - p_0}{E[s]}.$$

The average response time

$$W = \frac{N E[s]}{1 - p_0} - E[t].$$

TABLE 25

Steady State Formulas for D/D/c/K/K (Machine Repair with Multiple Repairmen) Queueing System

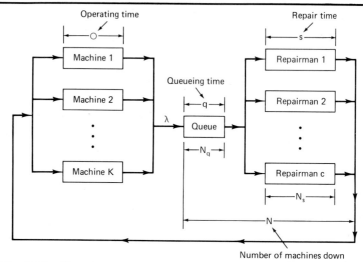

Fig. 5.2.9 Machine repair queueing system with *c* repairmen. (D/D/c/K/K queueing system.)

$E[\bigcirc] = 1/\alpha$ = operating time per machine (constant).

$E[s] = 1/\mu$ = repair time per machine by one repairman (constant).

$$\rho = \min \left\{ \frac{K}{c(1 + E[\bigcirc]/E[s])}, 1 \right\}.$$

$$\lambda = \frac{c\rho}{E[s]} = c\rho\mu.$$

$$W_{q} = E[q] = \frac{K}{\lambda} - E[\bigcirc] - E[s].$$

$W = W_{q} + E[s].$

$L = \lambda W$ = average number of machines down.

$L_{q} = \lambda W_{q}$ = average number of machines waiting for repair.

TABLE 26
Steady State Formulas for the "Straightforward Model"
of Boyse and Warn

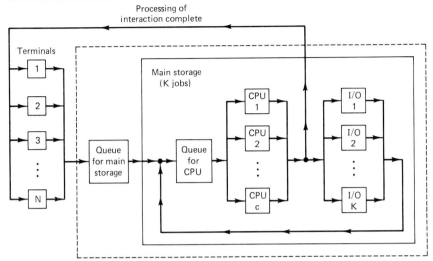

Fig. 6.3.3   The "straightforward model" of Boyse and Warn.

For the assumptions of the model see Section 6.3.3. The parameters are as follow.

$E[s]$ = average CPU usage interval between page faults.

m = average number of CPU intervals per interaction.

$mE[s]$ = average CPU time per interaction.

$E[\bigcirc]$ = average service time of an I/O (page) request.

K = constant multiprogramming level and thus number of parallel I/O servers.

N = number of active terminals.

$E[t] = 1/\alpha$ = average think time.

c = number of CPU's.

If both CPU and I/O servers have constant service time, then the CPU utilization is

$$\rho = \min \left\{ \frac{K}{c(1 + E[\bigcirc]/E[s])}, \, 1 \right\}.$$

If both the CPU service time and the I/O service time is exponential then use the appropriate machine repair equations (if single CPU system use the equations of Table 10; for multiple CPU's use Table 11) to get λ. Then the CPU utilization is calculated by

$$\rho = \lambda E[s]/c. \tag{6.3.10}$$

The average throughput λ_T is given by

$$\lambda_T = c\rho/mE[s]. \tag{6.3.19}$$

The average response time W is calculated by

$$W = \frac{NmE[s]}{c\rho} - E[t]. \tag{6.3.20}$$

TABLE 27

Steady State Equations of Central Server Model of Multiprogramming

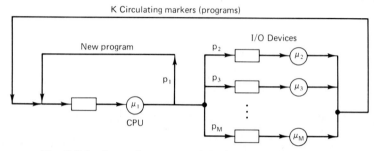

K Circulating markers (programs)

New program

I/O Devices

p_1 p_2 p_3 p_M

CPU

Fig. 6.3.4   Central server model of multiprogramming.

For the assumptions of the model see Section 6.3.4.
Calculate $G(0)$, $G(1)$, ..., $G(K)$ by Algorithm 6.3.1 (Buzen's Algorithm).
Then the server utilizations are given by

$$\rho_i = \begin{cases} G(K-1)/G(K) & i = 1 \\ \dfrac{\mu_1 \rho_1 p_i}{\mu_i} & i = 2, 3, \dots, M. \end{cases} \qquad (6.3.25)$$

The average throughput λ_T is given by

$$\lambda_T = \mu_1 \rho_1 p_1. \qquad (6.3.26)$$

If the central server model is the central processor model for the interactive computing system
of Fig. 6.3.1, then the average response time W is calculated by

$$W = \frac{N}{\lambda_T} - E[t] = \frac{N}{\mu_1 \rho_1 p_1} - E[t]. \qquad (6.3.27)$$

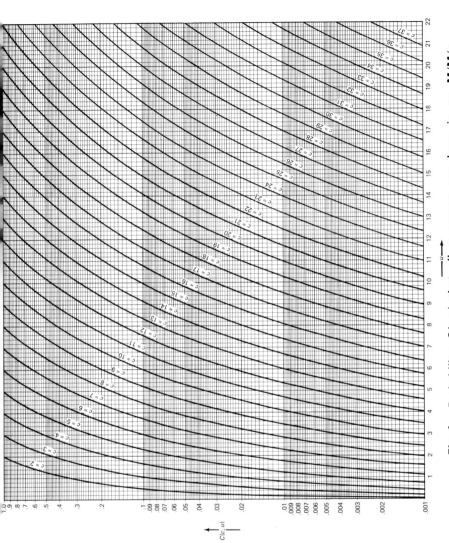

Fig. 1 Probability $C(c,u)$ that all c servers are busy in an M/M/c queueing system versus traffic intensity $u = \lambda E[s]$.

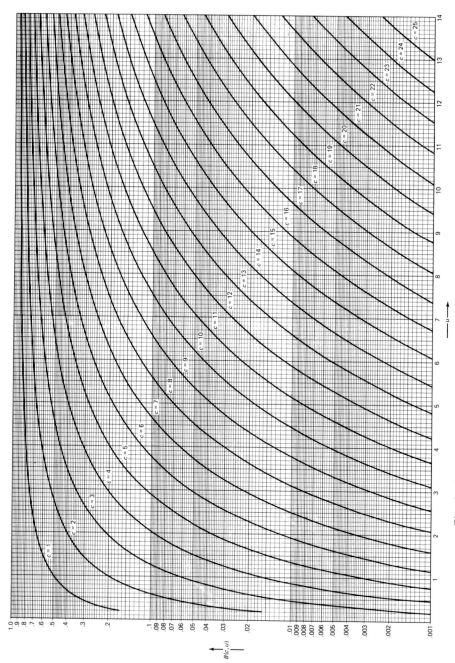

Fig. 2 Erlang's loss formula $B(c,u)$, the probability that all c servers are busy in an M/M/c loss system.

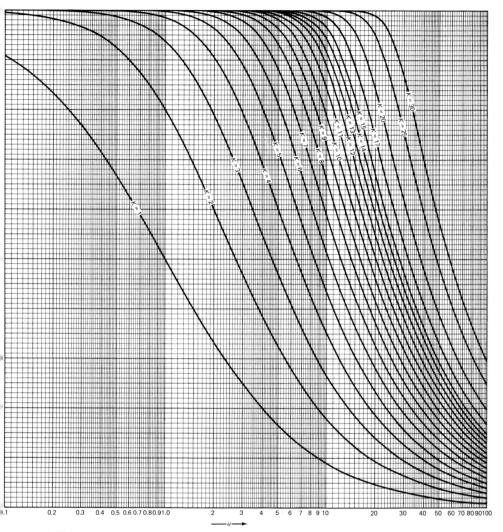

Fig. 3 Server utilization ρ as a function of K and $z = E[\bigcirc]/E[s] = \mu/\alpha$ for the M/M/1/K/K queueing system (machine repair).

NUMERICAL ANSWERS
TO SELECTED EXERCISES

Chapter Two

1. (a) 0.3645. (b) 0.4275. (c) 0.8585. (d) 0.0665.
5. $1 - 13/6^4 = 1283/1296$. **6.** (a) 1, 2, 3, 4. (b) $p(1) = 20/35$, $p(2) = 10/35$, $p(3) = 4/35$, $p(4) = 1/35$. **9.** 0.99978. **10.** mean = 3.95, standard deviation = 2.295. **12.** 94 bytes. **14.** Four. **15.** (a) No.
(b) 75/216, 15/216, 1/216. (c) $-\$0.0787$. **17.** (a) Method B. (b) Yes.

Chapter Three

1. (a) 0.3955. (b) 0.08789. **2.** (a) $3/16 = 0.1875$. (b) 0.10057.
(c) 0.5642. **4.** (a) 2.5. (b) 0.1296. **6.** 0.14197. Poisson approximation is
0.14288. **7.** 0.0902. **8.** 0.013475. Normal approximation 0.00944.
11. mean = 20, variance = 33.33. Approximation (a): mean = 20, variance = 40.
(b): mean = 20, variance = 34. **12.** (a) 0.77687. (b) 0.13534.

(c) 11.5 seconds.   **14.** (a) 8 seconds.   (b) 0.265.   (c) 0.242424.   (d) No.
17. Erlang-5 with $\lambda = 1/90$.   **19.** Normal approximation = 0.001035.
One-sided inequality ≤ 0.0868. Correct value = 0.00266.   **20.** Normal
approximation = 0.00944. Correct value = 0.01347.   **21.** (a) Gamma with
parameters 5 and λ.   (b) 0.4405.   **22.** 66,564.

Chapter Four

2. $EP(2,3) = 5/3$.   **3.** 138.63 hours, 110.06 hours, 105.84 hours, respectively.

Chapter Five

5. Cost per 8-hour day for I. M. Slow is $800, for I. M. Fast $384. Hire Fast fast.
6. (a) 1/3.   (b) 1.5.   (c) 0.063.   (d) 7.2 minutes.   **7.** 2.486028 minutes.
11. $W_q = 0.2984$ minutes. $W = 5.2984$ minutes.   **12.** (a) 22.22 hours.
(b) 49 minutes.   **13.** All times in minutes. Terminal 1: $W_q = 6.92$, $W = 36.92$,
$\pi_q(90) = 23.21$. Terminal 2: $W_q = 10.00$, $W = 40.00$, $\pi_q(90) = 36.35$. Terminal 3:
$W_q = 4.29$, $W = 34.29$, $\pi_q(90) = 7.65$. Terminal 4: $W_q = 13.64$, $W = 43.64$, $\pi_q(90) =$
49.72. Terminal 5: $W_q = 2.00$, $W = 32.00$, $\pi_q(90) = 0.00$.   (b) $W_q = 9.30$,
$W = 39.30$, $\pi_q(90) = 32.01$.   (c) 45.83.   **14.** (a) 0.7576.   (b) 0.2424.
(c) 65.45 minutes.   (d) 5.45.   (e) 0.43841.   **15.** (a) 3.   (b) 0.47368 minutes,
0.23684.   (c) 0.23684, 0.2105.   **16.** (a) B(10,5) = 0.018385.   (b) 4.908.
(c) B(9,5) = 0.037458.   **17.** (a) 13.62 customers/hour.   (b) B(12,7) =
0.027081.   (c) 6.81   **23.** 23.25 minutes.   **26.** All times in seconds.
Line 1: $W_q = 6$, $W = 8$, $\sigma_w = 10.2$, $L_q = 1.8$, $L = 2.4$, $\pi_w(90) = 21.26$ by Martin's
estimate. $\pi_w(90) = 2.585 \times 8 = 20.68$ by Table 5.3.2. Line 2: $W_q = 3$, $W = 5$,
$\sigma_w = 5$, $L_q = 0.9$, $L = 1.5$, $\pi_w(90) = 11.5$ by both methods. Line 3: $W_q = 2$, $W = 4$,
$\sigma_w = 3.127$, $L_q = 0.6$, $L = 1.2$. $\pi_w(90) = 8.065$ by Martin's estimate and $2.07 \times 4 =$
8.28 by Table 5.3.1. Line 4: $W_q = 1.5$, $W = 3.5$, $\sigma_w = 2.06$, $L_q = 0.45$, $L = 1.05$.
$\pi_q(90) = 6.18$ by Martin's estimate and $1.79 \times 3.5 = 6.27$ by Table 5.3.1.
29. 68.457 queries/second.   **30.** (a) By Fig. 3, $\rho = 0.86$, $\lambda = 0.43$ entries/second,
$E[q] = 7.77$ seconds. MACHΔREP gives $\rho = 0.8675402095$, $\lambda = 0.4337701048$
entries per second, $W_q = 7.161059$ seconds.   (b) MACHΔREP gives $\rho =$
0.9855909875, $\lambda = 0.4927954937$ entries per second, $W_q = 19.1695734$ seconds.
(c) MACHΔREP gives $\rho = 0.9997790557$, $\lambda = 0.4998895278$ entries per second,
$W_q = 38.022099$ seconds.   **36.** (a) 0.178046.   (b) 1.217 people.   (c) 0.0115.
37. All times seconds. For HOL: $W_{q_1} = 8.477$, $W_{q_2} = 60.55$, $W_q = 55.3427$,
$W_1 = 13.477$, $W_2 = 60.95$, $W = 56.2027$. For Preemptive-resume: $W_{q_1} = 3.333$,
$W_{q_2} = 60.95$, $W_q = 55.1883$, $W_1 = 8.333$, $W_2 = 61.35$, $W = 56.0483$.   **38.** All
times seconds. By heavy traffic approximation $E[q] = 0.986$, $E[w] = 1.0335$,
$\pi_q(90) = 2.2678$, $P[q > 1.5] = 0.218$. For M/M/1 system $E[q] = 0.9025$, $E[w] =$
0.95, $\pi_q(90) = 2.139$, $P[q > 1.5] = 0.196$. Allen–Cunneen approximation yields
$E[q] = 0.9250625$.   **39.** $0 \leq W_q \leq 3.2$. True $W_q = 2.7084$.   **40.** $4.0005 \leq W_q \leq$
20.35 seconds. Allen–Cunneen approximation yields $W_q = 18.1845$.

Chapter Six

1. $R_1/R_2 = (1 + C_s^2)/2 = 2.$ **2.** (a) $\rho_1 = 5/6$, $\rho_2 = 0.96.$ (b) 5 jobs.
(c) 24 jobs. (d) 29 jobs. (e) 174 seconds.   **3.** $\rho_1 = 0.8328$, $\rho_2 = 0.95938$,
average response time = 60.04 second, $\lambda_T = 0.16656$ jobs/second.
6. $\rho_1 = 0.87254$, $\rho_2 = 0.83006$, $W = 22.92$ seconds, $\lambda_T = 0.21814$ jobs/second.
7. $\rho = 0.9987$, $\lambda_T = 0.4933$ jobs/second, $W = 16.027$ seconds, $N^{**} = 12.48 \approx 12.$
8. (a) $\rho = 0.5493$, $\lambda_T = 0.5507$ jobs/second, $W = 14.16$ seconds, $N^{**} = 13.77 \approx 14.$
(b) $\rho = 0.8009$, $\lambda_T = 0.8029$ jobs/second, $W = 8.45$ seconds, $N^{**} = 20.07 \approx 20.$
(c) $\rho = 0.8455$, $\lambda_T = 0.8477$ jobs/second, $W = 7.8$ seconds, $N^{**} = 21.19 \approx 21.$
(d) $\rho = 0.981$, $\lambda_T = 0.9835$ jobs/second, $W = 6.17$ seconds, $N^{**} = 24.59 \approx 25.$
9. $p_0 = 0.1105$, $\rho = 0.7513$, $\lambda_T = 0.7532$ jobs/second, $W = 9.2771$ seconds,
$N^{**} = 18.8295 \approx 19.$

Chapter Seven

3. 19.96372, 1.01931.   **4.** $12.1506 \le \mu \le 13.6494$, $7.508 \le \sigma^2 \le 14.658.$
$10.440 \le \mu \le 15.360.$ **5.** $16.600 \le \mu \le 21.516$, $15.959 \le \sigma^2 \le 58.865.$
6. $0.34 \le \rho \le 0.715$, $0.285 \le p_0 \le 0.66$, $0.471 \le L \le 2.060$, $0.131 \le L_q \le 1.345.$

Chapter Eight

1. $z = -1.6172.$ Accept.   **2.** $n = 271.$ **3.** $n = 515.$ **4.** $t = 2.493.$
Reject.   **5.** Division A programmers are more productive.   **6.** Test does not
indicate Division A programmers more productive.   **7.** Cannot reject $\sigma^2 = 30.$
8. Variances equal.   **9.** Population is normal.   **10.** Response time is
$N(30, 5^2).$ **11.** Response time is normally distributed.   **12.** Line time not
exponential.   **13.** Population normally distributed.

INDEX